Pharmaceutical Salts and Co-crystals

RSC Drug Discovery Series

Titles in the Series:

1: Metabolism, Pharmacokinetics and Toxicity of Functional Groups: Impact of Chemical Building Blocks on ADMET
2: Emerging Drugs and Targets for Alzheimer's Disease; Volume 1: Beta-Amyloid, Tau Protein and Glucose Metabolism
3: Emerging Drugs and Targets for Alzheimer's Disease; Volume 2: Neuronal Plasticity, Neuronal Protection and Other Miscellaneous Strategies
4: Accounts in Drug Discovery: Case Studies in Medicinal Chemistry
5: New Frontiers in Chemical Biology: Enabling Drug Discovery
6: Animal Models for Neurodegenerative Disease
7: Neurodegeneration: Metallostasis and Proteostasis
8: G Protein-Coupled Receptors: From Structure to Function
9: Pharmaceutical Process Development: Current Chemical and Engineering Challenges
10: Extracellular and Intracellular Signaling
11: New Synthetic Technologies in Medicinal Chemistry
12: New Horizons in Predictive Toxicology: Current Status and Application
13: Drug Design Strategies: Quantitative Approaches
14: Neglected Diseases and Drug Discovery
15: Biomedical Imaging: The Chemistry of Labels, Probes and Contrast Agents
16: Pharmaceutical Salts and Co-crystals

How to obtain future titles on publication:
A standing order plan is available for this series. A standing order will bring delivery of each new volume immediately on publication.

For further information please contact:
Book Sales Department, Royal Society of Chemistry, Thomas Graham House, Science Park, Milton Road, Cambridge, CB4 0WF, UK
Telephone: +44 (0)1223 420066, Fax: +44 (0)1223 420247, Email: books@rsc.org
Visit our website at http://www.rsc.org/Shop/Books/

Pharmaceutical Salts and Co-crystals

Edited by

Johan Wouters
University of Namur, Namur, Belgium

Luc Quéré
UCB S.A., Brussels, Belgium

RSC Publishing

RSC Drug Discovery Series No. 16

ISBN: 978-1-84973-158-4
ISSN: 2041-3203

A catalogue record for this book is available from the British Library

Published by The Royal Society of Chemistry,
Thomas Graham House, Science Park, Milton Road,
Cambridge CB4 0WF, UK

Registered Charity Number 207890

For further information see our web site at www.rsc.org

Preface

Multi-component crystalline systems are not new, but only recently has the term *co-crystal* (or cocrystal) gained momentum in the glossary of the pharmaceutical world. In contrast to salts, formation of a co-crystal can be envisaged for any active pharmaceutical ingredient (API), regardless of the presence of acidic or basic groups, thus enlarging the pharmaceutical space associated with an already marketed drug or a drug candidate. Indeed, making pharmaceutical co-crystals allows modifications to be introduced to the crystalline form of an API, which in turn can alter its physicochemical properties, without compromising its intended biological activity. In this context, one excellent reason, among others, to investigate co-crystals is to increase the solubility or the dissolution properties of a poorly soluble API. For neutral APIs, co-crystals can potentially expand the number of solid forms in development, while for a free acid or free base, the approach of selecting of both salts and co-crystals can be used to improve the solubility profile.

By the same token, the study of pharmaceutical co-crystals in the context of crystal engineering continues to be an exciting and rewarding endeavour. It provides, among other things, the formalism of supramolecular synthons that allows us to describe and design these original multi-component entities. It is interesting to note that the reversible molecular interactions that hold together the API and a co-crystal former (or co-former) have the same nature (*i.e.* H bonds, Van der Waals interaction, metal coordination) than those that are found between the API and its biological target, making co-crystallization a general approach in the pharmaceutical arena.

In this exciting perspective, the challenge we tried to take up was to combine reports of the latest academic research and comprehensive overviews of basic principles, with more applied contributions from the industry. The result is a unique book, ideal for pharmaceutical development scientists and researchers

RSC Drug Discovery Series No. 16
Pharmaceutical Salts and Co-crystals
Edited by Johan Wouters and Luc Quéré

Published by the Royal Society of Chemistry, www.rsc.org

in crystal engineering and useful for both seasoned experts and graduate students.

As this book makes it clear, the study of pharmaceutical co-crystals opens up new opportunities for the pharmaceutical industry. Several chapters of this book have addressed essential aspects, such as the role of co-crystals in the pharmaceutical development continuum (N. Schultheiss and J.-O. Henck), solid forms and pharmacokinetics (N. Biswas), the process of co-crystallization and scale up issues (E. Gagnière, D. Mangin, S. Veesler and F. Puel), solid state analytical techniques and strategies for characterization (S. Reutzel-Edens) as well as opportunities and pitfalls when embarking on co-crystal search for broadening IP rights (M. Hoffmann and J. Lindeman).

The study of pharmaceutical co-crystals is also an exciting field of academic research in the context of crystal engineering. The fundamental aspects of cohesive forces and supramolecular chemistry have been reviewed in this book by A. Bond and G. Row. W. Jones and T. Friščić report on the mechanochemical methods utilized for the discovery and development of novel pharmaceutical solid forms, while co-crystal solubility and thermodynamic stability aspects have been assessed by L. Roy, M.P. Lipert and N. Rodriguez-Hornedo. The importance of phase diagrams in a co-crystal search (T. Rager and R. Hilfiker) and knowledge of heterogeneous equilibria addressed by G. Coquerel, will help the reader to view the concept of a co-crystal with the wide angle provided by the thermodynamic perspective.

From purely intermolecular recognition to size or polarity matches between molecules, a large number of factors influence the (co-)crystallisation outcome. The theoretical basis for its prediction is still under development and F. Leusen and J. Kendrick and F. Laszlo and T. Friščić have reviewed recent advances in the field.

The number and nature of potential non-toxic co-crystal formers (or co-formers) that can be co-crystallized with an API is numerous. The last chapter is an attempt to list promising ones. Even if each API must be examined and evaluated on a case-by-case basis in terms of molecular structure, we are confident that this listing could serve as a valuable companion for both screening and retrosynthetical design approaches to co-crystallization. The data provided should help chemical intuition to assist the process of selection of co-crystal formers.

Before finishing this preface, we would like to express our deep gratitude to the authors of the different chapters of this book who have generously and confidently accepted to share with all of us their valuable expertise. In particular, we thank Prof G.R. Desiraju for his continuous support and encouragement and for his enlightening introduction. We would also like to thank Prof S. Guccione, series editor, who gave us the opportunity to include this book in the RSC Drug Discovery series and the team of RSC for assistance and guidance.

Enjoy the reading,

Johan Wouters and Luc Quéré

Contents

RSC Drug Discovery Series No. 16
Pharmaceutical Salts and Co-crystals
Edited by Johan Wouters and Luc Quéré

Published by the Royal Society of Chemistry, www.rsc.org

CHAPTER 1

Pharmaceutical Salts and Co-crystals: Retrospect and Prospects

GAUTAM R. DESIRAJU

Solid State and Structural Chemistry Unit, Indian Institute of Science, Bangalore 560012, India

Every new field in chemistry needs a link to an application of commercial and practical use to sustain interest. Each such field generates a whole new set of ideas, paradigms and models. These concepts need to be tested in as wide a variety of forums as possible because their generality has to be proved. The industrial enterprise has always provided an excellent testing ground for new ideas in the chemical sciences. Many fundamental concepts took root because of an impetus from industry, the most spectacular ones being the discovery of stereochemistry by Pasteur, Haber's process for nitrogen fixation, and the birth of polymer chemistry starting with the production of synthetic rubber from isoprene. The subject of crystal engineering appeared in its modern manifestation in the late 1980s and early 1990s.[1,2] Two important branches of this subject emerged. The field of co-ordination polymers[2] quickly found its practical application in the gas absorption properties of metal-organic framework compounds.[3] The field of organic crystal engineering[1] found its practical application, a little later, in the area of pharmaceutical co-crystals and salts.[4]

RSC Drug Discovery Series No. 16
Pharmaceutical Salts and Co-crystals
Edited by Johan Wouters and Luc Quéré

Published by the Royal Society of Chemistry, www.rsc.org

The chapters in this book illustrate the tremendous growth in this area during the past decade.

Co-crystals have been around in the chemical literature ever since Wöhler described quinhydrone in 1844.[5] Pfeiffer's monumental work in 1922, *Organische Molekulverbindungen*[6] now has a worthy successor in Herbstein's two volume magnum opus, *Crystalline Molecular Complexes and Compounds*,[7] nearly a century later. What crystal engineering did was to provide a context wherein these compounds were given a name, *co-crystals*, however contentious it was. Secondly, it provided the formalism of the supramolecular synthon[8] with which these compounds could be described, designed and deconvoluted. Finally it defined a setting in which these compounds could be assessed, namely in the pharmaceutical industry. The chapters in this book cover all the above aspects of research.

One might ask "What's in a name?". But names are important in chemistry, and there is usually much disputation about chemical names especially in areas that are still evolving. Some of us have defended or questioned the very term *co-crystal*.[9,10] Coquerel, in his chapter in the present book, advises us in somewhat harsh terms when he says, "Therefore, rather than creating questionable terms and maintaining endless semantic debates with poor added values, the scientific community should concentrate more on the three long lasting problems...". Now, I would tend to both agree and disagree with this view. While no discussion about nomenclature should be endless and there are indeed real practical scientific issues that need to be addressed, a part of this discussion about naming something in chemistry is about refining one's scientific ideas. Names in incompletely developed subjects keep changing and evolving. In the end, a name that survives, survives. Pfeiffer's *Molekulverbindung* and Herbstein's *Molecular Compound* are one and the same, but they are separated by a century of structural chemistry, an enormous time period in science. Bond's definition of a co-crystal as a multi-component molecular crystal[10] has the great advantage of scientific accuracy with reasonable brevity but it lacks the pep and vigour of the term co-crystal. At the same time, one should not get carried away with style and ascribe scientific value to the term co-crystal. It has none. It is a bit like the term *pseudopolymorph*.[11] In a recent perspective article,[12] Bernstein levels criticism at this latter term, as being loose and lacking in scientific value. But the very same criticism can be applied to co-crystal, a term that the same author uses without flinching. My view is that both co-crystal and pseudopolymorph are of similar standing as far as the philosophy of nomenclature is concerned. Neither term has any real scientific value. Both terms are commonly used and in most cases there is little disagreement about whether or not a particular compound is a co-crystal or a pseudopolymorph. They are trivial names, as opposed to systematic names and, as I have written elsewhere,[13] the trivial name sometimes survives. Few call acetic acid ethanoic acid. We shall have to wait and see if the term co-crystal is still extant 100 years from now.

What about *pharmaceutical co-crystals*? Zaworotko and co-workers defined these as "co-crystals that are formed between a molecular or ionic active pharmaceutical ingredient (API) and a co-crystal former that is a solid under ambient conditions".[14] Bernstein, in the above-mentioned perspective article,[12] criticises this term too and asks rhetorically if we would call something a pharmaceutical polymorph. But this is begging the question because in calling something a pharmaceutical co-crystal, one is moving away from science and going towards patent litigation. As Hoffman and Lindeman have argued in their chapter, it does not matter that every inventor uses the same definition of what is a co-crystal or, for that matter, a pharmaceutical co-crystal. A patentee may choose to define a term in any way in a patent specification and the patentee's definition controls the meaning of the term in the claim. For example, a co-crystal may be defined to exclude solvates, salts and amorphs. Therefore, terms in the patent literature not only include certain examples and situations but also deliberately exclude others. The issue of nomenclature is therefore not just a strictly scientific matter but is also of legal concern. The term pharmaceutical co-crystal needs to be looked at from this perspective. Both scientists and patent attorneys need to approach this matter with caution and care.

A major reason for the popularity of pharmaceutical co-crystals in industry is that they lend themselves well to patent protection. They admirably satisfy the three criteria of patentability, namely novelty, non-obviousness and utility. A co-crystal almost always satisfies the novelty criterion because it is a new composition of matter. Non-obviousness is provided by the fact that the identification of the co-former is hardly ever routine, unlike say salt formation wherein an acid is obviously required to make a salt from a base. Utility is generally the only criterion that must be established but it is often easy to demonstrate—usually it is the lack of a particular attribute (solubility, bio-availability, dissolution profile, good shelf life) that has led to the identification of a pharmaceutical co-crystal. With respect to patentability, co-crystals offer opportunities *vis-à-vis* polymorphs. They are clearly new substances, problems of inherent anticipation[15] are not likely to arise so often and more of them can be made for any given API, expanding the pharmaceutical space around it and consequently the types of advantageous properties that may be accessed.

The issue of non-obviousness is particularly attractive: the design of a co-crystal using synthon theory has all the elements of design and strategy. In this sense, the matter is predictable and the choice of a co-former is neither arbitrary nor random. The necessary experimentation is executable and manageable. At the same time, there is no guarantee that every co-crystal that is designed retrosynthentically will actually be obtained in practice. This lack of inevitability strongly supports non-obviousness and in this respect co-crystals are unlike salts. There are other methods of design. Fábián and Friščić, in their chapter, discuss shape and polarity descriptors and identify the dihydric phenol orcinol as a promising co-former. Quite independently, in a high-throughput exercise, we have recently confirmed that orcinol is indeed an excellent co-former.[16] High-throughput methods are of obvious relevance. We have stated elsewhere, in a paper on the drug–drug co-crystal of lamivudine and zidovudine that a combination of logic-driven synthon-based design and high-throughput crystallisation is what might actually be needed.[17]

If crystal engineering required a concept known as the *supramolecular synthon*, pharmaceutical co-crystals needed the *heterosynthon*.[4] The synthon was sensed[18] before it was identified and named.[8] Similarly, we knew about heterosynthons long before Zaworotko coined the term in 2003. This structural unit was highlighted in the early 1990s as an example of molecular recognition in the very extensively studied melamine–barbituric acid molecular complex and related compounds.[19] Interestingly, this particular co-crystal is described in the chapter by Biswas on pharmacokinetics; it was the cause of lethal poisoning through kidney disease because its solubility is so much less than either of its constituents. Identifying the heterosynthon as a particular type of synthon was important because it helped to focus design strategies for pharmaceutical co-crystals. These ideas are elaborated by Bond in greater detail in his chapter.

A notable aspect of co-crystal research has been the development of high-throughput crystallisation screens. A review from the TransForm group appeared as early as 2004, with the phrase *high-throughput crystallisation* in the article title.[20] It was recognised that a large number of factors influence crystallisation outcome and that the theoretical basis for predicting such outcomes is poorly developed. Accordingly, there is a need for high-throughput crystallisation methods that sample variables such as temperature, solvent, concentration, additives, vessel design, time, heating and cooling rates, pH and mixing rates. Such research is also important for crystal engineering itself, in a general sense. A nagging worry in any experimental study is that all possible crystal forms of a single- or a multi-component system have not been isolated. Some of the conclusions we draw about structures and structure design could be biased by the fact that we are not dealing with a statistically significant number of examples. This is of even greater concern today, with our just emerging ideas about crystal energy landscapes and structural landscapes, and the notion that *a* crystal structure of a compound is just that, a data point. It is not *the* crystal structure of that compound. High-throughput crystallisation, accompanied by the related technique of high-throughput crystallography will go a long way to reduce these concerns.

A drug molecule is identified after a long and arduous process that begins with target identification, virtual screening, lead optimisation and process chemistry. Issues such as solubility, dissolution profiles and bioavailability normally come later in the drug design process. Issues pertaining to absorption, distribution, metabolism, excretion and toxicity (ADMET) are taken up even later. Often, drugs are acids or bases and it is possible to convert them easily into salts, wherein many of the above mentioned properties are favourable. However, current *in silico* strategies for lead molecule identification and optimisation are biased toward target binding and therefore lead more often than not to the development of lipophilic molecules as drugs.[21] These molecules cannot generally be converted into salts so easily. The chapters by Schultheiss and Henck, and by Biswas address the issue of whether solid form properties need to be taken into account earlier in the drug design continuum. The traditional approach has been to address problems of solubility, dissolution and bioavailability, and ADMET in general at the formulation stage. Schultheiss and Henck state that early identification of a pharmaceutical co-crystal has the potential to minimise the need for multiple changes of the solid form of an API during drug development, which in turn reduces costs directed towards *in vitro* and *in vivo* studies. Whether or not solid forms should be taken into account earlier in the drug development cycle, or whether problems relating to solubility and bioavailability should be addressed during formulation, is a complex corporate decision involving scientists, research managers and financial experts. However, the identification of the drug co-crystal as a legitimate extension of pharmaceutical space has certainly increased the flexibility of decision makers in this regard.

The great majority of co-crystals are constructed with strong hydrogen bonds and there is the possibility that the proton involved in the hydrogen bonding interaction is actually transferred from the donor (acid) to the acceptor (base) to form a salt. The salt to co-crystal continuum has been discussed extensively[22] and a common rule of thumb is it that when the difference in pK_a between the acid and the conjugate acid of the base is greater than 3 units, salt formation is expected. A number of chapters discuss these salt and co-crystal variations in structure. It is expected that salts are generally more soluble than co-crystals. Whether or not a salt is formed could have patent implications with respect to non-obviousness because salt formation could always be considered to be an operation that could be performed by a person skilled in the art. We found, for example, in a screen of the relatively strong acid saccharin with several basic APIs, that salt formation was the almost inevitable outcome (the solitary exception was piroxicam).[23]

Any crystal structure of a co-crystal or a salt appears to be reasonable in terms of hydrogen bonds and electrostatic interactions, if it is examined *post facto*. Predicting the interaction patterns for a given API-co-former combination *a priori* is quite another matter. Crystal structure prediction (CSP) is the most demanding type of crystal engineering because it seeks to predict fine details of crystal packing.[24] The addition of salts, co-crystals and hydrates (generally multi-component crystals) to the list of compounds given in the Cambridge Crystallographic Data Centre (CCDC) sponsored blind test for CSP has significantly added to the difficulty of this exercise. Gratifyingly it was found that in the latest (2010) blind test, good success was obtained in the CSP of both salts and hydrates. Leusen and Kendrick,[24] who were part of a group that made a number of accurate predictions, have reviewed this topic in their chapter. We found, in an earlier blind test, that it is necessary in CSP to know whether or not a particular binary crystal exists as a co-crystal or a salt before one embarks upon rigorous computation.[25]

CSP is closely allied to the subject of polymorphism. A frequently asked question is whether or not a pharmaceutical co-crystal is less likely to exhibit polymorphism when compared to the API itself.[4] In a more general context, are co-crystals less prone to polymorphism than single component crystals? The current consensus seems to be that the number of examples available at hand is far too small to permit any reliable answer to such questions. That a co-crystal is inherently less prone to polymorphism than is a single component crystal appears to be intuitive, but this hypothesis cannot have a direct answer because to prove the absence or lesser incidence of polymorphism is tantamount to proving the negative.[4] More specific is the matter of synthon polymorphism in a co-crystal. Various types of polymorphism are possible but the most direct manifestation of this phenomenon occurs when the primary synthons in the forms are different. Here, the data, although limited, seem to indicate a trend. Zaworotko and co-workers report that, of the 38 pairs of polymorphic organic co-crystals in the Cambridge Structural Database (CSD), polymorphism in 35 of these cases is a result of conformational flexibility and/or minor structural changes in the packing.[26] The hydrogen bonded heterosynthons persist. True synthon polymorphism, or a deep seated difference at the core of the structure, occurs only thrice. In this context, we isolated synthon polymorphs in a fourth example, namely the 1:2 co-crystals of 4,4′-bipyridine and 4-hydroxybenzoic acid.[27]

The study of pharmaceutical co-crystals in the context of crystal engineering continues to be an exciting and rewarding endeavour. Systematic crystal engineering of ternary co-crystals is still a nascent subject because handling three distinct chemical species in the design exercise is a daunting exercise.[28] New strategies in high-throughput crystallisation and the organisation of libraries of co-formers are expected to yield more co-crystals with advantageous properties.[29] A recent report states that a racemic compound can be converted into a racemic co-crystal which can show preferential enrichment, which is a symmetry breaking spontaneous enantiomeric resolution phenomenon, with big implications in the area of asymmetric synthesis.[30] A final sobering thought is that after 10 years of intense research, no current drug product is still marketed as a co-crystal. It is hoped that this will lead to a greater focus of research effort on the solution of practical problems, using cutting edge research at the frontiers of chemical knowledge.

Acknowledgement

I thank the Department of Science and Technology, New Delhi, for the award of a J. C. Bose fellowship.

References

References to chapters within this volume are not given here and are specified by author name(s) within the text.

1. G. R. Desiraju, *Crystal Engineering. The Design of Organic Solids*, Elsevier, Amsterdam, 1989.
2. B. F. Hoskins and R. Robson, *J. Am. Chem. Soc.*, 1989, **111**, 5962.
3. O. M. Yaghi, G. Li and H. Li, *Nature*, 1995, **378**, 703.
4. Ö. Almarsson and M. J. Zaworotko, *Chem. Commun.*, 2004, 1889.
5. F. Wöhler, *Annalen Chem. Pharm.*, **51**, 1844, 145.
6. P. Pfeiffer, *Organische Molekulverbindungen*, Verlag von Ferdinand Enke, Stuttgart, 1922.

7. F. H. Herbstein, *Crystalline Molecular Complexes and Compounds*, vols. 1 and 2, Oxford University Press, Oxford, 2005.
8. G. R. Desiraju, *Angew. Chem. Int. Ed. Engl.*, 1995, **34**, 2311.
9. G. R. Desiraju, *CrystEngComm*, 2003, **5**, 466; J. D. Dunitz, *CrystEngComm*, 2003, **5**, 506.
10. A. D. Bond, *CrystEngComm*, 2006, **8**, 833.
11. A. Nangia, *Cryst. Growth Des.*, 2006, **6**, 2.
12. J. Bernstein, *Cryst. Growth Des.*, 2005, **5**, 1661; J. Bernstein, *Cryst. Growth Des.*, 2011, **11**, 632.
13. G. R. Desiraju, *Angew. Chem. Int. Ed.*, 2011, **50**, 52.
14. P. Vishweshwar, J. A. McMahon, J. A. Bis and M. J. Zaworotko, *J. Pharm. Sci.*, 2006, **95**, 499.
15. A. V. Trask, *Mol. Pharmaceutics*, 2007, **4**, 301.
16. A. Mukherjee, P. J. Grobelny, T. S. Thakur and G. R. Desiraju, *Cryst. Growth Des.*, 2011, **11**, 000.
17. P. M. Bhatt, Y. Azim, T. S. Thakur and G. R. Desiraju, *Cryst. Growth Des.*, 2009, **9**, 951.
18. K. Biradha, C. V. K. M. Sharma, K. Panneerselvam, L. Shimoni, H. L. Carrell, D. E. Zacharias and G. R. Desiraju, *J. Chem. Soc., Chem. Commun.*, 1993, 1473.
19. J. A. Zerkowski, J. C. Macdonald, C. T. Seto, D. A. Wierda and G. M. Whitesides, *J. Am. Chem. Soc.*, 1994, **116**, 2382.
20. S. L. Morisette, Ö. Almarsson, M. L. Peterson, J. F. Remenar, M. J. Read, A. V. Lemmo, S. Ellis, M. J. Cima and C. R. Gardner, *Adv. Drug Delivery Rev.*, 2004, **56**, 275.
21. G. L. Amidon, H. Lennernäs, V. P. Shah and J. R. Crison, *Pharm. Res.*, 1995, **12**, 413.
22. S. L. Childs, G. P. Stahly and A. Park, *Mol. Pharmaceutics*, 2007, **4**, 323; C. B. Aakeröy, M. E. Fasulo and J. Desper, *Mol. Pharmaceutics*, 2007, **4**, 317; S. Mohamed, D. A. Tocher, M. Vickers, P. G. Karamertzanis and S. L. Price, *Cryst.Growth Des.*, 2009, **9**, 2881; V. R. Hathwar, R. Pal and T. N. Guru Row, *Cryst. Growth Des.*, 2010, **10,** 3306; R. Thakuria and A. Nangia, *CrystEngComm*, 2011, **13**, 1759; P. Grobelny, A. Mukherjee and G. R. Desiraju, *CrystEngComm*, 2011, **13**, 4358.
23. R. Banerjee, P. M. Bhatt, N. V. Ravindra and G. R. Desiraju, *Cryst. Growth Des.*, 2005, **5**, 2299.
24. G. R. Desiraju, *Angew. Chem. Int. Ed.*, 2007, **46**, 8342; M. A. Neumann, F. J. J. Leusen and J. Kendrick, *Angew. Chem. Int. Ed.*, 2008, **47**, 2427; P. Raiteri, R. Martonak and M. Parrinello, *Angew. Chem. Int. Ed.*, 2005, **44**, 3769; S. L. Price, *Acc. Chem. Res.*, 2009, **42**, 117.
25. T. S. Thakur and G. R. Desiraju, *Cryst. Growth Des.*, 2008, **8**, 4031.
26. H. D. Clarke, K. K. Arora, H. Bass, P. Kavuru, T. T. Ong, T. Pujari, L. Wojtas and M. J. Zaworotko, *Cryst. Growth Des.*, 2010, **10**, 2152.
27. A. Mukherjee and G. R. Desiraju, *Chem. Commun.*, 2011, **47**, 4090.
28. D. S. Reddy, D. C. Craig and G. R. Desiraju, *J. Chem. Soc., Chem. Commun.*, 1994, 1457; C. B. Aakeröy, A. M. Beatty and B. A. Helfrich,

Angew. Chem. Int. Ed., **40**, 2001, 3240; B. R. Bhogala, S. Basavoju and A. Nangia, *CrystEngComm*, 2005, **7**, 551; J. N. Moorthy, P. Natarajan and P. Venugopalan, *Chem. Commun.*, 2010, **46**, 3574.
29. N. Blagden, M. de Matas, P. T. Gavan and P. York, *Adv. Drug Delivery Rev.*, 2007, **59**, 617.
30. R. G. Gonnade, S. Iwama, Y. Mori, H. Takahashi, H. Tsue and R. Tamura, *Cryst. Growth. Des.*, 2011, **11**, 607.

CHAPTER 2

Fundamental Aspects of Salts and Co-crystals

ANDREW D. BOND

Department of Physics and Chemistry, University of Southern Denmark, Campusvej 55, 5230 Odense M, Denmark

2.1 Introduction

The purpose of this chapter is to provide some background information about fundamental aspects of salts and co-crystals and to describe some of the approaches that have been taken for strategic co-crystal or salt preparation. The chapter considers first the definitions of co-crystals and salts, which have been debated quite vigorously in the academic literature over the last several years. The focus then moves to intermolecular interactions and strategies for co-crystal or salt preparation, with an emphasis on structure rather than thermodynamics, since the latter is discussed later in this book. With regard to salts, the focus is on organic, molecular ions, as a specific complement to co-crystals. "Inorganic" salts such as halides or sulphates, although prolific in pharmaceutical products, are not the main interest. The examples presented within the chapter are intended to provide a coherent illustration of the concepts discussed and for this purpose the focus is placed on a specific selection of intermolecular interactions. This should not give the impression that these are the only types of interactions that can be employed in co-crystal or salt design strategies—a vast number of other examples are presented in other chapters of this book.

RSC Drug Discovery Series No. 16
Pharmaceutical Salts and Co-crystals
Edited by Johan Wouters and Luc Quéré

Published by the Royal Society of Chemistry, www.rsc.org

2.2 Definitions: Salts and Co-crystals

According to the *Compendium of Chemical Terminology* produced by the International Union of Pure and Applied Chemistry (the IUPAC "Gold Book"), a salt is a "chemical compound comprising an assembly of cations and anions".[1] Thus, a pharmaceutical salt comprises an active pharmaceutical ingredient (API) that is molecular and either cationic or anionic and a counterion that might be molecular or monatomic (*e.g.* a halide anion). The requirement for charge balance means that a salt must have a definite stoichiometry. The term co-crystal, which does not currently have any entry in the IUPAC Gold Book, is used to describe the analogous situation where neutral molecular components are present within a crystalline compound in a definite stoichiometric ratio. For pharmaceutical co-crystals, one molecular component should be an API and the second component is often referred to as the *co-crystal former*. If the intention is to apply the co-crystal as part of a drug product, the co-crystal former should be non-toxic, and the GRAS (Generally Regarded as Safe) list provides one source of potential compounds for this purpose.[2] Co-crystals need not be restricted to binary compounds and numerous ternary co-crystals have been strategically prepared.[3–5]

The distinction between a salt and a co-crystal based on charged or neutral molecular components is widely accepted. Difficulties have emerged, however, with further elaboration of the co-crystal definition, in particular with attempts to distinguish a co-crystal from a solvate. The term solvate carries a widely accepted meaning for chemists: a solvate is the outcome of a solution crystallisation process where solvent molecules become included in the crystalline product, either in stoichiometric or non-stoichiometric quantities. A typical attempt to distinguish a co-crystal from a solvate (taken from Almarsson and Zaworotko[6]) is therefore as follows: "the primary difference is the physical state of the isolated pure components: if one component is a liquid at room temperature, the crystals are referred to as solvates; if both components are solids at room temperature, the products are referred to as co-crystals".[†] One obvious problem with such a definition is the arbitrary choice of "room temperature" for assessing the state of the pure components, but this could easily be clarified by defining standard conditions as for other rigorous scientific definitions. A more fundamental problem comes from the arbitrary borderline introduced between obviously very similar chemical compounds. For example, molecules containing a carboxylic acid group form stoichiometric crystalline compounds in combination with N-containing bases such as pyridine or

[†]In the academic literature, the "solid-component" definition of a co-crystal is frequently accompanied by a reference to a 2005 review paper by Aakeröy and Salmon.[7] This is unfair, since these authors state very clearly in their paper that it is not their intention to apply any such definition. Specifically, they write: "The purpose of this article is not to propose new definitions or to weigh in on the current semantic/semiotic debate, but it will be necessary to delineate the scientific realm of this [review]". One criterion applied to limit the realm of the review is that "Only co-crystals made from reactants that are solids at ambient conditions will be included", with an accompanying footnote that states "One could also make a case for including materials such as those prepared through co-condensation (of liquids or gases) at reduced temperatures or elevated pressure".

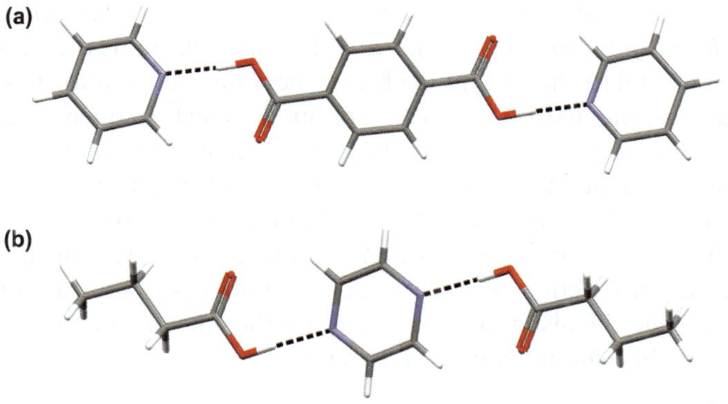

Figure 2.1 Hydrogen-bonded units within the co-crystals (a) terephthalic acid: pyridine (1:2)[8] and (b) butanoic acid:pyrazine (2:1).[10] The carboxylic acid–pyridine hydrogen-bond motif forms a common theme throughout this chapter.

pyrazine (Figure 2.1).[8–11] Since pyridine is a liquid under ambient conditions and pyrazine is a solid, any compound containing pyridine must be referred to as a solvate, while the comparable compound containing pyrazine would be a co-crystal. If the co-crystallisation were to be carried out with the *n*-alkyl carboxylic acids, for example, there is a further problem since short-chain acids are liquid under ambient conditions while longer chain acids are solids. Crystalline compounds of the short-chain acids with pyrazine would therefore be solvates while those with the longer-chain acids would be co-crystals, even though the compounds form an obviously continuous homologous series.[10,11] Compounds of the short-chain acids with pyridine would presumably be neither co-crystals nor solvates!

The problems associated with formulating a precise definition for a co-crystal are fundamental because they come from an attempt to impose a definite borderline on Nature's continuum. The borderline may come from a choice of physical state under a choice of conditions, or some definition of what is or is not a solvent, *etc*. The only way to avoid the issue is to generalise the definition so that it contains as few arbitrary distinctions as possible. Various proposals that have been given in the literature[8,12–14] amount to more or less the same thing: a co-crystal should refer to the outcome of any co-crystallisation process, which means that it should refer to any case where two (or more) chemical components crystallise together. In other words, it should apply to any *multi-component crystal*, where "component" has its meaning as in the phase rule. In this way, a solvate is a type of co-crystal. Of course, definitions that become too general are frequently of limited utility and some distinction is often imposed in a practical sense. For example, chemists distinguish between covalent and ionic bonding, but crucially with the understanding that these represent extremes of a continuum of bonding types. The same understanding and flexibility should be applied to the definition of a co-crystal. Practically, the term co-crystal will usually be understood as *a multi-component molecular crystal containing neutral molecules in a definite stoichiometric ratio*. For pharmaceutical co-crystals, the API and co-crystal former are quite likely to be solids under ambient conditions because this is the nature of the compounds most frequently applied as pharmaceuticals, but it should not be a requirement to define a co-crystal, pharmaceutical or otherwise. The term solvate will remain in common usage because it is so well established. It will continue to refer to multi-component crystals where one molecule is a commonly recognised crystallisation solvent. In the end, any distinction between a solvate and a co-crystal cannot be taken too literally—one chemist's solvent may be another chemist's co-crystal former! And we probably should not start to embark on definitions of "co-crystal solvates", *etc.*

Returning to the distinction between a co-crystal and a salt, it has been sensibly stated in the literature by Childs and co-workers that this must also be viewed as a continuum.[15] The extent of charge transfer in a crystal might be variable, particularly when this refers to proton transfer between an acid and base, and in some cases it might even be dependent on conditions, especially temperature.[16] Thus, salts and co-crystals should be viewed as the end points of a continuum, again with the understanding that there is no arbitrary borderline between them. This notion of a continuum essentially implies that a co-crystal *must* be stoichiometric, since the corresponding salt must have its anionic and cationic components present in a stoichiometric ratio. The continuum will mostly be implicit in everyday discussion (as it is when chemists talk about covalent or ionic bonding), but it should be consistently invoked where there are attempts to impose arbitrary borderlines.

2.3 Supramolecular Chemistry, Intermolecular Interactions and Crystal Packing

Using modern classifications, academic research involving salts and co-crystals, particularly targeted attempts to prepare them, falls within the realm of crystal engineering. This classification has emerged only within the last 20 or so years, but it is well established today. Of course, multi-component molecular crystals were identified and utilised much earlier, but the explosive growth of crystal engineering and the realisation that it is relevant to pharmaceutical materials is a relatively modern phenomenon. It is no coincidence that this development has accompanied the step-change in the rate and ease with which crystal structures can be determined owing to the introduction of CCD-based X-ray instruments in the early 1990s. The ready availability of molecular crystal structures, and the fact that the community has been especially good at collecting and distributing them by virtue of the Cambridge Structural Database (CSD),[17] has led to great progress in our ability to understand and to control the association of molecules in the crystalline state.

The close relationship between crystal engineering and supramolecular chemistry is often expressed by the quote that a crystal is a "supermolecule par excellence".[18] This reflects the fact that the intermolecular interactions responsible for non-covalent association of molecules in the solid state are the same interactions that operate in solution or in the gas phase. If any practical distinction is to be made between the disciplines of supramolecular chemistry and crystal engineering, it is that the latter deals with the preparation of crystalline materials, while the former deals principally with molecular complexes in solution or gas phases.[‡] One advantage of the crystalline state is that crystal structures provide an exceptionally detailed geometrical picture of intermolecular interactions through X-ray diffraction analysis. However, the requirement for a given molecule to interact with numerous near neighbours within a crystal means that intermolecular interactions between pairs of molecules will not necessarily be at their energetic or geometrical optima within crystal structures. For example, the dimeric O–H···O hydrogen-bonded pair formed between two formic acid molecules in the gas phase is replaced in the crystalline state by infinite hydrogen-bonded ribbons.[19] In other words, the influence of crystal packing can distort local intermolecular interactions so that the gas-phase preference (which is most readily computed) or indeed the situation in any one established crystal structure cannot be taken as an absolute indication of the way in which molecules will arrange themselves within other crystals.

A characteristic feature of the crystalline state is the drive for close-packing of molecules, which can be understood on the basis of geometrical factors outlined by Kitaigorodski.[20] The geometrical approach assumes essentially that all intermolecular interactions are isotropic and that the observed crystal structure is the one that makes the most efficient use of space. The supramolecular approach on the other hand is chemical-based and acknowledges that molecular recognition processes depend on specific, anisotropic interactions between molecules. For strategic co-crystal or salt formation, the geometrical approach alone is of little general utility because it is very difficult to predict the formation of multi-component crystals based solely on shape complementarity. The supramolecular approach, based on chemical complementarity, is very much more useful, but it must be considered in appropriate combination with the geometrical factors that are inherent in the crystalline state.

2.3.1 Supramolecular Synthons

The supramolecular synthon concept combines the chemical elements of molecular recognition with the geometrical requirements of crystal packing. The seminal 1995 review paper of Desiraju introduced the formal literature definition of the supramolecular synthon,[21] based on analogy with Corey's definition of synthons for covalent synthesis. The explicit link between covalent synthesis and crystal engineering is stressed by this analogy, and co-crystal or

‡There is also a fundamental difference between the nature of the crystallisation process and the formation of supramolecular complexes in solution or in the gas phase: the former is a kinetically controlled process, while the latter is controlled by thermodynamics.

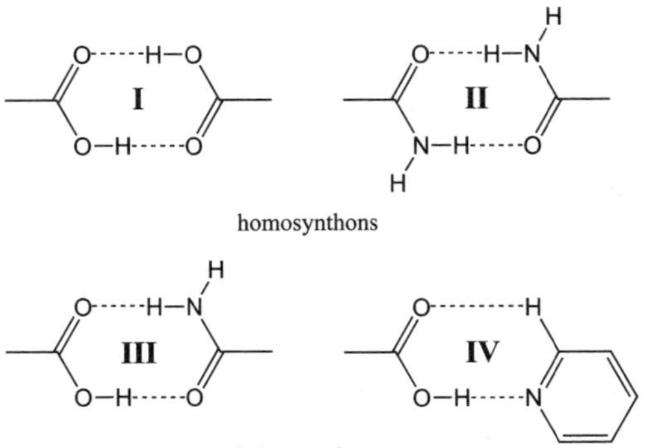

Figure 2.2 Some synthons referred to in this chapter. The term homosynthon is frequently used as the complement to heterosynthon.

salt formation is truly chemical synthesis since it produces new compounds in the crystalline state. Supramolecular synthons are defined as "spatial arrangements of intermolecular interactions" which occur frequently within crystal structures.[21] Some synthons that will be referred to consistently in this chapter are illustrated in Figure 2.2. Synthons are distinct from the interactions themselves and the definition does not make any assumption regarding the nature of the interactions or their energies. The concept is "purely probabilistic and concerned only with the frequencies of occurrence of subjectively chosen but hopefully representative patterns in crystal structures".[22] By taking such a pragmatic approach, elements of supramolecular recognition and close packing in crystals are both acknowledged and it is of no consequence whether crystallisation is under kinetic or thermodynamic control: either a particular synthon is observed in a crystal structure or it is not.[§] The most useful supramolecular synthons are those that are observed most frequently when they have the opportunity to occur—that is, if all of the functional groups necessary for formation of the synthon are present in the system. For strategic co-crystal or salt preparation, desirable synthons are those formed with a high frequency between different functional groups (commonly referred to as *heterosynthons*) when the required groups are present within the molecular components that are intended to be co-crystallised.

2.3.2 The Cambridge Structural Database

For application (and indeed the first realisation[23]) of the supramolecular synthon approach, one requirement is a comprehensive and accessible collection of molecular crystal structures that can be used to identify robust supramolecular

[§]However, given that crystallisation is kinetically controlled, supramolecular synthons can be referred to specifically as kinetically favourable patterns. They might also be thermodynamically favourable and the most robust supramolecular synthons are likely to be those that are simultaneously kinetically and thermodynamically favourable.

synthons. This function is fulfilled by the Cambridge Structural Database (CSD).[17] The importance of the CSD for crystal engineering can hardly be overstated and there are numerous reviews in the recent literature that describe its role within the research field.[24–28] In this chapter, it is relevant to describe some features of the CSD system that can be utilised to devise strategies for co-crystal or salt formation. In particular, the *Materials Mercury* module[29] facilitates searches of the CSD to identify supramolecular synthons involving particular combinations of functional groups (a *motif search*), or to identify crystal structures containing a particular geometrical arrangement of atoms (a *crystal packing feature*). The *Materials Mercury* module provides a straightforward interface for performing such searches on a practical timescale.

An example of a motif search might consider the carboxylic acid–amide synthon, denoted III in Figure 2.2, which is available as a pre-defined motif within *Materials Mercury*. A search of the entire CSD (Version 5.32, November 2010, 525095 entries) identifies this motif in 137 structures and provides a frequency of occurrence (defined as the number of instances of the motif divided by the number of structures that contain the two functional groups) of 47.2% (Figure 2.3). This statistic encompasses both single- and multi-component molecular crystals. If the purpose is to guide co-crystallisation strategies, the set of structures against which the search is performed can be limited to those containing more than one distinct molecular component and further limited to structures that have the acid and amide groups on different molecules. In this case, a search of 190 relevant structures identifies the motif in 109 cases, amounting to a frequency of 59.9%. On the basis of these results, synthon III can be expected to be realised within a co-crystal with a frequency of *ca.* 60% when molecules containing these two functional groups are co-crystallised. These results could be further analysed to identify particular circumstances where the synthon is *not* frequently realised. For example, 64 of the 190 structures contain nicotinamide or isonicotinamide (Figure 2.4) and synthon III is present in only 18 (28.1%) of these. A rapid screen of the 64 structures shows that all of them contain the acid–pyridine synthon denoted IV in Figure 2.2, while 44 (68.8%) contain the amide–amide synthon II.** Thus, a co-crystallisation strategy based on the expected formation of synthon III is likely to be significantly less robust if one of the molecules also contains a pyridyl group. This example is discussed further later in the chapter.

Motif searches, defined by specific interactions between particular functional groups, are conceptually based on the supramolecular aspects of crystal packing. They permit examination of *potential* crystal packing features on the basis of the chemical groups present in the molecules before any (co-)crystallisation is carried out. For an established crystal structure,

**The two structures that do not contain either synthon II or synthon III have CSD refcodes MELYEI[30] and SODDOF.[31] The first of these is a hydrate, in which a water molecule donates hydrogen bonds to the carbonyl group of the amide and to the carbonyl group of one carboxyl unit. The second structure contains salicylic acid, which accepts one hydrogen bond from the amide NH_2 group at its OH group, while the amide molecules form a catemeric N–H···O hydrogen-bond motif.

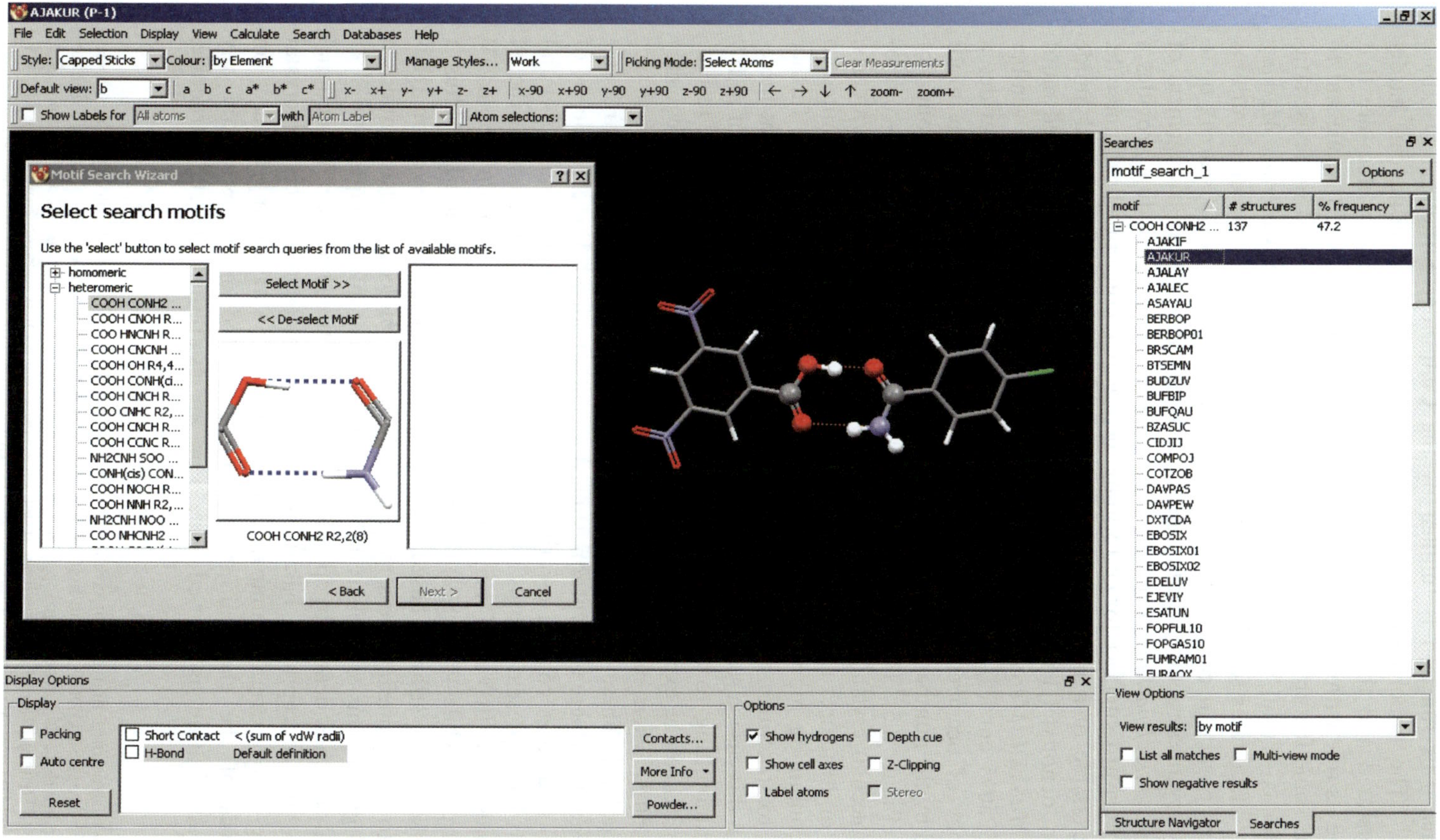

Figure 2.3 Screenshot taken from the *Materials Mercury* program[29] during a motif search for synthon III. The selection of pre-defined motifs is visible on the left, one structure with the identified motif highlighted is visible in the middle and the full list of search results is visible on the right, with accompanying frequency of observation for the motif. Reproduced by permission of the Cambridge Crystallographic Data Centre.

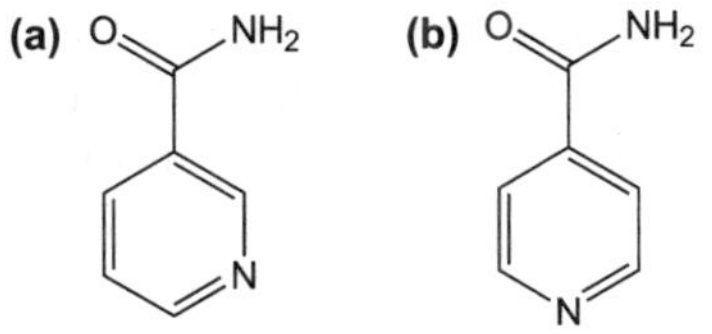

Figure 2.4 Chemical structures of (a) nicotinamide and (b) isonicotinamide.

Materials Mercury also allows convenient searches of the CSD to identify the frequency of any particular observed packing feature, incorporating geometrical information as well as chemical information. This provides an opportunity to establish new synthons by defining a particular geometrical arrangement of atoms observed in a given crystal structure and asking how often that arrangement is observed in other crystal structures. Of course, selecting, for example, an acid–amide dimer using its constituent atoms within an established crystal structure will achieve the same results as the motif search described previously. The distinction is that the motif search implicitly defines default geometrical criteria so that the search can be performed solely by specification of the chemical features.

2.3.3 Supramolecular Yield

The previous discussion about the frequency of observation of synthons within co-crystal structures does not necessarily reflect the practical robustness of a particular heterosynthon, because it is based only on the frequency of occurrence within established co-crystals containing the functional groups necessary to construct the synthon. It could be that two molecular components containing a selected combination of functional groups crystallise separately rather than together. A more meaningful practical statistic might consider the frequency with which co-crystallisation occurs to give a structure containing the targeted heterosynthon compared to all co-crystallisation attempts using suitable molecular pairs. Of course, this can only ever be a practical statistic because the number of suitable molecular pairs is essentially infinite and the co-crystallisation outcome is frequently dependent on the practical conditions, for example the choice of solution crystallisation or mechanical grinding (see Chapter 9).

Aside from these fundamental limitations, it is difficult to establish any general guidelines because negative co-crystallisation results are reported less consistently in the academic literature and they are very much harder to collate in a systematic way. Statistics based on known crystal structures (and possibly robust theoretical predictions of hypothetical crystal structures) therefore provide the best practical guideline that is available: frequent observation of a particular heterosynthon in the CSD establishes it to be viable within co-crystals and shows that co-crystallisation of molecules containing the constituent functional groups frequently can compete with separate crystallisation of the two molecular components. Thus, synthon frequency established from the CSD often provides an adequate indication of an anticipated co-crystallisation outcome and forms the basis of most co-crystal or salt design strategies.

In the literature, Aakeröy and co-workers have frequently referred to supramolecular yield as a measure of the "experimental regime where a particular synthon prevails despite competition from other non-covalent forces".[7] It is not clear whether this concept is (or can be) defined in any rigorous way. It does not seem to be analogous to the yield in covalent synthesis, since the latter refers to the outcome of a particular reaction with respect to a particular product rather than the frequency with which a particular reaction type occurs compared to all chemical circumstances where it has the potential to occur. It is also not clear whether the supramolecular yield refers solely to the frequency with which a synthon prevails within characterised crystals (in which case it is just an elegant label for everything that has been discussed so far) or whether it also involves failed co-crystallisation attempts (in which case it is dependent on the number and variety of co-crystallisation attempts that are made). Therefore, "supramolecular yield" should probably be viewed solely in practical terms: if ten attempted co-crystallisations based on a particular targeted heterosynthon under particular experimental conditions are successful ten times, that can be said to be a high-yielding supramolecular reaction that is useful practically.

2.4 Design Strategies for Salts and Co-crystals

2.4.1 Synthon Hierarchy and Synthon Interference

In order to implement targeted co-crystal or salt design, one of the necessary lines of investigation over the last decade or so has been to examine experimentally the robustness of particular heterosynthons in the presence of other functional groups that could potentially form alternative synthons. This has been presented variously in terms of establishing a *synthon hierarchy* or investigating the nature of *synthon interference*. This work can be viewed in general to populate the CSD with the necessary data for meaningful statistical assessment of synthon frequency within a broad range of compounds. It frequently provides surprises and it has also provided some examples of inspired design. In particular, the targeted realisation of ternary co-crystals (Section 2.4.2) falls into the latter category. The number of studies that could potentially be mentioned in this section is enormous and the few examples that are given are necessarily limited. They are deliberately selected from some of the most prominent groups in the academic literature (in order to highlight who these groups have been) and they are intended to represent the types of studies that have commonly been carried out. They are also restricted principally to a limited set of compounds and synthons (specifically, those in Figure 2.2) in an effort to illustrate how cumulative knowledge of a particular class of compounds can lead to effective co-crystal or salt design strategies.

The first selected example is taken from the research of Zaworotko and co-workers, involving the hydroxyl group as a hydrogen-bond donor and either pyridine or a cyano group as a hydrogen-bond acceptor.[32] In 2007, analysis of the CSD showed that hydroxyl groups form hydrogen bonds to pyridine in more than 99% of crystal structures (single-component or multi-component) containing the two groups and no other conventional hydrogen-bonding groups. The corresponding statistic for the occurrence of hydroxyl–cyano hydrogen bonds was *ca.* 77%, indicating that both heterosynthons compete favourably with the competing hydroxyl–hydroxyl homosynthon.

The practical utility of this observation for targeted co-crystal formation was then proven by synthesis of 17 co-crystals using molecules containing various permutations of OH, pyridine and CN functional groups (Figure 2.5). Each co-crystal contains the hydroxyl–pyridine synthon, establishing this firmly to be at the top of the synthon hierarchy: OH–pyridine > OH–cyano > OH–OH. Other notable features of this study were the application of a variety of preparation techniques to explore as comprehensively as possible whether co-crystallisation can be achieved, together with the fact that the authors document failed co-crystallisation trials: all attempts to co-crystallise 3-hydroxypyridine or 5-hydroxyisoquinoline (Figure 2.6) with any cyano-containing molecule failed

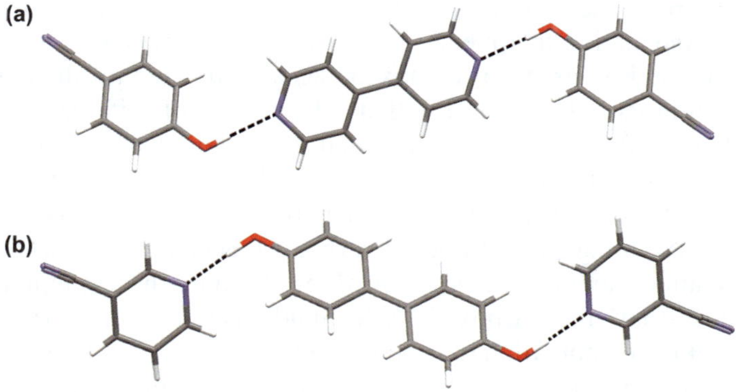

Figure 2.5 Hydrogen-bonded units within the co-crystals (a) 4-cyanophenol:4,4′-bipyridine (2:1) and (b) 4,4′-biphenol:3-cyanopyridine (1:2), established by OH–pyridine synthons.[32]

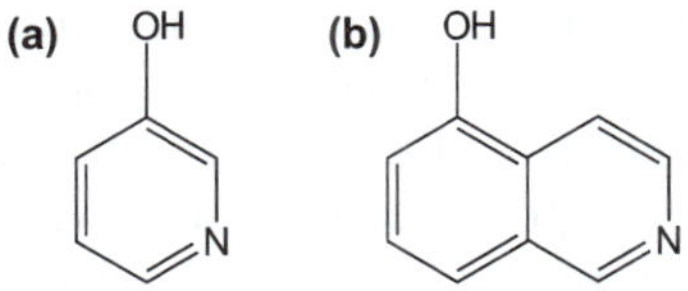

Figure 2.6 Chemical structures of (a) 3-hydroxypyridine and (b) 5-hydroxyisoquinoline. Both molecules contain OH and pyridine groups and they form OH–pyridine synthons in their individual crystal structures.

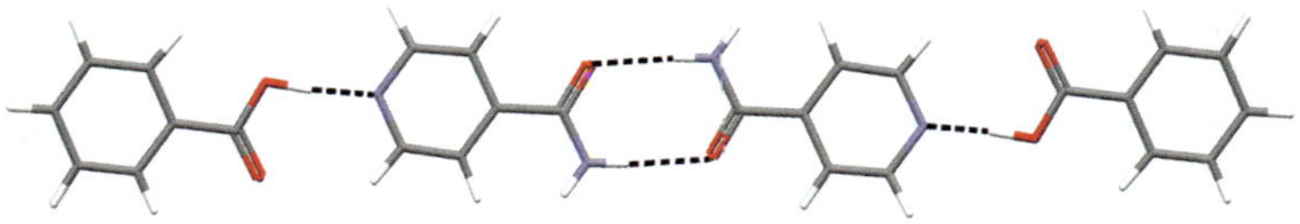

Figure 2.7 Hydrogen-bonded unit within the co-crystal benzoic acid:isonicotinamide (1:1), containing synthons II and IV.[3]

to produce any co-crystal. This was rationalised by the fact that these molecules contain both OH and pyridine groups in a single molecule, so that the single-component crystal should be favoured according to the established synthon hierarchy.

A second illustrative example is related to the carboxylic acid–nicotinamide/isonicotinamide co-crystals introduced in Section 2.3.2. In 2001, Aakeröy and co-workers reported a 1:1 co-crystal of isonicotinamide and benzoic acid.[3] It contains molecules of isonicotinamide joined by homosynthon II, with carboxylic acid molecules forming hydrogen bonds to the pyridine groups of each isonicotinamide to construct heterosynthon IV (Figure 2.7). This is consistent with a hierarchy of interactions where the best hydrogen-bond donor (COOH) combines with the best acceptor (pyridine N), leaving the second best donor and acceptor ($CONH_2$) to hydrogen bond to each other. In seeking to establish the general robustness of this pattern, the same researchers subsequently reported a systematic experimental study of twelve co-crystals containing isonicotinamide and a variety of carboxylic acids.[33] In the introduction to that paper, the authors note specifically that "the paucity of suitable crystal structures containing all three groups means that we are not able to establish reliable patterns of behaviour based on existing crystallographic data", which alludes to the role of these types of studies for populating the CSD. All twelve co-crystals were found to contain the carboxylic acid–pyridine heterosynthon IV and the amide–amide homosynthon II, establishing this combination to be more robust than the carboxylic acid–amide heterosynthon III and also showing that the combination of synthons II and IV is robust in the presence of other functional groups in the various carboxylic acid molecules.

In a subsequent study, Nangia and co-workers utilised the same combination of synthons to prepare co-crystals of isonicotinamide with a series of five α,ω-alkanedicarboxylic acids.[34] The expected 2:1 co-crystals, comprising isonicotinamide dimers linked into ribbons by diacid molecules, were formed in all five cases (Figure 2.8). However, 1:1 co-crystals including heterosynthons III and IV were also identified for two of them (glutaric and adipic acid; Figure 2.9). Thus, the robustness of heterosynthon IV is emphasised, but the carboxylic acid–amide heterosynthon III is shown to compete with the amide–amide synthon II to produce co-crystals with differing stoichiometry. These observations were rationalised on the basis of (gas-phase) energy calculations for the isolated molecules and for the hydrogen-bonded pairs: the energy of the synthon is defined as the energy of the hydrogen-bonded pair minus the sum of the energies of the two isolated molecules. The calculations

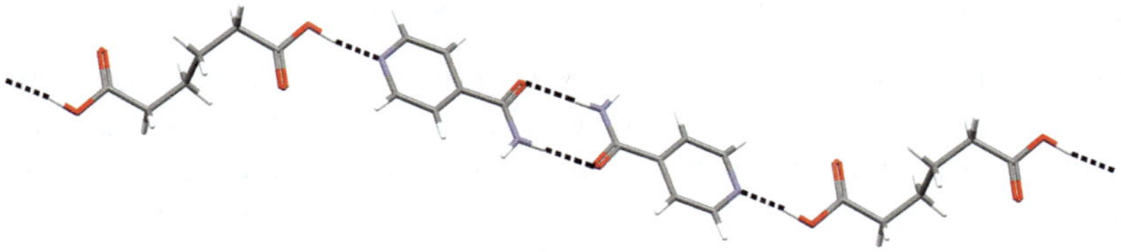

Figure 2.8 Fragment of a hydrogen-bonded ribbon within the co-crystal adipic acid:isonicotinamide (1:2), containing synthons II and IV.[34]

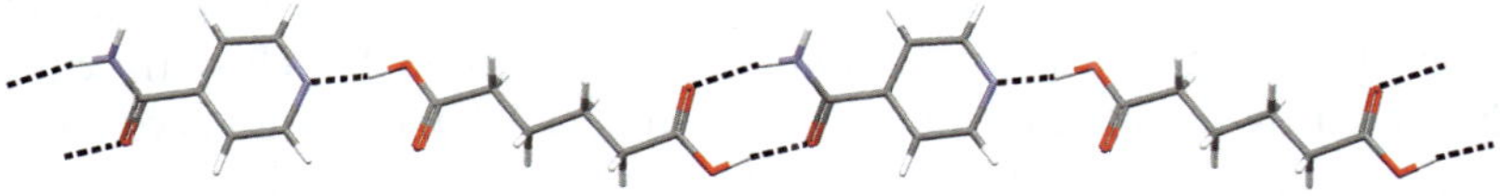

Figure 2.9 Fragment of a hydrogen-bonded ribbon within the co-crystal adipic acid:isonicotinamide (1:1), containing synthons III and IV.[34]

suggest that synthon IV is significantly more stabilising than synthon III for strong carboxylic acids, but that the difference is much less for weaker carboxylic acids. This is consistent with the fact that synthon III is observed for glutaric and adipic acid, since these are the weakest acids amongst the set. This study provides an illustration of how the synthon hierarchy might be modulated in specific chemical circumstances and how this might influence the resulting co-crystal stoichiometry.

A final illustrative example within the same theme comes from Seaton and co-workers,[35] who reconsidered the 1:1 co-crystal of isonicotinamide with benzoic acid (Figure 2.7) and showed also that a 2:1 co-crystal can be formed, containing synthons III and IV, comparable to the 1:1 co-crystals described by Nangia for glutaric and adipic acid (Figure 2.9). Only the 1:1 co-crystal was obtained if the co-crystallisation was carried out in ethanol solution, while either co-crystal stoichiometry could be obtained from aqueous or methanol solution by varying the initial composition. Energy calculations for the molecular pairs in this case considered the influence of solvent, and the relative energies of synthons II, III and IV showed some dependence on the solvent system involved. However, the differences were relatively small and do not account for the dominance of the 1:1 co-crystal from ethanol. More significant are the ternary phase diagrams, which show that the regions of growth of the 2:1 co-crystal are restricted compared to those of the 1:1 co-crystal, so that the 2:1 co-crystal can be obtained only from solutions with a relatively high level of benzoic acid (mole fraction >0.9), and generally as a mixture with other phases (either the 1:1 co-crystal or the individual components). This reminder of the thermodynamic aspects of co-crystallisation (to be taken up in earnest in later chapters of this book) illustrates that structural factors alone are not always sufficient to account for co-crystallisation outcomes.

2.4.2 Ternary and Quaternary Co-crystals

Apparently the first deliberately designed ternary co-crystals came from the work of Aakeröy with isonicotinamide.[3] The key to the design strategy is that isonicotinamide offers two distinct functional groups with different hydrogen-bond acceptor strengths. The pyridine N atom is the strongest hydrogen-bond acceptor and it therefore should combine preferentially with the strongest available hydrogen-bond donor. In the isonicotinamide–benzoic acid 1:1 co-crystal, the strongest donor is the carboxylic acid, which leaves the amide groups of isonicotinamide to combine with each other by default. To prepare a ternary co-crystal requires two different carboxylic acid molecules, one of which should form synthon IV, and the second of which should form an acid–amide synthon III in preference to the amide–amide synthon II. This strategy was implemented by targeted formation from three 1:1:1 ternary co-crystals using 3,5-dinitrobenzoic acid as the strong hydrogen-bond donor and either 3-methylbenzoic acid, 4-(dimethylamino)benzoic acid or 4-hydroxy-3-methoxycinnamic acid as the weaker hydrogen-bond donor (Figure 2.10). With reference to the supramolecular synthon approach, the strategy goes beyond the general probabilistic nature of synthon occurrence by tuning the expected interaction

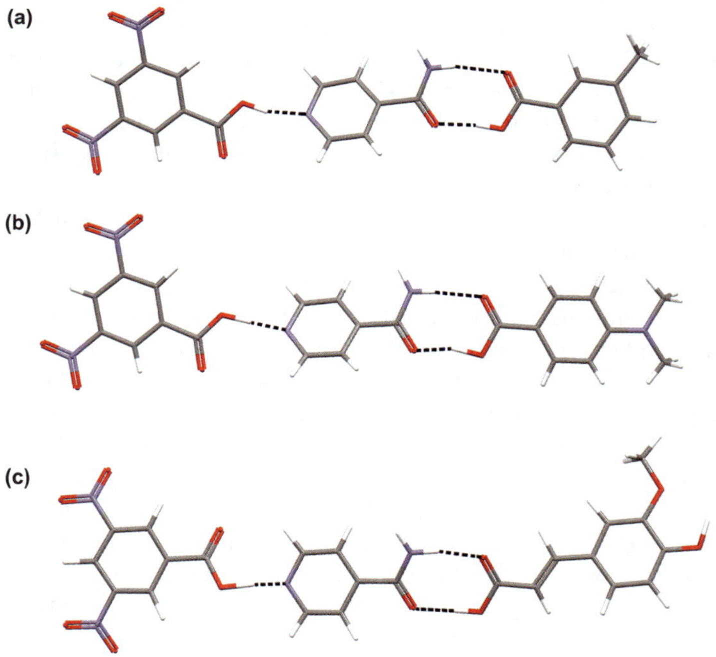

Figure 2.10 Hydrogen-bonded units within the ternary co-crystals (a) 3,5-dinitrobenzoic acid:isonicotinamide:3-methylbenzoic acid (1:1:1), (b) 3,5-dinitrobenzoic acid:isonicotinamide:4-(*N*,*N*-dimethylamino)benzoic acid (1:1:1), and (c) 3,5-dinitrobenzoic acid:isonicotinamide:4-hydroxy-3-methoxycinnamic acid (1:1:1), incorporating isonicotinamide and carboxylic acids with different hydrogen-bond donor strengths.[3]

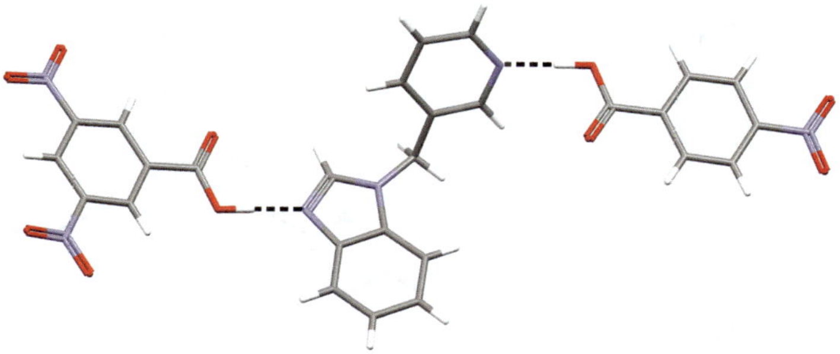

Figure 2.11 Hydrogen-bonded unit within the ternary co-crystal 3,5-dinitrobenzoic acid:4-nitrobenzoic acid:1-((3-pyridyl)methyl)benzimidazole (1:1:1), containing a multi-functional base combined with two different carboxylic acids.[4]

strengths in such a way as to control the synthon probabilities within this specific chemical system. Another ternary system designed on the same principles involves a multi-functional base incorporating pyridine and benzimidazole N atoms (Figure 2.11), which act as hydrogen-bond acceptors with different strengths.[4] It should be noted that the basis of the control in these cases is thermodynamic, which need not necessarily be the origin of the frequent occurrence of a particular synthon. In other words, this elegant hierarchical design approach could still very well be thwarted by uncontrollable crystallisation kinetics.

Ternary co-crystals generated from a multi-functional acid molecule combined with two different bases have also been realised by Nangia and co-workers.[5] Co-crystallisation of 1,3,5-cyclohexane tricarboxylic acid with 4,4′-bipyridine and 4,4′-bipyridine-*N*-oxide results in a ternary co-crystal sustained through synthon IV and O–H···O hydrogen bonds to the O atom of the *N*-oxide (Figure 2.12). Water molecules are also included in the structure, so it is actually a four-component system, although inclusion of water obviously was not an inherent part of the design strategy.

A similar approach incorporating the same tricarboxylic acid molecule and two different types of bipyridine molecules has provided several other ternary co-crystals.[36] In the latter system, several of the co-crystals also incorporate molecules of the crystallisation solvents (Figure 2.13), including *ortho*- and *para*-xylene, *ortho*- and *para*-dichlorobenzene, methoxybenzene (anisole) and benzyl alcohol, and they are referred to explicitly as quaternary co-crystals. Again, incorporation of the fourth molecular component is not based on any specific design strategy and it bears all of the attributes that would be expected for a "solvate" or a "clathrate": the solvent molecules do not form any specific directional interactions within the crystal; they display crystallographic disorder and the seven reported examples are isostructural (although the crystal containing benzyl alcohol apparently has one crystallographic axis doubled compared to the other cases). Aside from the semantic discussion in Section 2.2, it is clear that the level of genuine synthetic control in all of these cases is

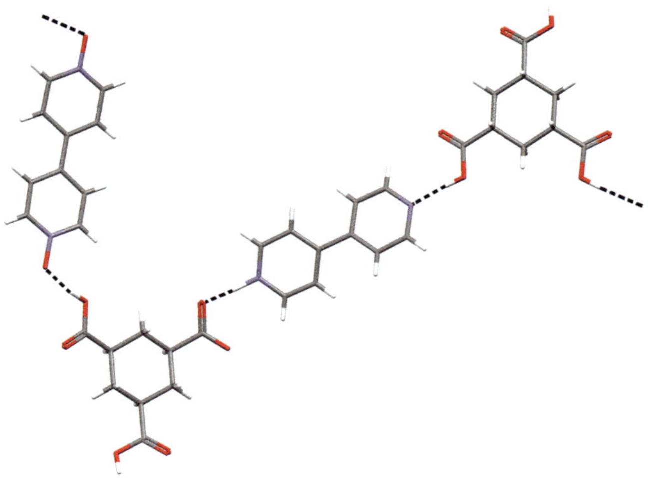

Figure 2.12 Hydrogen-bonded unit within the hydrated ternary co-crystal cyclohexane-1,3,5-tricarboxylic acid:4,4′-bipyridine: 4,4′-bipyridine-*N*,*N*′-dioxide (2:1:1) dihydrate.[5] In the crystallographic model, proton transfer is considered to occur for one O–H···N hydrogen bond.

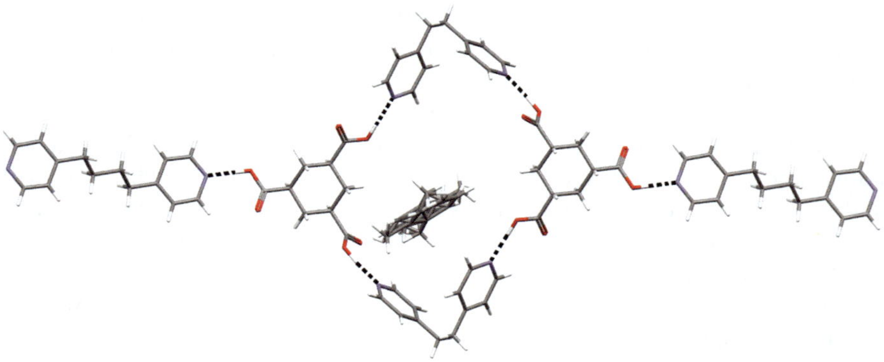

Figure 2.13 Hydrogen-bonded unit within the quaternary co-crystal cyclohexane-1,3,5-tricarboxylic acid:1,2-bis(4-pyridyl)ethane:1,4-bis(4-pyridyl)butane:*ortho*-xylene (2:2:1:1).[36] The *ortho*-xylene molecules are disordered in cavities formed by the hydrogen-bonded components.

associated with targeted design of the ternary system and this appears to date to be the limit to which the hierarchical synthon approach has been successfully applied.

2.4.3 Salts or Co-crystals?

For co-crystal design strategies based on strong hydrogen bonds (as the vast majority are), there is a possibility that the proton involved in the hydrogen-bonding interaction could be transferred from the donor (acid) to the acceptor (base) to form a salt. To predict the final ionisation states of the molecules in a multi-component crystal, *i.e.* to establish whether a co-crystal or a salt might be formed, the acid ionisation constant, pK_a, is the measure that is most commonly considered.[37] Although pK_a values are properly defined only in solution under equilibrium conditions, salt formation in the solid state is established by experiment to occur reliably when the difference in pK_a is sufficiently large. A common guideline is the "rule of three", which states that a salt should be expected when the difference in pK_a between the acid and the conjugate acid of the base is greater than 3 units. Refinements to this rule have considered that a "grey area" can exist, and that the lower limit required for reliable expectation of a co-crystal should be a pK_a difference of 2.[38] These empirical rules must be treated with some caution, however, since it is well established that proton transfer behaviour in molecular crystals is influenced by the crystalline environment and temperature.[39,40]

To provide one practical example of the complexity of the situation, proton transfer in the carboxylic acid–pyridine synthon has been investigated systematically by Tocher and co-workers through co-crystallisation of pyridine or 4-dimethylaminopyridine (DMAP) with five different carboxylic acids (Table 2.1).[9] Since the pK_a value for the conjugate acid of DMAP ($pK_a = 9.70$) is considerably larger than that of any of the carboxylic acids, salts should be expected for all crystals containing DMAP and this was indeed found to be the case. However, two different crystal forms were established for the fumarate salt, both of which also contained neutral fumaric acid molecules.

Table 2.1 Summary of the results obtained for co-crystallisation of pyridine or 4-dimethylaminopyridine (DMAP) with five different carboxylic acids.[9]

	Pyridine (py) *(pK_a for Hpy = 5.14)*	*4-Dimethylaminopyridine (DMAP)* *(pK_a for HDMAP = 9.70)*
Maleic acid		
$pK_a(1) = 1.92$	acid:py (1:2) co-crystal[a]	$acid^-:DMAP^+$ (1:1) salt
$pK_a(2) = 6.23$		
Fumaric acid		
$pK_a(1) = 3.02$		$acid:acid^{2-}:DMAP^+$ (1:1:2) salt
$pK_a(2) = 4.38$		$acid:acid^{2-}:DMAP^+$ (2:1:2) salt
Phthalic acid		
$pK_a(1) = 2.98$	$acid^-:py^+$ (1:1) salt	$acid^-:DMAP^+$ (1:1) salt
$pK_a(2) = 5.14$		
Isophthalic acid		
$pK_a(1) = 3.46$	acid:py (1:1) disordered	$acid^{2-}:DMAP^+$ (1:2) salt
$pK_a(2) = 4.46$		
Terephthalic acid		
$pK_a(1) = 3.41$	acid:py (1:2) co-crystal	$acid^{2-}:DMAP^+$ (1:2) salt
$pK_a(2) = 4.82$		

[a]Maleic acid and fumaric acid provide the same fumaric acid:pyridine (1:2) co-crystal due to isomerisation of maleic acid in solution.

For pyridine ($pK_a = 5.14$), the expectation on the basis of pK_a values is much less clear and the resulting multi-component crystals were characterised as three co-crystals, one salt and one structure including disordered H atoms.[8] Thus, even this limited grid demonstrates the difficulties associated with predicting salt or co-crystal formation.

Analysis of the resulting crystal structures revealed no significant distinction between the molecular structures in the salts compared to the co-crystals and the pyridine–carboxylic acid synthons could be considered to operate equally effectively in a structure-determining sense whether they are formally ionised or not. Of course, this is a good illustration of the continuum between salts and co-crystals and the authors specifically questioned whether the positions of the protons mattered at all in these types of structures. One significant difference was found in the computed lattice energies of the structures: on energy minimisation, the majority of the crystallographically ordered structures were reproduced significantly less well if the location of the proton on the acid or base was chosen incorrectly (*i.e.* an established salt structure was modelled as an ordered co-crystal, or vice versa). Thus, for computational modelling of salts and co-crystals (at least using the common force field-based approaches), the assignment as a salt or co-crystal can significantly influence the results.

2.5 Concluding Remarks

This chapter has concentrated on fundamental aspects of molecular recognition and crystal packing, which can be applied to strategic preparation of pharmaceutical co-crystals and salts. Thermodynamic factors associated with the formation of multi-component crystals have not been specifically considered since these are addressed later in this book. Although co-crystallisation must ultimately be controlled by the rules of thermodynamic and/or kinetics, the structural approach described in this chapter, based on the frequency of observation of heterosynthons within known crystal structures, often suffices to produce the intended outcome. The academic literature now contains a great many successful examples of rational co-crystallisations, including specifically designed ternary co-crystals, only a very small number of which have been based on any "proper" consideration of thermodynamics. This is a testament to the probabilistic nature of the supramolecular synthon approach and also to the effectiveness of the Cambridge Structural Database as a tool for identifying robust supramolecular synthons in crystal structures. For a given API, containing a certain combination of functional groups that might be targeted for heterosynthon formation, a specific and comprehensive CSD survey can be completed in a matter of hours (at most) and a small number of probable heterosynthons can be identified. This in turn can provide a relatively small list of (pharmaceutically acceptable) co-crystal formers that can be applied within targeted experimental trials. Since the supramolecular synthon approach is probabilistic in its nature, experimental trials planned in this way will always be genuine "trials" and some of them should be expected not to produce the intended co-crystal or salt outcome. If the targeted heterosynthon is sufficiently robust, however, *some* of these trials should be expected to provide the targeted multi-component crystal. With careful choice of the co-crystal formers, the level of targeted control should be quite high for binary systems, and in specific cases might be extendable to ternary systems.

References

1. *IUPAC Compendium of Chemical Terminology*, ed. A. D. McNaught and A. Wilkinson, Royal Society of Chemistry, Cambridge, 2nd edn,1997.
2. http://www.fda.gov/Food/FoodIngredientsPackaging/GenerallyRecognized-asSafeGRAS/default.htm.
3. C. B. Aakeröy, A. M. Beatty and B. A. Helfrich, *Angew Chem. Int. Ed.*, 2001, **40**, 3240.
4. C. B. Aakeröy, J. Desper and J. F. Urbina, *Chem. Commun.*, 2005, 2820.
5. B. R. Bhogala, S. Basavoju and A. Nangia, *CrystEngComm*, 2005, **7**, 551.
6. Ö. Almarsson and M. J. Zaworotko, *Chem. Commun.*, 2004, 1889.
7. C. B. Aakeröy and D. J. Salmon, *CrystEngComm*, 2005, **7**, 439.
8. S. H. Dale, M. R. J. Elsegood, M. Hemmings and A. L. Wilkinson, *CrystEngComm*, 2004, **6**, 207.
9. S. Mohamed, D. A. Tocher, M. Vickers, P. G. Karamertzanis and S. L. Price, *Cryst. Growth Des.*, 2009, **9**, 2881.
10. A. D. Bond, *Chem. Commun.*, 2003, 250.
11. A. D. Bond, *CrystEngComm*, 2006, **8**, 333.
12. G. R. Desiraju, *CrystEngComm*, 2003, **5**, 466.
13. J. D. Dunitz, *CrystEngComm*, 2003, **5**, 506.
14. A. D. Bond, *CrystEngComm*, 2007, **9**, 833.
15. S. L. Childs, G. P. Stahly and A. Park, *Mol. Pharmaceutics*, 2007, **4**, 323.
16. A. Parkin, C. C. Seaton, N. Blagden and C. C. Wilson, *Cryst. Growth. Des.*, 2007, **7**, 531.
17. F. H. Allen, *Acta Cryst. Sect. B*, 2002, **58**, 380.
18. J. D. Dunitz, *Pure Appl. Chem.*, 1991, **63**, 177.
19. I. Nahringbauer, *Acta Cryst. Sect. B*, 1978, **34**, 315.
20. A. I. Kitaigorodski, *Molecular Crystals and Molecules*, Academic Press, New York, 1973.
21. G. R. Desiraju, *Angew. Chem. Int. Ed.*, 1995, **34**, 2311.
22. G. R. Desiraju, *Angew. Chem. Int. Ed.*, 2007, **46**, 8342.
23. G. R. Desiraju, *Nature*, 2004, **431**, 25.
24. F. H. Allen and W. D. S. Motherwell, *Acta Cryst.*, 2002, **B58**, 407.
25. A. Nangia, *CrystEngComm*, 2002, **4**, 93.
26. J. A. Chisholm, E. Pidcock, J. van de Streek, L. Infantes, W. D. S. Motherwell and F. H. Allen, *CrystEngComm*, 2006, **8**, 11.
27. A. D. Bond in *Organic Crystal Engineering*, ed. E. R. T. Tiekink, J. J. Vittal and M. J. Zaworotko, Wiley, Chichester, 2010, p. 1.
28. C. R. Groom and F. H. Allen, *Future Medicinal Chemistry*, 2010, **2**, 933.

29. C. F. Macrae, I. J. Bruno, J. A. Chisholm, P. R. Edgington, P. McCabe, E. Pidcock, L. Rodriguez-Monge, R. Taylor, J. van de Streek and P. A. Wood, *J. Appl. Cryst.*, 2008, **41**, 466.
30. H. Aghabozorg, M. Ghadermazi and J. A. Gharamalek, *Acta Cryst. Sect. E*, 2006, **62**, o3445.
31. D. J. Berry, C. C. Seaton, W. Clegg, R. W. Harrington, S. J. Coles, P. N. Horton, M. B. Hursthouse, R. Storey, W. Jones, T. Friščić and N. Blagden, *Cryst. Growth. Des.*, 2008, **8**, 1697.
32. J. A. Bis, P. Vishweshwar, D. Weyna and M. J. Zaworotko, *Mol. Pharmaceutics*, 2007, 4, 401.
33. C. B. Aakeröy, A. M. Beatty and B. A. Helfrich, *J. Am. Chem. Soc.*, 2002, **124**, 14425.
34. P. Vishweshwar, A. Nangia and V. M. Lynch, *Cryst. Growth Des.*, 2003, **3**, 783.
35. C. C. Seaton, A. Parkin, C. C. Wilson and N. Blagden, *Cryst. Growth Des.*, 2009, **9**, 47.
36. B. R. Bhogala and A. Nangia, *New J. Chem.*, 2008, **32**, 800.
37. Z. J. Li, Y. Abramov, J. Bordner, J. Leonard, A. Medek and A. V. Trask, *J. Am. Chem. Soc.*, 2006, **128**, 8199.
38. Q. Tong and G. Whitesall, *Pharma. Rev. Tech.*, 1988, **3**, 215.
39. G. P. Stahly, *Cryst. Growth Des.*, 2007, **7**, 1007.
40. T. Steiner, I. Majerz and C. C. Wilson, *Angew. Chem. Int. Ed.*, 2001, **40**, 2651.

CHAPTER 3

Role of Fluorine in Weak Interactions in Co-crystals

SEETHA LEKSHMI SUNIL,[a] SUSANTA K. NAYAK,[a] VENKATESHA R. HATHWAR,[a] DEEPAK CHOPRA[b] AND TAYUR N. GURU ROW[a]*

[a] Solid State and Structural Chemistry Unit, Indian Institute of Science, Bangalore 560012, India; [b] Department of Chemistry, Indian Institute of Science Education and Research, Bhopal, 462023, India

3.1 Introduction

According to Jack Dunitz "Weak interactions are not only important in modern supramolecular chemistry, they hold the organic world together and are responsible for the very existence of liquids and solids; a characteristic feature of weak interactions is the phenomenon of enthalpy–entropy compensation".[1] Such weak interactions are the predominant factors in the formation of salts and co-crystals and it is obvious that a clear understanding of these interactions, in terms of their geometry and energy, is crucial in the design of new pharmaceutical compounds. Hydrogen bonding is the most sought after interaction and its relevance in this context is well explored.

Fluorine occupies a unique position in the periodic table and forms the strongest known single bonds with boron, carbon, silicon, and hydrogen. With carbon the bond energy increases with the degree of fluorination, a feature not found in other halogen atoms.[2] It is also observed that

RSC Drug Discovery Series No. 16
Pharmaceutical Salts and Co-crystals
Edited by Johan Wouters and Luc Quéré

Published by the Royal Society of Chemistry, www.rsc.org

fluorination increases the strength of C–C single bonds while decreasing the strength of the double bonds. Several explorations involving fluorinated organic compounds indicate that pharmacokinetic properties are optimized with the addition of fluorine which has resulted in the formation of a significantly large number of fluorinated drugs in clinical use. Further, physicochemical properties such as metabolic stability, lipophilicity, pK_a can be engineered to arrive at better drug availability and enhancement of activity. Introduction of a fluorine atom to the drug fragment increases the lipophilicity, which in turn improves the penetration through lipid membranes and tissues, consequently favoring the intracellular delivery of the drug. The substitution of hydrogen by fluorine generally does not influence steric effects although their van der Waals radii (1.20 and 1.47 Å, respectively) differ to a significant extent. In fact, multiple fluorination of aromatic rings leads to changes in quadrupole moment which significantly influences intermolecular interactions involving fluorine. Among the numerous marketed pharmaceuticals in the world, more than 150 drug compounds are fluorinated.[3] The relatively huge number of drugs compared with other halogen-containing pharmaceuticals is interesting, particularly the fact that organochlorine and organobromine compounds are found abundantly in natural products. It is noteworthy that the activity of modern crop protection products has increased to a great extent with the introduction of fluorine-substituted moieties. In addition, there has been considerable increase in the number of commercial products containing mixed halogens.[4]

The isosteric substitution of hydrogen by a fluorine atom produces geometrically analogous compounds and this feature satisfies steric requirements at enzyme receptor sites. In this context, the development of new and active drugs based on "structure based drug design, (SBDD)" has induced a new era in drug discovery and generated considerable interest.[5] Intermolecular interactions of varying strengths such as van der Waals, electrostatic, hydrogen bonds *etc.* are essentially involved in the binding of a drug to its receptor. The C–F bond is longer, has an opposite bond polarity and is stronger than the C–H bond, which makes the molecule more resistant towards metabolic degradation. All these observations suggest that an understanding of the nature of the interactions involving organic fluorine is essential. In this chapter, a brief discussion of the requirements of the basic tools used in the supramolecular assembly of fluorine-containing compounds and the importance of quantifying interactions involving organic fluorine is highlighted in terms of the results obtained from the analysis of the charge density distribution.

Understanding of intermolecular interactions in molecular crystals in terms of "synthons" forms the basis of "crystal engineering".[6,7] Crystallization of organic compounds can lead to different forms of solids such as polymorphs, hydrates, solvates, salts, and co-crystals.[8] These different forms of solid exhibit variable physico-chemical properties such as stability, crystal shape, compressibility, and density without changing its intrinsic chemical structure.[9,10] In the pharmaceutical industry, these various solid forms of active pharmaceutical ingredients (APIs) can be used to improve physico-chemical properties and thereby produce significant improvements in the API's performance as a final deliverable drug in the market.[11–13]

3.1.1 Co-crystals and Salts

It is well known that if a given compound in its ionic form is crystallized with a counter ion, the resulting molecular complex could result in salt formation.[14,15] Depending on the extent of polarization, the two components could either remain as a molecular complex or become a salt. Co-crystallization offers an option that has enormous potential to provide new stable structures that may improve the properties of an API. To date, a universal and agreeable definition of what constitutes a "co-crystal" is still unavailable and several controversial views on this term have been expressed by several scientists.[16,17] It is generally understood that a co-crystal is nothing but a multi-component solid form made up of neutral molecules at ambient conditions, whereas a salt is made of any pair of ionized molecules. In both cases it becomes essential to correlate a definite identification of the hydrogen atom position by crystallography with other physico-chemical parameters.

Salt formation is one of the primary solid state approaches used to modify the physical properties of APIs and it is estimated that over half of the medicines available in the market are administered as salts. However, co-crystals offer a different pathway, where any API, regardless of acidic, basic, or ionizable groups, could potentially be co-crystallized with pharmaceutically acceptable compounds. This aspect helps complement existing methods by reintroducing molecules that have limited pharmaceutical profiles based on their non-ionizable functional groups. In addition, the number of potentially non-toxic co-crystal formers (or co-formers) that can be incorporated into a co-crystalline reaction is numerous. Co-crystallization improves mechanical properties, for example, layered co-crystal forms of non-ionizable API paracetamol have been generated through liquid assisted grinding. The resulting crystal structure is composed of corrugated hydrogen-bonded layers of molecules which allows for improved tableting process by preventing chipping and disintegration.[14–21]

A widely accepted working definition in industry is that a "pharmaceutical co-crystal" is formed between a molecular or ionic API and a co-crystal former that is solid under ambient conditions.[17] Since a significantly large number of drugs contain "organic fluorine", it is important to gain insights into the nature of intermolecular interactions induced by fluorine such as C–H...F–C, N–H...F–C, O–H...F–C along with halogen bonds like F...F, F...X (X = any other halogen), F...O and F...N. Use of high resolution X-ray diffraction data based charge density analysis provides the ideal tool for quantifying such interactions.[22,41,66,75,76] Subsequent sections will describe the nature of intra- and intermolecular interactions with special reference to halogens, opening up debate on the special element fluorine. This is followed by recent results of interactions involving fluorine from our group both from the structural and charge density perspective.

3.2 Role of Intra- and Intermolecular Interactions

In crystal engineering, intra- and intermolecular interactions play a pivotal role as structure directing parameters. Interactions like van der Waals, coulombic, steric repulsions, strong and/or weak hydrogen bonds lead to the arrangement of molecules in a crystal.[23] A crystal structure hence is the manifestation of a supermolecule, "a supermolecule par excellence".[24] Hydrogen bonds are the extensively studied and applied in crystal engineering. Based on their geometrical, energetic, thermodynamic and functional nature, hydrogen bonds are categorized as either strong and weak. Charge density studies provide quantitative information on the entire range of weak and strong interactions based on both experimental and theoretical models. Although the energy associated with weak interactions is small, their geometrical preference and cooperative action can significantly contribute towards the conformation and packing of organic and biological structures. Extensive studies related to interactions involving π systems, such as π···π, C–H···π, O–H···π, N–H···π, C–H···X, C–X···π, shows that these interactions can be employed as structure-directing tools in both chemically and biologically important systems. For example, attractive interactions between π systems along with hydrogen bonds provide stability in the formation of the DNA structure, packing modes in aromatic compounds, and intercalation of drugs and host–guest systems. Similarly C–X···π (X = F, Cl, Br, I) and anion···π are found to display directional preferences and participate in organizing molecules in the crystal lattice.[25] Among the C–H···X (F, Cl, Br and I) interactions, C–H···F interactions contribute significantly to crystal packing.[26] Based on statistical analysis using the Cambridge Structural Database (CSD), it has been demonstrated that C–H···Cl interactions have similar characteristics to C–H···O hydrogen bonds.[27] Halogen···halogen (X···X) interactions have also been a study of interest in recent times.[28]

3.2.1 Halogen···Halogen Interactions

Interactions involving halogens of the type X···X (where, X = F, Cl, Br and I) can be considered as donor–acceptor, secondary, electron-transfer, highest occupied molecular orbital (HOMO) and lowest unoccupied molecular orbital (LUMO) interactions.[28–32] The electrostatic potential on the surfaces of the halogen atom is highly anisotropic and potential values vary according to the nature of the halogen atom and follow the sequence Cl < Br < I.[33] The intermolecular distance between the two halogen atoms represents the equilibrium between the attractive and repulsive forces, which have a contribution from electrostatic, charge transfer, polarization, dispersive, and exchange components.[34] The strengths of these contacts decrease in the following order I···I > Br···Br > Cl···Cl as evidenced by crystallographic and theoretical studies. Also, the relative strength of these interactions decreases as a function of the hybridization of the *ipso* carbon atom in the following order $sp^2 > sp > sp^3$ and is reinforced by the presence of a substituent electronegative atom.[35]

In recent literature, it has been demonstrated that halogen···halogen interactions have a structure-directing influence both in the presence and absence of stronger intermolecular interactions.[36] It has also been demonstrated that substitution of halogens steers the molecules in the crystalline lattice to afford a favorable orientation for topochemical photodimerization.[37] X···X interactions can be put into two categories, (1) homo or symmetrical (*e.g.* F···F, Cl···Cl, Br···Br, and I···I) and (2) hetero or unsymmetrical (*e.g.* F···Cl, F···Br, F···I, and Br···Cl *etc.*) interactions. The geometrical features of halogen···halogen interactions have been extensively studied and a well-defined classification has emerged.[28,36,38] Figure 3.1 gives the details of such a classification as Type I and Type II geometries.

Type I interactions can be referred as '*cis*' and '*trans*' according to the orientation which prefers either a crystallographic inversion center or a 2-fold symmetry and are essentially van der Waals type and repulsive in character. Type II interactions are commonly associated with crystallographic screw axes and glide planes and are attractive in nature due to polarization charges. This may be understood based on a model that assigns a positive polarization in the polar region of the halogen atom and a negative polarization in its equatorial region.[28c,36,39] *Ab initio* calculations on organic chlorine compounds by Price, and Nyburg and Wong-Ng point out that the decrease in repulsion and increase in attraction in Cl···Cl contacts is attributable to the anisotropy around the chlorine resulting in "polar flattening effect".[39a,40] Charge density of electron density measurements clearly indicate the existence of polar flattening and demonstrate that the Type II contacts are indeed attractive in nature.[41] Type II contacts are preferred from Cl to I as the polarization effect increases. Inter- or hetero-halogen···halogen interactions have been utilized in the supramolecular assembly of two dimensional (2D) layers in *bis*-haloamides through co-crystal

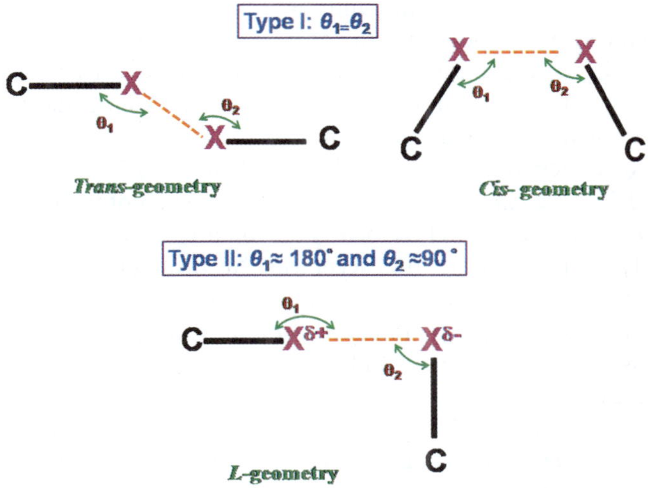

Figure 3.1 Geometrical classification of a halogen···halogen interaction as Type I and Type II.

formation. The preferred hetero-halogen interaction is Type II with the heavier atom being polarized $\delta^{(+)}$.[36]

Another well known interaction involving halogen atoms is halogen bonding, with the general representation C–D···X–Y, wherein intermolecular non-covalent interactions exist between the halogen atom and atoms containing lone pairs (mostly D = N or O).[42] The D···X distance is less than the sum of the van der Waals radii of D and X. The stronger the interaction, the shorter the D···X distance and the more linear is the angle D···X–Y. *Ab initio* calculations performed on covalently bonded halogen atoms prove that both electrostatics and charge transfer play an important role in halogen bonding.[43]

3.3 Debate on Organic Fluorine

Fluorine with its high electronegativity, relatively small size and very low polarizability of the bound and three non-bonding electron pairs forms bonds very differently from other elements in the second row of the periodic table.[44] In the book, *The Nature of the Chemical Bond*, Pauling stated that "Only the most electronegative atoms should form hydrogen bonds and the strength of the bond should increase with increase in the electronegativity of the two bonded atoms".[45] Accordingly, it is assumed that being the most electronegative element, fluorine should form the strongest hydrogen bonds. The hydrogen bonding potential in the hydrogen fluoride ion provides evidence for this observation. Subsequently, Pauling also observed that fluorine atoms attached to carbon are not strong enough to act as proton acceptors in the formation of hydrogen bonds, an observation which he specially pointed out in the later edition of his book.[46]

Jack Dunitz has recognized organic fluorine to be very special and has classified it as an "odd man out" based on the extent of polarizability.[47] Organic fluorine-containing compounds have been found to show differences in physico-chemical properties from the corresponding hydrogen analogs. Since electronegativity is the measure of the propensity to attract electrons, the covalently bonded organic fluorine is an extremely weak base and thus is a poor proton acceptor to form hydrogen bonds.

Analysis of the crystal structures obtained from the CSD shows that covalently bound fluorine barely acts as a hydrogen bond acceptor. In contrast to the usual acceptors, like carbonyl oxygen and aromatic nitrogen, C–F groups are exceptionally weak hydrogen bond acceptors. According to Dunitz and Taylor, the extremely weak hydrogen bonding capability of the covalently bound F can be attributed to its low proton affinity (low basicity, low-lying lone pair orbitals and tightness of its electron shell) and its inability to alter this by intramolecular electron delocalization or an intermolecular cooperative effect.[48]

On the other side of the debate, a recent review, highlights that C–F bond is significantly polarized, has unusual stability (strength) and interacts electrostatically with its environment.[44] In addition, the principles which direct the conformational preferences of C–F bonds in organic molecules are rationalized in terms of the gauche and anomeric effect. In general the introduction of fluorine atom(s) into organic molecules causes key changes in their physicochemical properties, the chemical reactivity and biological activity compared to their non-fluorinated analogues.[49–54] Reichenbächer *et al.* have systematically analyzed intermolecular interactions involving the C–F bond in the context of its structure-directing ability and concluded that indeed both F···F and C–H···F–C interactions are dominant in many structures, including pharmaceutical compounds.[26a] The replacement of hydrogen by a fluorine atom is often regarded as isosteric substitution.[55] A mono-fluorinated analog is geometrically very similar to its parent molecule and hence satisfies the steric requirements at enzyme receptor sites.[56,57] The formation of intermolecular O–H···F–C and N–H···F–C hydrogen bridges is important in binding fluorinated compounds to enzyme active sites which bring significant changes in enzyme–ligand binding affinity along with the changes in pharmaco-kinetic properties.[58,59]

The observation that fluorine does not readily accept hydrogen bonding in an organic environment has been explored in recent times on a variety of model compounds experimentally as well as in theoretical considerations.[60,61] Desiraju and Parthasarathy, based on a CSD analysis, predicted that the presence of organic fluorine offers a topologically different packing arrangement from those of other heavier halogen atoms.[62] The introduction of fluorophilicity and fluorophobicity remains only a concept and so far no quantification of these terms is available.[63] It has been clearly established that the introduction of a fluorine atom leads to a modification of the packing motif compared to that of the non-fluorinated counterpart.[64] Though interactions involving fluorine atoms (namely C–H···F–C, C–F···F–C and C–F···π) are weak, they also contribute to the stability of the crystal structure, especially in the absence of other strong interactions like hydrogen bonds and weak interactions like π···π or C–H···π. Further there are also indicators of the formation of C–F···F–C contacts which have been explored in terms of experimental and theoretical charge density studies using high resolution X-ray data.[65,66] More recently, the versatility of such weak interactions involving organic fluorine acting cooperatively in the presence of strong hydrogen bonds has also been demonstrated.[67]

3.3.1 Topological Analysis of the Electron Density of Compounds Containing Organic Fluorine: Insights from Experimental and Theoretical Electron Density Analysis

Evaluation of intermolecular interactions in terms of charge density distributions using high resolution X-ray data has opened up new vistas in quantifying a variety of intra- and intermolecular interactions beyond the criteria of mere geometry.[28,68] These results have also been substantiated by theoretical methods and all topological factors associated with bonding features have been understood using the atoms in molecules (AIM) approach.[69]

Crystal structures of Cl_2, Br_2 and I_2 suggest deviations from normal van der Waals type structures.[70] Potential energy models indicated the addition of anisotropic terms and in this context two possible models were generated. The first model is based on a postulate that anisotropic non-bonded radii are associated with the Cl atom in the chlorine molecule in the solid state.[40] The other model proposes that halogen atoms are weakly bonded in molecular crystals with an estimated energy for Cl···Cl interactions about 3% of the energy of the Cl–Cl covalent bond.[71]

Charge density studies in solid molecular chlorine depicted the closed shell nature of Cl···Cl interactions with a (3, –1) bond critical point.[72] A charge density-based analysis of Type II contacts showed that these interactions are attractive.[41b] The following three model compounds, 2-chloro-3-quinolinyl methanol (VCL1), 2-chloro-3-hydroxypyridine (VCL2) and 2-chloro-3-chloromethyl-8-methylquinoline (VCL3) (Figure 3.2) representing the geometries of Type I and Type II contacts were subjected to both experimental and theoretical charge density studies.[73] The results clearly show that Type I contacts support the decreased repulsion model while Type II clearly follow the attractive model as seen from three-dimensional (3D) deformation density and Laplacian maps (Figures 3.3 and 3.4). This classification is indeed essential to

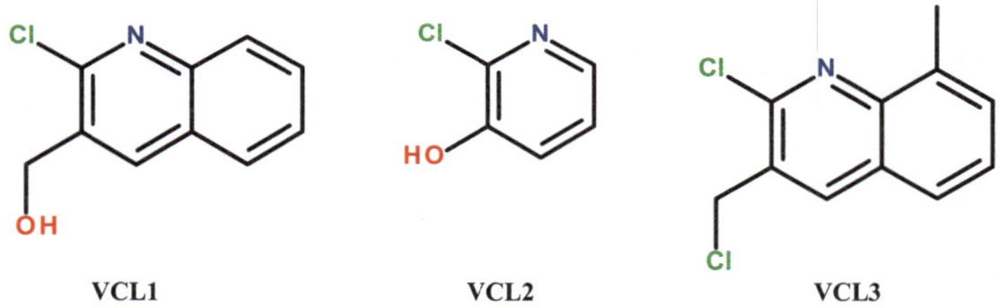

Figure 3.2 Schematic diagram of (a) 2-chloro-3-quinolinyl methanol (VCL1) (b) 2-chloro-3-hydroxypyridine (VCL2) and (c) 2-chloro-3-chloromethyl-8-methylquinoline (VCL3).

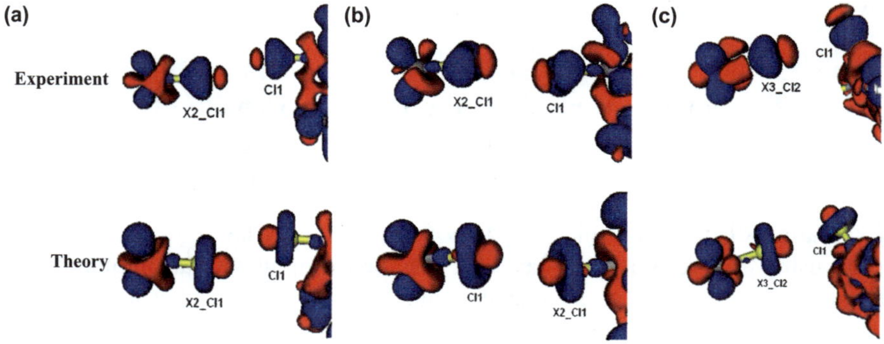

Figure 3.3 3D static deformation density maps from experimental and theoretical charge density calculations for Cl···Cl intermolecular interactions of the compounds (a) VCL1, (b) VCL2 and (c) VCL3. Blue and red colors represent positive and negative values, respectively. The $\Delta\rho(r)$ iso-surfaces are drawn at ± 0.01 eÅ^{-3}.

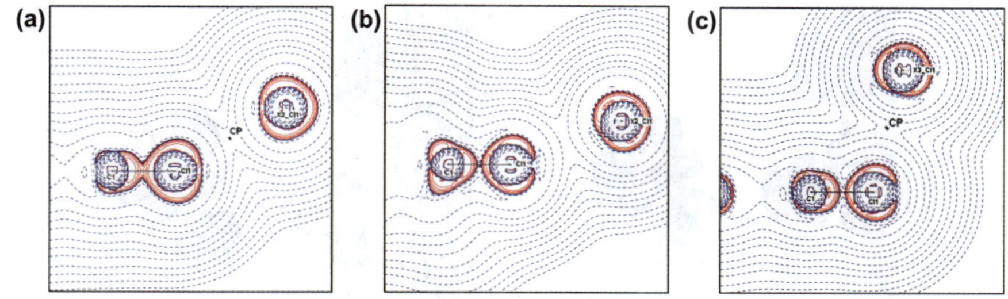

Figure 3.4 Laplacian maps from experimental charge density analysis with (3, –1) critical points (CP) for compounds (a) VCL1, (b) VCL2 and (c) VCL3. The contours are drawn on the logarithmic scale.

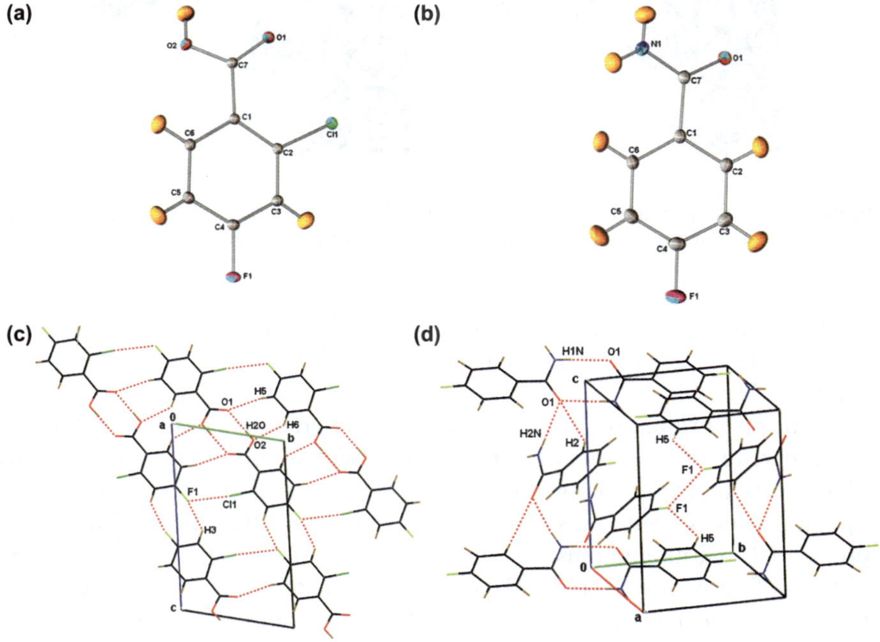

Figure 3.5 Thermal ellipsoids diagrams (a) and (b) and the intermolecular interactions (c) and (d) present in the crystal structure of 2-chloro-4-fluorobenzoic acid and 4-fluorobenzamide, respectively.

extend understanding of interactions involving fluorine either in the context of hetero halogen contacts or as F···F interactions.

In an electron density (ED) study using multipolar refinement, Bach *et al.*[70] have analyzed the intermolecular interactions of the type C–F···O and C–F···F–C associated with pentafluorobenzoic acid at 110 K. These studies highlight the discrepancies in the experimental and theoretical values of the Laplacian at the bond critical points of C–F and examined the energetic disadvantage of the F···F interaction. Bianchi *et al.*[71] have shown that in the (*E*)-1,2-bis(4-pyridyl)ethylene

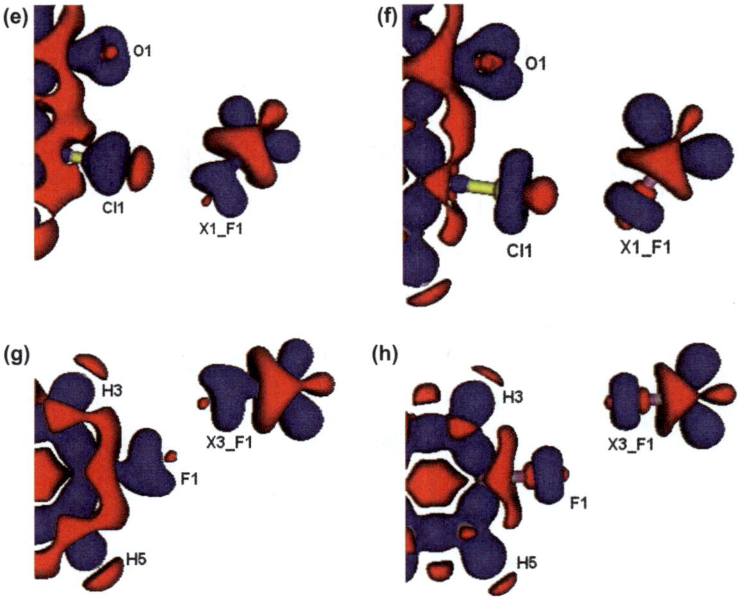

Figure 3.6 Experimental (e), (f) and theoretical (g), (h) 3D deformation density maps of 2-chloro-4-fluorobenzoic acid and 4-fluorobenzamide, respectively.

complex with 1,4-diiodotetrafluorobenzene at 90 K, F···F, C···F and C···C intermolecular interactions are appreciably noticeable and underpin the crystal packing.

Quantitative evaluation of Cl···F and F···F interactions has brought about the comparison of hetero-halogen interactions with homo-halogen interactions and suggests that even though F···F contacts occur ubiquitously, the extent of polarization is arguably small, but correlates with the observations made from the Cl···Cl interactions.[76] Figures 3.5 and 3.6 show the two molecules, 2-chloro-4-fluorobenzoic acid and 4-fluorobenzamide, which were subjected to detailed experimental and theoretical topological evaluation. The results clearly depict the extent of polarization in Cl···F and F···F intermolecular interactions.

3.4 Conclusions

Fluorinated compounds have attained much attention as drugs and agrochemicals and the structural diversity associated with these compounds is incredible. The ever increasing amount of structural information that is available, together with theoretical studies, indicate that interactions involving "organic fluorine" are indeed a stabilizing force and not a consequence of close packing. Studies pertaining to the deformation of the electron density and the experimental electrostatic potential in the region of the C–H and C–F bond regime can provide a quantitative perception of the versatility and usefulness of these interactions. These inputs will serve in the engineering of several novel pharmaceutical co-crystals, particularly with the ever increasing number of fluorinated organic molecules proving to be potential drugs, by exploiting the rather weak yet significant polarization available as "organic fluorine".

Acknowledgement

TNG thanks Department of Science and Technology for the award of the J. C. Bose fellowship and SS thanks UGC for the award of the Dr. D. S. Kothari post-doctoral fellowship. SKN thanks Council of Scientific and Industrial Research, Department of Science and Technology and VRH thanks IISc for a fellowship. DC thanks DST-Fast Track scheme for a research grant. We thank Professors K. Venkatesan, G. R. Desiraju and K. Nagarajan for their interest and constant encouragement in our attempts to understand fluorine interactions. Professor Nagarajan has also been a constant source for providing fluorinated organic compounds of our interest and choice. We wish to express our deepest thanks to all the members of our group who have contributed to the work on fluorine interactions and in particular we record the efforts of Drs. Angshuman Ray Choudhury, Usha K. Urs, M. D. Prasanna, Ishu Saraogi and Vijay Thiruvenkatam. We thank Indian Institute of Science XRD facility for the provision of all data collection instruments.

References

1. J. D. Dunitz, *Mol. Cryst. Liq. Cryst.*, 1996, **279**, 209.
2. B. E. Smart, in *Chemistry of Organic Fluorine Compounds*, ed. M. Hudlicky and A. E. Parlath, ACS Monograph 187, ACS, Washington DC, 1995, p. 979.
3. J. Bégué and D. Bonnet-Delpon, *J. Fluorine Chem.*, 2006, **127**, 992.
4. P. Jeschke, *ChemBioChem*, 2004, **5**, 570.
5. H. W. Kim, P. Rossi, R. K. Shoemaker and S. G. DiMagno, *J. Am. Chem. Soc.*, 1998, **120**, 9082.
6. (a) G. R. Desiraju, *Crystal Engineering. The Design of Organic Solids*, Elsevier, Amesterdam, 1989; (b) G. R. Desiraju, *Angew. Chem. Int. Ed. Engl.*, 1995, **34**, 2311; (c) D. D. MacNicol, F. Toda and R. Bishop, *Comprehensive Supramolecular Chemistry, Solid State Supramolecular Chemistry: Crystal Engineering*, Pergamon, Oxford, UK, 1996.
7. (a) J. W. Steed and J. L. Atwood, *Supramolecular Chemistry*, Wiley, Chichester, 2000, p. 389; (b) G. R. Desiraju, *The Crystal as a Supramolecular Entity: Perspective in Supramolecular Chemistry*, ed. G. R. Desiraju, Wiley, Chichester, 1995.
8. N. Schultheiss and A. Newman, *Cryst. Growth Des.*, 2009, **9**, 2950.
9. (a) S. Datta and D. J. Grant, *Nat. Rev. Drug Discov.*, 2004, **3**, 42; (b) L. F. Huang and W. Q. Tong, *Adv. Drug Deliver. Rev.*, 2004, **56**, 321;

(c) A. Ballabh, D. R. Trivedi, P. Dastidar, P. K Ghosh, A. Pramanik and V. G. Kumar, *Cryst. Growth Des.*, 2006, **6**, 1591.
10. (a) B. Rodriguez-Spong, C. P. Price, A. Jayasankar, A. J. Matzger and N. Rodriguez-Hornedo, *Adv. Drug Deliver. Rev.*, 2004, **56**, 241; (b) R. J. Bastin, M. J. Bowker and B. J. Slater, *Org. Process Res. Dev.*, 2000, **4**, 427.
11. A. V. Trask, W. D. S. Motherwell and W. Jones, *Mater. Res. Bull.*, 2006, **31**, 874.
12. (a) J. Haleblian and W. McCrone, *J. Pharm. Sci.*, 1969, **58**, 911; (b) D. Singhal and W. Curatolo, *Adv. Drug Deliver. Rev.*, 2004, **56**, 335.
13. (a) T. Friščić and W. Jones, *Faraday Discuss.*, 2007, **136**, 167; (b) T. Friščić, L. Fábián, J. C. Burley, D. G. Reid, M. J Duer and W. Jones, *Chem. Commun.*, 2008, 1644.
14. (a) S. L. Childs, G. P. Stahly and A. Park, *Mol. Pharmaceutics*, 2007, **4**, 323; (b) A. Ballabh, D. R. Trivedi, P. Dastidar and E. Suresh, *CrystEngComm*, 2002, **4**, 135.
15. (a) D. R. Trivedi and P. Dastidar, *Cryst. Growth Des.*, 2006, **6**, 1022; (b) S. Roy, G. Mahata and K. Biradha, *Cryst. Growth Des.*, 2009, **9**, 5006.
16. (a) A. D. Bond, *CrystEngComm,* 2007, **9**, 833; (b) A. Parkin, C. J. Gilmore and C. C. Wilson, *Z. Kristallogr.*, 2008, **223**, 430.
17. (a) C. B. Aakeröy and D. J. Salmon, *CrystEngComm,* 2005, **7**, 439; (b) M. J. Zaworotko, *Cryst. Growth Des.*, 2007, **7**, 4; (c) N. Rodriguez-Hornedo, *Mol. Pharmaceutics*, 2007, **4**, 299.
18. A. V. Trask and W. Jones, *Top. Curr. Chem.*, 2004, **254**, 41.
19. V. R. Pedireddi, W. Jones, A. P. Chorlton and R. Docherty, *Chem. Commun.*, 1996, 987.
20. M. D. Eddleston and W. Jones, *Cryst. Growth Des.*, 2010, **10**, 365.
21. S. Karki, T. Friščić, L. Fábián, P. R. Laity, G. M. Day and W. Jones, *Adv. Mater.*, 2009, **21**, 3905.
22. (a) P. Coppens, *X-ray Charge Densities and Chemical Bonding,* Oxford University Press, Oxford, U.K, 1997; (b) T. S. Koritsanszky and P. Coppens, *Chem. Rev.*, 2001, **101**, 1583.
23. G. R. Desiraju and T. Steiner, *The Weak Hydrogen Bond,* Oxford University Press, 1999.
24. J. D. Dunitz, *Pure Appl. Chem.*, 1991, **63**, 177.
25. (a) F. Cozzi, M. Cinquini, R. Annuziata, T. Dwyer and J. S. Siegel, *J. Am. Chem. Soc.*, 1992, **114**, 5729; (b) I. Saraogi, V. G. Vijay, S. Das, K. Sekhar and T. N. Guru Row, *Cryst. Eng.*, 2003, **6**, 69; (c) A. Robertazzi, F. Krull, E.-W. Knapp and G. Gamez, *CrystEngComm*, 2011, **13**, 3293.
26. (a) K. Reichenbächer, H. I. Süss and J. Hulliger, *Chem. Soc. Rev.*, 2005, **34**, 22; (b) A. Schwarzer, W. Seichter, E. Weber, H. S. Evans, M. Losada and J. Hulliger, *CrystEngComm*, 2004, **6**, 567; (c) T. S. Thakur, M. T. Kirchner, D. Bläser, R. Boese and G. R. Desiraju, *CrystEngComm*, 2010, **12**, 2079; (d) D. Chopra and T. N. Guru Row, *CrystEngComm*, 2011, **13**, 2175.
27. C. B. Aakeröy, T. A. Evans, K. R. Seddon and I. Pálinkó, *New J. Chem.*, 1999, **23**, 145.

28. (a) N. Ramasubbu, R. Parathasarathy and P. Murray-Rust, *J. Am. Chem. Soc.*, 1986, **108**, 4308; (b) E. Bosch and C. L. Barnes, *Cryst. Growth Des.*, 2002, **2**, 299; (c) B. K. Saha, A. Nangia and J. F. Nicoud, *Cryst. Growth Des.*, 2006, **6**, 1278; (d) G. Cavallo, P. Metrangolo, T. Pilati, G. Resnati, M. Sansotera and G. Terraneo, *Chem. Soc. Rev.*, 2010, **39**, 3772; (e) S. Samai and K. Biradha, *CrystEngComm,* 2009, **48**, 3838.
29. J. Xu, W.-L. Wang, T. Lin, Z. Sun and Y.-H. Lai, *Supramol. Chem.*, 2008, **20**, 723.
30. H. A. Bent, *Chem. Rev.*, 1968, **68**, 587.
31. R. E. Rosenfield, R. Parthasarathy and J. D. Duntiz, *J. Am. Chem. Soc.*, 1977, **99**, 4860.
32. T. N. Guru Row and R. Parthasarathy, *J. Am. Chem. Soc.*, 1981, **103**, 477.
33. M. Podsiadlo and A. Katrusiak, *J. Phys. Chem. B*, 2008, **112**, 5355.
34. G. M. Espallargas, L. Brammer, D. R. Allan, C. R. Pulham, N. Robertson and J. E. Warren, *J. Am. Chem. Soc.*, 2008, **130**, 9058.
35. F. F. Awwadi, R. D. Willett, K. A. Peterson and B. Twamley, *Chem. Eur. J.*, 2006, **12**, 8952.
36. (a) V. R. Pedireddi, D. S. Reddy, B. S. Goud, D. C. Craig, A. D. Rae and G. R. Desiraju, *J. Chem. Soc. Perkin Trans. 2*, 1994, **2**, 2353; (b) S. Samai and K. Biradha, *CrystEngComm,* 2009, **11**, 482; (c) S. K. Nayak, S. J. Prathap and T. N. Guru Row, *J. Mol. Struct.*, 2009, **935**, 156. (e) S. K. Nayak, M. K. Reddy and T. N. Guru Row, *Cryst. Growth Des.*, 2011, **11**, 1578.
37. (a) P. Venugopalan, T. B. Rao and K. Venkatesan, K. *J. Chem. Soc. Perkin. Trans. 2,* 1991, 981; (b) K. Vishnumurthy, T. N. Guru Row and K. Venkatesan, *Tetrahedron,* 1999, **55**, 4095.
38. E. A. Medlycott, K. A. Udachin and G. S. Hanan, *Dalton Trans.*, 2007, 430.
39. (a) S. L. Price, A. J. Stone, J. Lucas, R. S. Rowland and A. E. Thornley, *J. Am. Chem. Soc.*, 1994, **116**, 4910; (b) J. A. R. P. Sarma and G. R. Desiraju, *Acc. Chem. Res.*, 1986, **19**, 222.
40. (a) S. C. Nyburg and W. Wong-Ng, *Inorg. Chem.*, 1979, **181**, 2790; (b) S. C. Nyburg and W. Wong-Ng, *Proc. R. Soc. London Ser. A*, 1979, **367**, 29.
41. (a) P. Garc, S. Dahaoui, C. Katan, M. Souhassou and C. Lecomte. *Faraday Discuss.*, 2007, **135**, 217; (b) T. T. T. Bui, S. Dahaoui, C. Lacomte, G. R. Desiraju and E. Espinosa, *Angew. Chem. Int. Ed.*, 2009, **48**, 3838.
42. (a) P. Metrangolo and G. Resnati, *Chem. Eur. J.*, 2001, **7**, 2511; (b) E. Corradi, S. V. Meille, M. T. Messina, P. Metrangolo and G. Resnati, *Angew. Chem., Int. Ed.*, 2000, **39**, 1782; (c) Q. L. Chu, Z. M. Wang, Q. C. Huang, C. H. Yan and S. Z. Zhu, *J. Am. Chem. Soc.*, 2001, **123**, 11069.
43. (a) J-W. Zou, Y-J. Jiang, M. Guo, G-X. Hu, B. Zhang, H-C. Liu and Q-S. Yu, *Chem. Eur. J.*, 2005, **11**, 7400; (b) Z. Wang, B. Zheng, X. Yu, X. Li and Yi. Pinggui, *J. Chem. Phys.*, 2010, **132**, 164104.
44. D. O'Hagan, *Chem. Soc. Rev.*, 2008, **37**, 308.

45. L. Pauling, *Nature of the Chemical Bond*, Cornell University Press, Ithaca, NY, 2nd edn, 1939, p. 286.
46. L. Pauling, *Nature of the Chemical Bond*, Cornell University Press, Ithaca, NY, 3rd edn, 1960, p. 464.
47. J. D. Dunitz, *ChemBioChem.*, 2004, **5**, 614.
48. J. D. Dunitz and R. Taylor, *Chem. Eur. J.*, 1997, **3**, 89.
49. (a) R. K. Filler, *Biomedical Effects of Fluorine Chemistry*, Kodama Ltd, Elsevier, Amsterdam, 1982; (b) R. Filler, Y. Kobayashi and L. M. Yagupolskii, *Organofluorine Compounds in Medicinal Chemistry and Biomedical Applications Studies in Organic Chemistry*, Elsevier, Amsterdam, 1993, vol. 48; (c) K. Reichenbächer, H. I. Suss and J. Hulliger, *Chem. Soc. Rev.*, 2005, **34**, 22.
50. U. Grob, S. Rudiger, B. Bassner, H. Hagemann and J. C. Tatlow, Methods of organic chemistry (Houben–Weyl), in *Organo-Fluorine Compounds*, Thieme, Studdgart, 4th edn, 1999, vol. E10a, p.18.
51. C. G. Beuguin, *Physical and Structural Size of Fluorine in Fluoro-Organic Compounds in Enantiocontrolled Synthesis of Fluoro-Organic Compounds*, ed. V. A. Soloshonok, Wiley, Chichester, 1999, p. 601.
52. T. Hiyama, *Organofluorine Compounds, Chemistry and Applications*, Springer, Berlin, 2000.
53. (a) R. E. Banks, B. E. Smart and J. C. Tatlow, *Organofluorine Chemistry, Principles and Commercial Applications*, Plenum Press, New York, 1994, p. 57.
54. (a) B. E. Smart, *J. Fluorine Chem.*, 2001, **109**, 3; (b) H. W. Roesky, *Nature Chem.*, 2010, **2**, 240.
55. G. A. Patani and E. J. LaVoie, *Chem. Rev.*, 1996, **96**, 3147.
56. I. Ojima, J. T. McCarthy and J. T. Welch, *Biomedical Frontiers of Fluorine Chemistry*, ACS Symposium Series 639, ACS, Washington, 2000.
57. P. V. Ramachandran, *Asymmetric Fluoro-Organic Chemistry: Synthesis, Applications and Future Directions*, ACS Symposium Series 639, ACS, Washington, 2000.
58. (a) J. L. Alderfer and A. V. Eliseev, *J. Org. Chem.*, 1997, **62**, 8225; (b) I. P. Street, C. Armstrong and S. G. Withers, *Biochemistry*, 1986, **25**, 6021; (c) L. H. Takahashi, R. Radhakrishnan, R. E. RosenfieldJr, F. E. Meyer, Jr. and D. A. Trainor, *J. Am. Chem. Soc.*, 1989, **111**, 3368; (d) R. H. Abeles and T. A. Alston, *J. Biol. Chem.*, 1990, **265**, 16705; (e) J. D. McCarter, M. J. Adam and S. G. Withers, *J. Biol. Chem.*, 1992, **286**, 721; (f) C. Mattos, B. Rasmussen, X. Ding, G. A. Petsko and D. Ringe, *Nat. Struct. Biol.*, 1994, **1**, 55.
59. (a) J. A. Olsen, D. W. Banner, P. Seiler, B. Wagner, T. Tschopp, U. Obst-Sander, M. Kansy, K. Muller and F. Diederich, *ChemBioChem*, 2004, **5**, 666; (b) E. T. Koul, J. C. Morales and K. M. Guckian, *Angew. Chem., Int. Ed.*, 2000, **39**, 990.
60. (a) L. Shimoni and J. P. Glusker, *Struct. Chem.*, 1994, **5**, 383; (b) J. A. K. Howard, V. J. Hoy, D. O'Hagan and G. T. Smith, *Tetrahedron*, 1996, **52**,

12613; (c) F. Emmerling, I. Orgzall, B. Dietzel, B. W. Schulz, G. Reck and B. Schulz, *J. Mol. Struct.*, 2007, **832**, 124.
61. (a) I. Y. Bagryanskaya, Y. V. Gatilova, A. M. Maksimov, V. E. Platonov and A. V. Zibarev, *J. Fluorine Chem.*, 2005, **126**, 1281; (b) A. R. Choudhury, K. Nagarajan and T. N. Guru Row, *Cryst. Growth Des.*, 2004, **4**, 47; (c) S. Kawahara, S. Tsuzuki and T. Uchimaru, *J. Phys. Chem. A*, 2004, **108**, 6744; (d) T. V. Rybalova and I. Y. Bagryanskaya, *J. Struct. Chem.*, 2009, **50**, 741; (e) A. Schwarzer, P. Bombicz and E. Weber, *J. Fluorine Chem.*, 2010, **131**, 345.
62. G. R. Desiraju and R. Parthasarathy, *J. Am. Chem. Soc.*, 1989, **111**, 8725.
63. J. Gladysz and D. P. Curren, *Tetrahedron*, 2002, **38**, 3823.
64. (a) M. D. Prasanna and T. N. Guru Row, *J. Mol. Struct.*, 2001, **562**, 55; (b) M. D. Prasanna and T. N. Guru Row, *CrystEngComm*, 2000, **2**, 134.
65. D. Chopra, K. Nagarajan and T. N. Guru Row, *Cryst. Growth Des.*, 2005, **5**, 1035.
66. D. Chopra, T. S. Cameron, J. D. Ferrara and T. N. Guru Row, *J. Phys. Chem. A*, 2006, **110**, 10465.
67. (a) D. Chopra and T. N. Guru Row, *CrystEngComm*, 2008, **10**, 54; (b) D. Chopra and T. N. Guru Row, *Cryst. Growth Des.*, 2005, **5**, 1679; (c) D. Chopra and T. N. Guru Row, *Cryst. Growth Des.*, 2006, **6**, 1267.
68. V. R. Hathwar, S. M. Roopan, R. Subhashini, F. N. Khan and T. N. Guru Row, *J. Chem. Sci.*, 2010, **122**, 677.
69. (a) R. F. W. Bader, *Atoms in Molecules–A Quantum Theory*, Clarendon, Oxford, 1990; (b) R. F. W. Bader. *J. Phys. Chem. A*, 1998, **102**, 7314.
70. (a) E. D. Stevens, *Mol. Phys.*, 1979, **37**, 27; (b) S. H. Walmsley and A. Anderson, *Mol. Phys.*, 1964, **7**, 411; (c) P. T. Wong and E. Whalley, *Can. J. Phys.*, 1972, **50**, 1856.
71. D. E. Williams and L. Y. Hsu, *Acta Crystallogr., Sect. A*, 1985, **41**, 296.
72. V. G. Tsirelson, P. F. Zou, T. H. Tang and R. F. W. Bader, *Acta Crystallogr., Sect. A*, 1995, **51**, 143.
73. V. R. Hathwar and T. N. Guru Row, *J. Phys. Chem. A*, 2010, **114**, 13434.
74. A. Bach, D. Lentz and P. Luger, *J. Phys. Chem. A*, 2001, **105**, 7405.
75. R. Bianchi, A. Forni and T. Pilati, *Chem.-Eur. J.*, 2003, **9**, 1631.
76. V. R. Hathwar and T. N. Guru Row, *Cryst. Growth Des.*, 2011, **11**, 1338.

CHAPTER 4

Polymorph Prediction of Small Organic Molecules, Co-crystals and Salts

FRANK J. J. LEUSEN AND JOHN KENDRICK

School of Life Sciences, University of Bradford, Richmond Road, Bradford, BD7 1DP, United Kingdom

4.1 Introduction

Crystal structure prediction is regarded by some as the holy grail of crystal engineering because reliable and accurate prediction of the polymorphs that a compound can crystallise in would allow the design of organic materials with specific properties from first principles. This chapter provides an overview of the current status of crystal structure prediction of small organic molecules in general and focuses on the specific issues encountered in the prediction of co-crystal and salt structures. Both the global optimisation problem of searching for all possible crystal structures of a compound and the problem of calculating accurate lattice energies in order to rank potential crystal structures according to stability are discussed. A number of illustrative examples are presented, including an overview of the Cambridge Crystallographic Data Centre's blind tests in crystal structure prediction, as well as some examples for co-crystals, solvates and salts.

4.1.1 Relevance of Crystal Structure Prediction

The solid state properties of organic materials not only depend on the molecular structure but also on the crystal structure. The colour of pigments, the

RSC Drug Discovery Series No. 16
Pharmaceutical Salts and Co-crystals
Edited by Johan Wouters and Luc Quéré

Published by the Royal Society of Chemistry, www.rsc.org

solubility and dissolution rate of pharmaceuticals, the melting point and texture of chocolate, the vapour pressure of agrochemicals and the performance characteristics of explosives are all influenced by the way in which the molecules are packed together in the solid state. Understanding, controlling and ultimately designing solid state properties requires detailed knowledge of crystal structure at the atomic level. This knowledge enables the simulation of a range of solid state properties including; crystal morphology, colour, mechanical properties, stability and diffraction patterns. The design of organic materials with desired properties from first principles is arguably the ultimate goal of crystal engineering and requires the ability to predict reliably and accurately crystal structures from molecular structure alone. Once the crystal structure is available, solid state properties can be simulated or predicted.

The phenomenon of polymorphism, which is the ability of a compound to crystallise in more than one distinct crystal structure, complicates matters further, as the physical and chemical properties of different polymorphs of a compound may vary significantly.[1] Polymorphism arises from a combination of the thermodynamic and the kinetic nature of the crystallisation process and different polymorphs of a compound can be observed owing to variations in the conditions of a crystallisation experiment, such as temperature, solvent and pressure. It may cause a problem, for instance, if a new polymorph is found after a product has gone to market, but polymorphism may also present opportunities to optimise the properties of a product or to extend its patent protection.

In computational chemistry, the general compromise is that the more accurate the simulation, the more computational effort is required. The simulation of a pathway is much more demanding and time consuming, and often less accurate, than the simulation of the start and end points of a process. The prediction of crystallisation behaviour from first principles involves the simulation of crystal nucleation as well as crystal growth, considering factors such as temperature and super-saturation. Various approaches have been developed to simulate these processes using computational chemistry tools,[2,3] but the timescales of both the nucleation and the growth step are too large to be simulated without assuming some prior knowledge of the crystal structure or reducing the accuracy of the calculations. Successful crystal structure prediction approaches therefore consider only the end points of the crystallisation process, that is, the various ways in which a compound can pack in the solid state. All possible crystal packing alternatives are generated and ranked according to calculated lattice energy. It is assumed that low-energy structures are likely to be observed experimentally. Since stability differences between polymorphs are generally very small, it is of paramount importance that the lattice energy calculation method used is as accurate as possible.

4.1.2 Historic Overview

The field of crystal packing simulations developed in parallel with advances made in modern computational hardware. Pioneering work was done by

Kitaigorodskii[4] and Williams[5] in the 1950s and 1960s. An early example of the determination of a crystal structure using molecular packing analysis was reported in 1966, where the structure of dibenzoylmethane was solved by using the non-bonded distances between rigid molecules as a quality measure and refining the packing using X-ray diffraction data.[5] To automate the procedure, software was developed to minimise non-bonded interatomic close contacts and to assist in the determination of the crystal structures of simple organic molecules.[6–8] More refined potential energy functions were introduced to consider interatomic forces in molecules and crystals and applied to study the molecular and solid state structures of a variety of organic materials.[9–11] These early studies required experimental information, usually unit cell dimensions derived from X-ray diffraction data, to limit the number of degrees of freedom of the simulations to a level manageable by the computational hardware available at the time. Despite further developments of potential energy functions to represent conformational strain as well as van der Waals and Coulomb interactions, by the late 1980s it was still not possible to predict the crystal structure of a simple organic molecule from first principles, which was considered 'one of the continuing scandals in physical sciences'.[12]

An editorial in *Nature*[12] led to renewed activity and several approaches to crystal structure prediction were developed in the 1990s, some of which are still in use today. These continuing developments are made possible by the increasing power of modern computers. Some approaches start with small clusters of molecules, which are then extended in three dimensions to form crystals.[13,14] It is assumed that strong intermolecular interactions within a small group of molecules, such as hydrogen bonds, dominate the crystal packing. This method is computationally efficient, but will provide unsatisfactory results in cases where stable molecular clusters can be generated that do not correspond to the solid state structures observed experimentally.[15] Other approaches in crystal structure prediction developed in the 1990s apply lattice symmetry throughout the calculations,[16] whilst the majority of methods use full space group symmetry.[17–21] These methods and their subsequent developments will not be discussed in detail here as they have been comprehensively reviewed in the literature.[22–25]

In order to assess progress in the development and application of crystal structure prediction tools, a series of blind tests have been organised by the Cambridge Crystallographic Data Centre. These tests, which took place in 1999, 2001, 2004, 2007 and 2010, involve predicting the crystal packing of a small number of molecules with varying complexity, for which the crystal structures have been determined experimentally. The experimental data is not available to participants, who are invited to predict up to three crystal structures for each compound. The experimental structures are released after the predictions have been submitted and the results are discussed at a workshop involving the organisers and the participants. Since these blind tests constitute an independent and reliable assessment of the capabilities and limitations of crystal structure prediction, they will be discussed in more detail in Section 4.3.1 of this chapter.

4.2 Theory of Crystal Structure Prediction

Successful crystal structure prediction requires two essential ingredients. One of these is a method for accurately ranking the relative stabilities of different crystal packing alternatives available to the target compound and the other is a method for generating a complete list of all stable and meta stable polymorphs. Both these elements are discussed below.

4.2.1 Lattice Energy Calculations

There are many ways to calculate the approximate stability of a crystal structure through computational chemistry approaches. The most important of these are discussed here, after an introduction to the thermodynamic considerations of the stability of a co-crystal.

4.2.1.1 Thermodynamic Considerations

The most stable crystal form of a compound at a given temperature and pressure is that structure which has the lowest free energy. The stability of a co-crystal ($A_a B_b$), with a molecules of A and b molecules of B in its formula unit, relative to the single-molecule crystalline components A and B can be calculated by considering the process below:

$$aA + bB \rightarrow A_aB_b \tag{4.1}$$

At any given temperature and pressure the free energy change is:

$$\Delta G^{\text{cocrystal}} = G(A_aB_b) - aG(A) - bG(B) \tag{4.2}$$

If $\Delta G^{\text{cocrystal}}$ is negative, the reaction is spontaneous and the co-crystal is more stable than its constituent single-molecule crystals. The free energy of a crystal at a temperature T can be written as a sum of enthalpy and lattice vibrational entropy (S^{vib}).

$$G = H - TS^{\text{vib}} \tag{4.3}$$

Since the volume change between a co-crystal and its constituent single-molecule crystals is very small, it can be assumed that the enthalpy (H) changes are the same as the internal energy (E) changes. Thus the expression for the free energy of the co-crystal relative to its single components becomes:

$$\Delta G^{\text{cocrystal}} = E(A_aB_b) - aE(A) - bE(B) - T(S^{\text{vib}}(A_aB_b) - aS^{\text{vib}}(A) - bS^{\text{vib}}(B)) \tag{4.4}$$

The internal energies are defined as the energies of the minima on the potential energy surface of the molecular crystals, including a quantum correction for zero-point energies due to vibration. Since the vibrational modes associated with the molecular vibrations in the co-crystal and the single-molecule crystals are likely to be very similar, the so called 0 K approximation is often used and in addition zero-point energy corrections may be neglected in the calculation of internal energies:

$$\Delta G^{\text{cocrystal}} \approx E(A_aB_b) - aE(A) - bE(B) \tag{4.5}$$

The lattice energies of the single-molecule crystals and co-crystal are assumed to be those for the most stable forms of each. To predict the stability of a given co-crystal it is therefore necessary to perform crystal structure prediction on each molecule independently and on the co-crystal itself. If it is necessary to predict which stoichiometry of co-crystal is the most stable, it is also necessary to perform crystal structure prediction on crystals with each possible stoichiometry. Since the cost and difficulty of a crystal structure prediction calculation increases considerably with the number of independent molecules in the asymmetric unit, this becomes a very hard problem which few have ventured to tackle. The role of crystal structure prediction is to identify the lowest energy structures that are possible for both co-crystal and its single-molecule crystalline components. The prediction of stoichiometry for a co-crystal requires consideration of all 'dissociation' processes available for a given number of molecules in the asymmetric unit, for example:

$$A_aB_b \rightarrow A_{a-i}B_{b-j} + iA + jB \tag{4.6}$$

Following van Eijck and Kroon,[26] Z'' will be used to designate the number of molecules $(a+b)$ in the asymmetric unit, as opposed to Z' which will be used to designate the number of formula units in the asymmetric unit. Typically all reactions consistent with a given Z'' must be considered and the most exothermic product gives the predicted stoichiometry.

There are various theoretical methods for calculating the lattice energy of a molecular crystal. It is important to note that the accuracy of these methods is key to the correct prediction of co-crystal stability. Some verification of the accuracy of the method for calculating the lattice energy is possible using experimental information. If the method, when used in crystal structure prediction, fails to predict the known polymorphs of the molecular crystals as low energy structures, then it is clear that, unless there is some fortuitous cancellation of errors, the method will not be able to predict co-crystal stability reliably.

For the case where a solvate is being formed, the thermodynamic considerations are slightly different. However, as the free energy associated with the phase change from solid to liquid at the melting point is zero, it is sometimes assumed that an approximation to the stability of a solvate can be obtained using the energy of the most stable crystalline form of the solvent in equations (4.4) or (4.5).[27]

The potential energy landscape has become an important concept in crystal structure prediction. The concept highlights the relationships between the energy of the structure and its geometry. An accurate landscape not only implies accurate energetics, but also accurate geometries. Nature strives towards low energy and if we are able to calculate the potential energy landscape with sufficient accuracy we will be able to predict co-crystal and polymorph stability. The next sections summarise some of the methods available for the calculation of the potential energy landscape, in particular the methods of molecular and quantum mechanics.

4.2.1.2 Molecular Mechanics

The molecular mechanics method represents the lattice energy of a crystal with a set of simple analytical functions representing different interactions between bonded and non-bonded atoms, as shown schematically in Figure 4.1.

Many early crystal structure predictions were performed assuming rigid molecules, in which case the relative lattice energies are determined only by non-bonded interactions. One of the most common non-bonded potentials which is used in crystal structure prediction to describe van der Waals dispersive interactions between atoms *i* and *j* and which prevents overlap of atoms at short distances is the Buckingham or exponential-6 potential:

$$V_{ij}{}^{vdw}(r_{ij}) = A_{ij}\exp(-B_{ij}r_{ij}) - C_{ij}r_{ij}{}^{-6} \tag{4.7}$$

Here the distance between the atomic centres is given by r_{ij} and A_{ij}, B_{ij} and C_{ij} are parameters which need to be provided. Williams and others[28,29] developed interaction parameters suitable for lattice energy calculations by fitting the

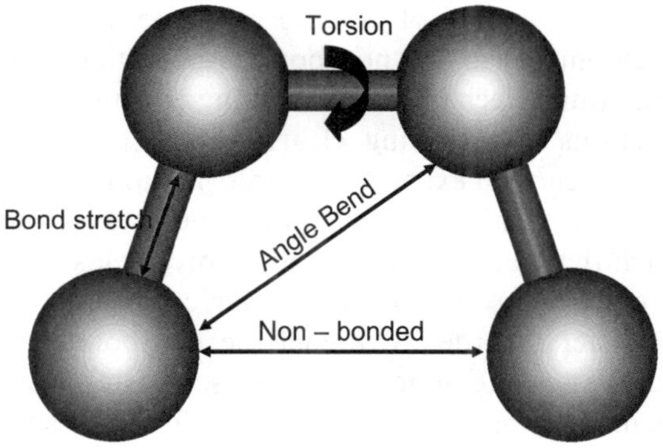

Figure 4.1 Components of the energy in molecular mechanics calculations.

parameters A, B and C to experimental information. An important aspect of this early work was the recognition that the self interaction parameters for a given atom type were valid in other molecules and crystals. It was also found that combination rules could be used to generate the interaction parameters between two different atom types from knowledge of the self interaction parameters only. This reduced the difficulty of the fitting of parameters substantially and encouraged transferability.

In their work and in many other force fields, a potential is used to describe the interactions between charge distributions based on Coulomb's law, where the partial charges on the atom centres, q_i and q_j, are derived from analysis of the electrostatic potential around the molecule calculated using molecular quantum mechanics. Such charges are sometimes called potential derived charges:[30]

$$V_{ij}^{electrostatic}(r_{ij}) \propto q_i q_j r_{ij}^{-1} \tag{4.8}$$

Because of the long-range nature of the electrostatic interaction it is necessary to use special methods, such as that due to Ewald,[31] to ensure convergence of the infinite sums over the lattice.

Whilst non-bonded potentials based on this approach were fairly successful, it soon became clear that more accurate non-bonded potentials were needed. One of the most significant developments was the use of a more sophisticated model for the electrostatic interactions between charge distributions based on a distributed multipole approximation of the two charge distributions.[32] Such an approach can represent electrostatic interactions in a crystal using the gas phase molecular charge distribution to almost any degree of accuracy required. It does not, however, easily describe effects such as induction or polarisation where the response of the charge distribution to its environment is important, nor does the method cope easily with flexible molecules. There has been progress in both these areas and programs are available which are capable of using such sophisticated non-bonded force fields in crystal structure prediction.[33] As well as developments in the accurate description of the charge interactions, there are some atom types for which it is necessary to introduce anisotropic short range repulsion. The crystal structures of the chlorobenzenes are an example of this.[34]

Alongside developments in the non-bonded potentials, the ever increasing importance of quantum mechanics in providing accurate benchmark data on molecular interactions is becoming clear. As the potentials become more complex, there is not enough experimental information for reliable fitting. The accuracy of molecular quantum mechanical calculations for small molecule interactions is such that complex non-bonded force fields can be developed. A note of caution must be given here; molecular calculations usually assume a dimer or small cluster in order to model the non-bonded interactions. The transferability of these interactions to the solid state environment is an assumption which must be continually tested. This is especially true when developing potentials for salts, where the anion is particularly sensitive to its environment.

Bonded interactions include the bond stretch, angle bend and torsion potentials shown in Figure 4.1. There are many force fields available which document parameters for different atom types and which use different functional forms for the interactions. Examples include Compass,[35] Dreiding,[36] Charmm,[37] Amber[38] and MM3.[39–41] The most common functional forms are simple harmonic potentials for bond stretch and angle bending, with parameters for the force constant and the minimum energy geometry of the bond or angle. Cosine expansions are often used for periodic torsional potentials. The examples given below are illustrative of the type of functions that can be used. The parameters k, w, n, r^0, θ^0 and ϕ^0 are specific for a given set of atom types defined in a particular force field:

$$\begin{aligned} V_{ij}^{\text{bond}}(r_{ij}) &= 1/2k_{ij}(r_{ij} - r_{ij}^0)^2 \\ V_{ijk}^{\text{angle}}(\theta_{ijk}) &= 1/2k_{ijk}(\theta_{ijk} - \theta_{ijk}^0)^2 \\ V_{ijkl}^{\text{torsion}}(\phi_{ijkl}) &= 1/2w_{jk}[1 - \cos(n_{jk}(\phi_{ijkl} - \phi_{jk}^0))] \end{aligned} \tag{4.9}$$

Because non-bonded interactions occur within a molecule as well as between molecules, the bonded and non-bonded parameters are not independent and must often be derived together.

The general emphasis in force field development is towards transferrable force fields, where the functional form and the values of associated parameters can be used in a wide variety of molecules and crystals. As the parameters are developed empirically, transferability implies a degree of reliability and confidence that the parameters will work for crystals for which they were not specifically parameterised. In a recent development of the so-called tailor-made force field,[42] it was pointed out that for the specific case of crystal structure prediction, the force field does not need to be transferable and that in fact there are some important advantages to having a force field derived specifically for the molecule of interest. Given sufficiently accurate information from quantum mechanical calculations, the tailor-made force field can be obtained by fitting to the quantum mechanical potential energy surface. Neumann[42] defined a number of quantum mechanical data sets which represented both the non-bonded and bonded interactions in the crystal. The parameters of the force field were then optimised to fit these data sets. The quantum mechanical method chosen for the calculations was the DFT(d) method which will be described below.

4.2.1.3 Quantum Mechanics

From the time-independent, non-relativistic Schrödinger equation, knowledge of the molecular structure allows the calculation of the wavefunction and energy of the molecule:

$$\widehat{H}\Psi = E\Psi \tag{4.10}$$

Here $\hat{H}$ represents the Hamiltonian operator for the system, describing the electron–nuclear interaction, the electron kinetic energy and the electron–electron interaction. Ψ is the N-electron wavefunction for the system and E is its energy. The starting point for any calculations of the electronic structure of a molecule is the solution of Schrödinger's equation using a quantum chemistry package such as GAMESS-UK,[43] Dalton,[44] Orca[45] or Gaussian.[46] The only input required is an initial geometry. The program can determine the nearest geometry, which will be a (possibly local) minimum energy structure, together with the electronic wavefunction or electronic density, from which all manner of molecular properties can be calculated. Often molecular calculations are referred to as gas phase calculations, although this is not strictly correct. Calculations of a molecule or cluster of molecules are isolated from all environmental influences; there is neither temperature nor pressure. A short description of the various algorithms used to solve the quantum mechanical molecular electronic structure problem is given below. Quantum mechanics has limited application to large molecular systems because the computational effort required to solve the equations increases rapidly with molecular size.

True *ab initio* methods usually begin with the Hartee–Fock approach[47] and have the unique property that the calculation can always be improved, as long as there is sufficient computational resource available. An improvement to a calculation can be made in one of two ways; the basis set of functions used to describe the electron density can be expanded or a more sophisticated method can be used to account for the correlated motion of the electrons. The most accurate methods used routinely are those based on coupled cluster theory and in particular the CCSD(T) method (which indicates a coupled cluster method including single and double excitations with triple excitation treated perturbatively) is capable of very accurate calculations as long as a sufficiently large basis set is employed. The second order Moller Plesset perturbation theory method (MP2) is not as accurate but may be used on much larger molecules than CCSD(T) methods.

Density functional theory (DFT) is a relative newcomer to the field of computational chemistry. Rather than calculating a wavefunction, density functional methods calculate the electron density directly. Kohn and Sham[48] provided the theoretical foundations for the approach. Instead of describing electron correlation through the wavefunction, correlation is introduced as a potential. The method assumes a functional form for the contribution of the electron density to the electron correlation and to the electron exchange energies. The functional may be local (local density approximation functionals), in which case it depends only on the value of the electron density in space. More recent and more accurate functionals use a non-local potential[49] and they depend on both the values of the electron density and its derivatives (gradient corrected functionals). DFT has proved to be as computationally efficient as Hartree–Fock methods whilst providing results with reliability as good as some of the post Hartree–Fock methods such as MP2. This combination of features has meant that it has become the method of choice for quantum calculations on larger molecules. However it is not possible to improve a DFT calculation consistently as the ideal functional to describe electron correlation is not known. A drawback of using molecular quantum mechanical calculations of all types for the calculation of lattice energies is that such calculations do not include the effect of the environment. One recent approach that includes these effects is to perform the quantum mechanical calculation as though it were embedded in a polarisable medium with an effective dielectric constant.[50] The charge densities and the energies obtained in a medium with a dielectric constant of 3.0 appear to represent the crystalline environment quite well and can be used to estimate the induction and polarisation energies of a molecule in the solid state.

Based on the same underlying principles as the molecular-based quantum methods, solid state quantum mechanics represents bulk material using periodic boundary conditions. The imposition of these boundary conditions means that it becomes possible to expand the electron density in periodic functions such as plane waves, as an alternative to the atom-based functions used in the molecular case. The efficiency of the calculations is enhanced by the use of pseudo-potentials to represent the core electrons and to make the electron density as smooth as possible near the nucleus, hence reducing the complexity of the plane wave expansion of the electron density. Because of the number of choices available for pseudo potentials, basis sets and whether calculations are done in real or reciprocal space, there are many choices of software for performing solid state quantum mechanical calculations. A few examples which have been used in crystal structure prediction include; the Vienna ab initio Simulation Package (VASP),[51,52] CASTEP[53] and CRYSTAL.[54] A wider ranging introduction to the area can be found in the references.[55,56]

A major drawback to nearly all the DFT functionals is that they are unable to describe the non-local dispersive interaction. A straightforward approach (DFT(d)) to correcting this was developed, in the case of molecular calculations by Grimme,[57] and in the case of periodic calculations by Neumann and Perrin.[58] Their approaches differ in detail but broadly speaking an empirical molecular mechanics type correction is added to the energies and gradients of all pair-wise interactions. The interaction has to be damped at short range to represent the effect of the overlapping of electron distributions on the dispersive interaction. Neumann and Perrin derived atomic parameters which depend on the atom hybridisation by determining C_6 coefficients from sum over states spectroscopic data and the parameters associated with the damping factor were determined by fitting to minimise the error in the prediction of the unit cell dimensions of representative low temperature crystal structures.

Gavezzotti developed a method for calculating lattice energy by integrating over the molecular electron density in the crystal.[59–61] The molecular electron density is typically taken from a molecular quantum mechanical calculation, although it is not restricted in this way as the density could come from solid state calculations too. The Coulomb interaction between the molecules is calculated by a numerical integration over the tabulated electron densities. The repulsion between molecules is calculated from the overlap between the electron densities and the dispersion interaction is calculated from a semi-empirical London-type formula. Molecular polarisation is also accounted for by assuming a linear polarisability model and calculating the electric field at each charge density point. The method is attractive in that it provides a breakdown of the interactions between molecules in the crystal into physically relevant terms. Care is needed, as neither the geometry of the lattice nor the electron density is determined in a self consistent manner.

4.2.1.4 Molecular Dynamics and Monte Carlo Methods

In the methods discussed above, both quantum mechanics and molecular mechanics calculate the total energy and its gradients as a function of the nuclear coordinates. Geometry optimisation using these methods finds the nearest local minimum energy configuration. Molecular dynamics and Monte Carlo methods provide two useful but distinct roles. On the one hand they provide algorithms for the rigorous calculation of thermodynamic properties. On the other hand, they can be used as convenient computational tools to explore the conformational and packing energy landscape of a crystal. Molecular dynamics calculations integrate Newton equations of motion forward in time, so that the dynamical behaviour of the system can be predicted. The calculations can be performed in a number of statistical mechanical ensembles, including NVT, NVE and NPT, where N, V, E, P and T indicate ensembles in which the number of atoms, the volume, the energy, the pressure or the temperature, respectively, are constant.

Whereas molecular dynamics follows the dynamical evolution of a system over time, Monte Carlo methods average over the phase space of the system. The averaging is usually done using a Metropolis sampling procedure. Given a starting point, a 'move' is attempted. As long as it is reversible, the move can be almost any change in the system, a change in volume, a change in pressure, a swap of two molecules in a crystal, or a rotation around a bond. It is this freedom of choice of move which allows the method to explore the phase space of the system rapidly. After the move, the energy of the new system is calculated. If the energy has gone down, the new state is accepted. If the energy has gone up, there is still a probability of it being accepted, which is determined by a Boltzmann probability, P, which is a function of the change in energy and the temperature, T:

$$P = \exp\left(\frac{E_{\mathrm{old}} - E_{\mathrm{new}}}{k_{\mathrm{B}}T}\right) \tag{4.11}$$

Boltzmann's constant is represented by k_{B}. It is more difficult to use Monte Carlo methods to investigate dynamic events as there is no intrinsic concept of time but an ensemble average over the generated states of the system should give the same equilibrium thermodynamic properties as molecular dynamics.

4.2.2 Searching for Potential Structures

The prediction of crystal structures involves the identification of minima on the potential energy landscape, regardless of the method used to calculate this landscape. Owing to the many degrees of freedom involved, this process is not trivial, in particular for co-crystals and salts. This section provides an overview of search methods used in crystal structure prediction and the most important pitfalls are discussed.

4.2.2.1 Conformational Analysis

A conformational search in the gas phase is usually the first step in any crystal structure prediction study. Conformational searching identifies low energy local minima on the molecular potential energy landscape. The degrees of freedom explored are usually only the torsional angles of the molecule as these lead to relatively small energy changes compared with other molecular degrees of freedom. The lowest energy structure found is not necessarily that which will give rise to the optimum packing in the crystal structure, so it is necessary to find all low energy conformers, as any of these may give rise to low energy crystal packings.

The simplest form of conformational search is a grid scan. Here each degree of freedom is explored in turn over a range of values which systematically samples that variable. The energies of the molecule at all geometries are calculated. Other degrees of freedom can be optimised, but all those being explored must be constrained to their value on the grid. Taking the example of two torsion angles, a typical conformational scan explores each angle from 0° to 350° in steps of 10°. This results in 36^2 energy calculations or constrained geometry optimisations, the results of which can be shown on a contour map of energy, with the x and y axes defining the geometry of the molecule (see Figure 4.2 for an illustration[62]). A grid scan is thorough and provides a clear picture of the roughness of the molecular potential energy surface. Although the number of calculations increases dramatically with the number of degrees of freedom, it is almost always possible to choose subsets of the degrees of freedom which are independent of each other and consider them separately. It may also be possible to reduce the search range for certain degrees of freedom due to symmetry or to increase the step size for less important degrees of freedom. For those cases where it is not possible to reduce the number of conformations to be considered to a manageable level, the potential energy landscape can be explored by choosing random values for the degrees of freedom or using global optimisation methods. However, with these approaches it is more difficult to know when to stop searching.

4.2.2.2 Space Groups and Degrees of Freedom

Crystal structure prediction requires algorithms which can search the potential energy surface of the crystal for all local minima. The search must explore the

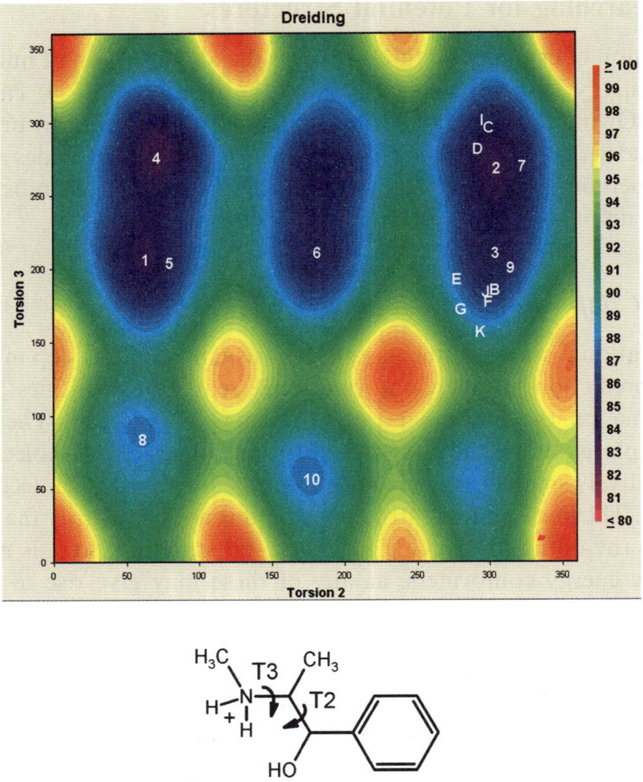

Figure 4.2 Top: Contour plot of conformational energy as a function of varying two torsion angles in ephedrinium.[62] Energies were calculated with the Dreiding force field and Gasteiger charges. The numbers indicate calculated local minima and the letters indicate experimentally observed conformations. Reproduced from Gourlay *et al.*[62] with permission from Elsevier. Bottom: Molecular structure of ephedrinium, showing the torsion angles 2 (indicated by T2) and 3 (T3) considered in the conformational analysis.

conformational freedom of the molecule as well as the packings of molecules within a unit cell, whose dimensions and shape can vary. Because of the interest in polymorphism and for operational reasons, crystal structure prediction aims to find not only the global minimum of a given potential energy surface but all of the minima within a specified energy window above the global minimum. In part this is recognition of the fact that the global minimum structure may not correspond to the experimental structure, but also because it is common to use screening methods, where less accurate methods are used initially to calculate candidate structures before refining their geometries, relative energies and their rank with a more accurate method.

An analysis of the Cambridge Structural Database (CSD)[63] for the space group frequency distribution of crystals is shown in Table 4.1. The 13 most common space groups, which account for over 90% of crystal structures in the database are shown, along with the number of degrees of freedom which must be considered when searching for packing alternatives of a crystal containing rigid molecules. Most of the entries refer to a single molecule in the asymmetric unit. When additional molecules are considered, a further six degrees of freedom are created for each new molecule, as can be seen from the $P1$ and $P\bar{1}$ entries in Table 4.1.

The earliest applications of crystal structure prediction were for rigid molecules and methods were developed which allowed the packing between molecules to be searched, making use of the symmetry of a particular space group. There are several advantages to doing this. The first is that it is not necessary to search all 230 space groups. As more than 90% of all crystal structures in the CSD are found in the 13 most common space groups, the experimental structure may be found with a high level of confidence by searching only in the most common space groups. Secondly, if a complete search is needed, this may be done most efficiently by considering more structures in the most common space groups rather than in the less common ones, weighting the effort of a search according to the likelihood of finding the structure in a particular space group. The final advantage is that, as can be seen from Table 4.1, the number of degrees of freedom that must be considered

Table 4.1 Frequency distribution of the most common space groups and their degrees of freedom.[a]

				Degrees of Freedom			
Space Group	*% Occurrence*[b]	*Crystal System*	*Z*[c]	*Cell*	*Trans.*	*Rot.*	*Total*
$P2_1/c$	35.1	Monoclinic	4	4	3	3	10
$P\bar{1}$	23.4	Triclinic	2	6	3	3	12
			4	6	6	6	18
$C2/c$	8.1	Monoclinic	8	4	3	3	10
$P2_12_12_1$	7.8	Orthorhombic	4	3	3	3	9
$P2_1$	5.4	Monoclinic	2	4	2	3	9
$Pbca$	3.5	Orthorhombic	8	3	3	3	9
$Pna2_1$	1.4	Orthorhombic	4	3	2	3	8
$Pnma$	1.2	Orthorhombic	8	3	3	3	9
Cc	1.1	Monoclinic	4	4	1	3	8
$P1$	1.0	Triclinic	1	6	0	3	9
			2	6	3	6	15
			4	6	9	12	27
$Pbcn$	0.9	Orthorhombic	8	3	3	3	9
$C2$	0.8	Monoclinic	4	4	2	3	9
$Pca2_1$	0.7	Orthorhombic	4	3	2	3	8

[a]The degrees of freedom refer to rigid molecules, with both translational (Trans.) and rotational (Rot.) degrees of freedom in a unit cell whose size and shape (Cell) also contribute to the total number of degrees of freedom.
[b]The percentage refers to data for all occurrences of the space groups in the CSD version 5.31 with updates up until May 2010.
[c]Z is the number of molecules in the unit cell.

when searching the potential energy landscape is fewer per molecule when space group symmetry is taken into account.

Consider the packing of four molecules in a unit cell. In the case of no symmetry (*P*1), there are 27 degrees of freedom to be taken into account, but if the search is done in the most common space group, $P2_1/c$, this reduces to 10. The importance of this reduction in complexity is made clearer by considering a grid scan of the rigid body packings in the two space groups. Let us assume that all the molecular rotational degrees of freedom are explored systematically from 0° to 350° in 10° steps, requiring 36 calculations for each rotational degree of freedom. This results in $36^{12}(4.7 \times 10^{18})$ and 36^4 (1.7×10^6) evaluations of the lattice energy in the *P*1 and $P2_1/c$ space groups, respectively, considering only the rigid body rotational degrees of freedom. Clearly there are great efficiency savings to be had by employing space group symmetry where possible. Furthermore, if it is assumed that the calculation of the lattice energy in *P*1 symmetry can be performed with 1 s of elapsed time, a grid scan of the rotational degrees of freedom would take 1.5×10^{11} years. This admittedly simplistic calculation leads to the conclusion that whilst a grid scan approach to determining the low energy structures is thorough and systematic, it is not an efficient method for crystal structure prediction.

The distribution of space groups is quite different for systems with more than one molecule in the formula unit than for the overall case. Cruz-Cabeza *et al.* published an analysis of the distribution of space groups for the case of solvates.[64] They found that 81% of solvates with an achiral main component crystallise in the most common space groups ($P\bar{1}$, $P2_1/c$ and $C2/c$) compared to about 65% for the overall statistics. The work demonstrates that the complexity of the search problem can be reduced by the judicious use of experimental information on the space group distribution of different crystal types.

4.2.2.3 Search Methods

Early implementations of search algorithms resulted in codes such as MOLPAK[65,66] and UPACK.[26,67] MOLPAK uses a repulsion only intermolecular potential and predicts candidate $Z' = 1$ crystal packings based on a number of co-ordination geometries observed experimentally in the most common space groups. UPACK was originally based on a grid search method[67] for exploring the potential energy landscape. Later developments[26] incorporated a random search method, allowed more than one molecule in the asymmetric unit and included the ability to search the torsional angles of the molecule using the same random search approach.

Of course when dealing with flexible molecules it is necessary to include further degrees of freedom. There are different approaches that can be taken to treat flexible molecules. Some methods treat a representative of each conformation of the molecule as a rigid entity. The search becomes more complex because many molecules may be needed to represent the molecular flexibility. Other methods treat the torsion angles which can give rise to the different conformations as additional degrees of freedom.

The methods originally developed in the Materials Studio Polymorph Predictor[21,68] are based on rigid body packing but include sophisticated global search methods using a simulated annealing algorithm. During the search, a random perturbation of the structure is made and the new structure is accepted or rejected on the basis of a Monte Carlo criterion. If the energy goes down the new structure is always accepted, if the energy goes up the new structure may be accepted, according to a probability calculated from the ratio between the energy increase and the expected thermal fluctuations in energy at the current system temperature (see equation (4.11)). By controlling the system temperature, the user controls the extent of exploration of the energy landscape. Typically the temperature is increased quickly to a maximum and then decreased slowly resulting in a set of low energy structures. Such calculations need to be repeated several times for different space groups and the results for different molecular conformations and space groups brought together into a single set of unique structures.

The methods developed in GRACE[69] use a Monte Carlo parallel tempering algorithm[70] to explore the potential energy landscape in all 230 space groups and include the important torsion angles as degrees of freedom with which to explore the low energy barriers connecting energy minima. The Crystal Predictor package[71,72] performs searches of both the packing in the unit cell and the important torsion angles in 59 of the most common space groups. Both GRACE and Crystal Predictor are implemented to make use of parallel computing, in recognition of the computer resources which can be needed to explore the potential energy landscape.

4.2.2.4 Clustering

After generating candidate structures using the search algorithms outlined above, it is necessary to search the structures for those which are unique. Because a unit cell for the same structure may be chosen in so many different ways, it is necessary to ensure that only one unique structure of those which are equivalent is carried forward for further investigation. This step, often referred to as the clustering step, requires the comparison between structures, but without imposing any symmetry constraints in the comparison. The COMPACK algorithm[73] is commonly used. It builds a cluster of nearest neighbour molecules around a selected molecule, matches the molecules in the two structures being compared and then computes the root mean squared (RMS) deviation between the atoms in the matching molecules. Another implementation of a clustering algorithm has been described in detail[74] and has been implemented for triclinic, monoclinic and orthorhombic systems.

Methods of clustering are implemented in most packages devoted to crystal structure prediction.[68,69] For flexible molecules it is always helpful to minimise the energy of the candidate structures without any constraints, before performing clustering. This ensures that the molecular geometries in equivalent structures are as similar as possible, given the accuracy of the minimisation.

Although this aspect of crystal structure prediction is not the subject of a great degree of literature, it is not a trivial problem to decide whether two crystal structures are the same and care has to be taken that valid structures are not thrown away during this step.

4.2.2.5 Stability Ranking

The candidate structures selected during the initial search and after clustering are usually selected on the basis of an initial molecular mechanical force field calculation and sometimes molecular flexibility has not been fully integrated into the search. The accuracy of such calculations for relative energies and for the prediction of geometries is not sufficient for reliable crystal structure prediction. However, the sheer number of structures which are considered during the search precludes the use of more accurate methods at that stage. It is therefore necessary to perform more sophisticated calculations on the low energy candidate structures after the search has completed in order to be more confident about the accuracy of the final lattice energy. One approach[75] is to use calculations involving a mixture of molecular mechanics for the dispersive interactions, quantum mechanics for the charge–charge interactions and quantum mechanics for the calculation of the molecular strain energy which results from crystal packing. Another approach is to use fully periodic quantum mechanical calculations.[58] At this stage of the crystal structure prediction it is also possible to include additional corrections to the energy calculation, in particular, the inclusion of zero-point energies and entropic contributions to the free energy. For harmonic motion, the corrections can be calculated from calculations of the phonon spectra of the crystals.[76] However, such calculations are generally restricted to molecular mechanics force fields. Although these calculations are also possible using solid state DFT methods, they are very time consuming.

4.2.2.6 Summary

A general outline of the procedure for crystal structure prediction is shown in Figure 4.3. The outline applies to the majority of methods which have been used for the crystal structure prediction of organic molecular crystals. In the field of crystal structure prediction for inorganic materials the problem has been addressed in different ways.[77] Methods such as the USPEX approach[78,79] have been developed which ignore space group symmetry and concentrate on the global optimisation of a large unit cell using techniques such as the evolutionary algorithms. These algorithms borrow concepts from the way nature optimises function through selection, recombination and mutation. Another approach which has been taken for crystal structure prediction is that of metadynamics.[80,81] Here, molecular dynamics calculations are used to find and explore those important degrees of freedom that lead to structural change.

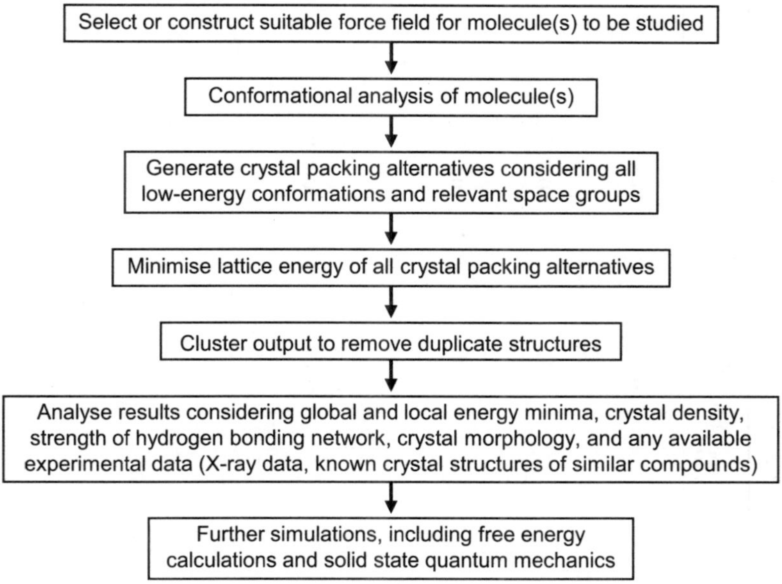

Figure 4.3 Schematic overview of the general procedure used in crystal structure prediction.

4.3 Application Examples

In this section, a number of examples of crystal structure prediction of small organic molecules, co-crystals, solvates and salts will be presented.

4.3.1 Blind Tests of Crystal Structure Prediction

A series of blind tests of crystal structure prediction has been organised by the Cambridge Crystallographic Data Centre to evaluate the development of crystal structure prediction tools on an ongoing basis. So far, the tests have taken place in 1999,[82] 2001,[83] 2004,[84] 2007[85] and 2010. The results of the 2010 blind test will not be discussed here as they have not yet been published. In each test, the molecular structures of three or four compounds are provided to a number of invited participants who have up to six months to submit a maximum of three predicted crystal structures for each compound. The experimental crystal structures have been determined, but are kept hidden from the participants. After the predictions have been submitted to an independent referee, the experimental structures are disclosed and compared to the various predicted structures. The participants then gather at a workshop to discuss the results and a joint publication is produced.

Table 4.2 summarises the results obtained in the first four blind tests, showing the molecular structure of each of the target compounds, their blind test number, the number of participants attempting each target and the number

Table 4.2 Summary of blind test results and results obtained with the DFT(d) method.

Blind Test Year and Molecule	*Molecular structure*	*No. Correct predictions/ No. Participants*	*Rank with DFT(d) Method*	*ΔE with DFT(d) (kcal mol^{-1})*	*RMS Geometric Deviation (Å)*
1999-I	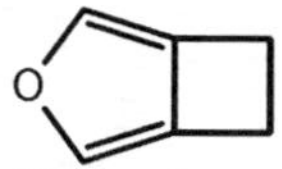 Two polymorphs	A: 4/11 B: 0/11	1 2	0.00 0.11	0.062 0.100
1999-II	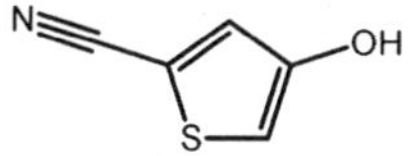	1/8	2	0.01	0.489
1999-III	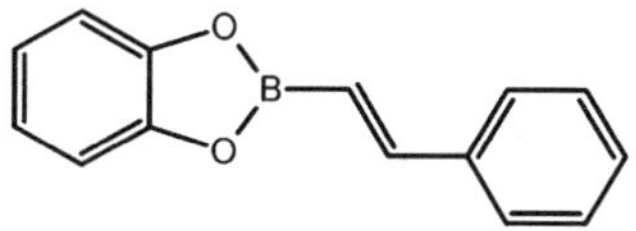	1/11	1	0.00	0.078
2001-IV	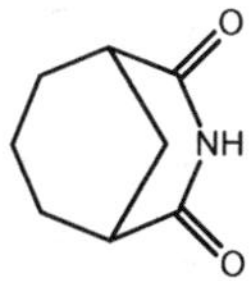 Second polymorph discovered after blind test	A: 2/15 B: not part of blind test	1 2	0.00 0.13	0.083 0.090
2001-V	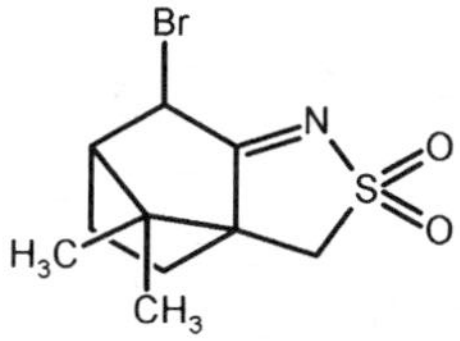	4/15	4	0.36	0.096

2001-VI	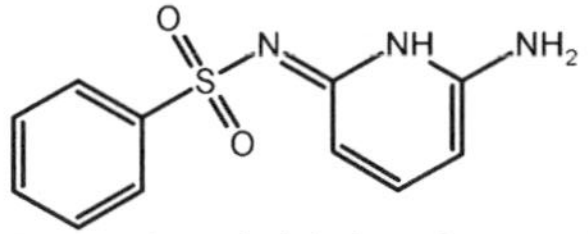 Second and third polymorphs discovered after blind test	A: 0/11 B: not part of blind test C: not part of blind test	1 2 3	0.00 0.21 0.22	0.113 0.093 0.091
1999-VII		1/6	1	0.00	0.074
2004-VIII	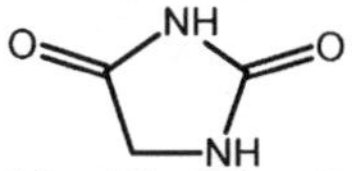 Not blind (partial structure discovered midway blind test)	4/15	1	0.00	0.183
2004-IX	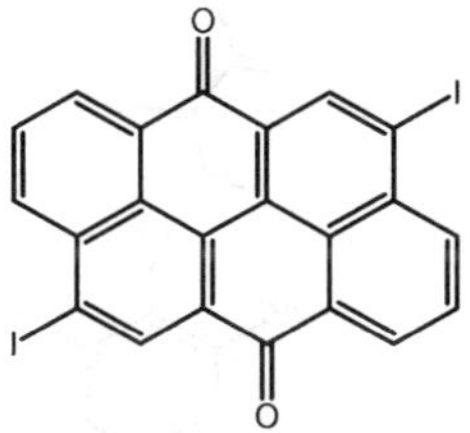	1/15	1	0.00	0.132
2004-X	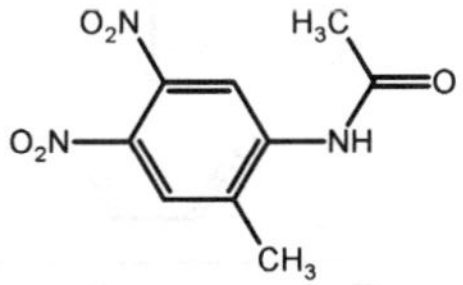	0/15	1	0.00	0.064
2004-XI	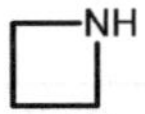 $Z' = 2$	0/18	1	0.00	0.107

Table 4.2 (*Continued*)

Blind Test Year and Molecule	*Molecular structure*	*No. Correct predictions/ No. Participants*	*Rank with DFT(d) Method*	*ΔE with DFT(d) (kcal mol^{-1})*	*RMS Geometric Deviation (Å)*
2007-XII	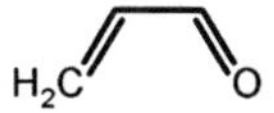	4/13	1	0.00	0.127
2007-XIII	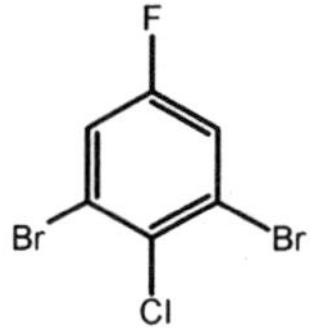	4/14	1	0.00	0.082
2007-XIV	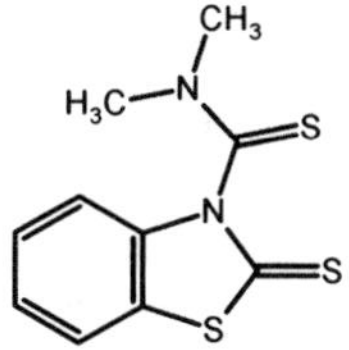	3/12	1	0.00	0.130
2007-XV	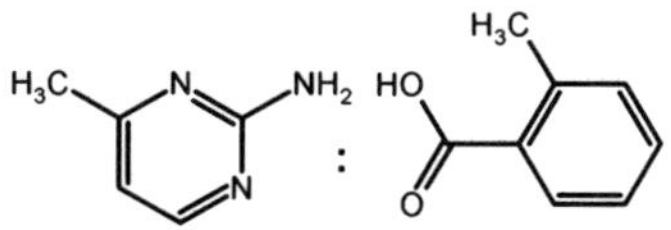 Co-crystal	2/12	1	0.00	0.075

of successful predictions. In the 2007 blind test, one method predicted each of the four target crystal structures correctly as the lowest-energy structure.[86] This solid state DFT(d) approach,[58] discussed in Section 4.2.1.3, was retrospectively applied to the structures of the first three blind tests. For each target compound, the experimental structure, as well as all the predictions made by all the participants, were recalculated.[87] The results of this study are reported in the last three columns of Table 4.2. The RMS geometric deviations between the experimental structures and the DFT(d) simulated structures were calculated using the crystal packing similarity tool in Mercury CSD 2.0 with a 16-molecule comparison (ignoring all hydrogen atoms).[73]

Unbeknown to the participants of the first blind test, the crystal structures of two polymorphs of molecule I had been determined, a $P2_1/c$ form and a meta-stable *Pbca* form.[82] The latter form was predicted correctly by four participants, but the stable $P2_1/c$ form was not found by any of the participants as one of their three predictions. Subsequent studies using molecular mechanics calculations to compute the lattice energies of the two polymorphs more accurately failed to obtain a satisfactory result for the stable form.[88,89] Using solid state quantum mechanics, however, it was found that only about 0.1 kcal mol^{-1} separates the calculated lattice energies of the two polymorphs, indicating that either form may be observed experimentally.[87]

For molecule II, whose crystal structure was predicted correctly by one of the 1999 blind test participants,[82] the DFT(d) method finds the experimentally observed structure to be the second most stable crystal packing alternative with an energy difference of only 0.01 kcal mol^{-1} compared to the lowest energy structure.[87] Such a small energy difference falls within the numerical error of the calculations. The DFT(d) optimised structure shows a geometric deviation of nearly 0.5 Å in comparison to the experimental structure, which is a much larger discrepancy than the values found for any of the other structures (around 0.1 Å). So far, the reason for this large deviation remains unclear despite efforts to unravel the problem.

A second polymorph of molecule IV was discovered after the second blind test.[90] Since the new polymorph has two independent molecules in the asymmetric unit, it was not predicted by any of the participants because it was indicated at the start of the blind test that the structure would have just one molecule in the asymmetric unit. Two participants predicted the first polymorph correctly.[83] The DFT(d) method ranks the two polymorphs as the first and second lowest energy structures among all the structures predicted by all the participants.[87]

The crystal structure of molecule V was predicted correctly by four participants in the 2001 blind test,[83] but the DFT(d) method ranks it as the fourth most stable packing alternative with a relative lattice energy of 0.36 kcal mol^{-1}. The DFT(d) lowest energy structure has a significantly higher density than any of the other low energy structures (including the experimental structure) and the results of a series of lattice energy minimisations as a function of pressure showed that its relative stability increases substantially in comparison to all other packing alternatives with increasing pressure.[87] It is therefore predicted that a new polymorph of molecule V may be obtained and that the discovery of this new polymorph may be facilitated by conducting a crystallisation experiment under pressure.

After the second blind test, two additional polymorphs have been discovered for molecule VI.[91,92] Neither the original polymorph nor either of the two additional polymorphs (one of which has two independent molecules in the asymmetric unit) were predicted by any of the participants.[83] The failure to predict these crystal structures has led to speculations in the literature that the observed polymorphs are kinetically favoured and could not be found in crystal structure prediction studies because such studies can only locate thermodynamically favoured forms.[91–95] However, a recent crystal structure prediction study using the DFT(d) method found all three experimentally observed polymorphs as the three most stable crystal packing alternatives, in the correct stability order.[96] These results, summarised in Table 4.2, indicate that previous failures to predict the crystal forms of molecule VI correctly did not arise because of the kinetic nature of the crystallisation process, but because of the inaccuracy of the molecular mechanics approaches used. The data reported in Table 4.2 shows that a purely thermodynamic methodology can predict the likely outcome of a crystallisation experiment, but it is a prerequisite that an accurate lattice energy calculation tool is used.

The crystal structure of molecule VIII of the third blind test was predicted correctly by four participants, but these predictions were not considered as blind because it was discovered halfway through the test that some experimental information on the structure (space group, unit cell dimensions and hydrogen bonding network) had been presented in a poster at the British Crystallographic Association meeting in March 2002.[84,97,98] This information would make it straightforward to construct the crystal structure. As a substitute, molecule XI was introduced, but none of the participants predicted its crystal structure correctly.[84] This failure may be associated with the fact that this structure has two independent molecules in the asymmetric unit and because it was introduced halfway through the blind test, many participants did not consider the $Z' = 2$ possibility due to time constraints. As discussed in Section 4.2.2.2, structures with more than one molecule in the asymmetric unit are considerably more difficult to predict than $Z' = 1$ crystal structures. From a computational point of view, there is little difference between co-crystals and structures with two of the same molecule in the asymmetric unit. The experimental structures of both molecule VIII and molecule XI were ranked correctly by the DFT(d) method as the most stable packing alternatives among the sets of predicted structures.[87]

Molecule IX was not considered in the re-evaluation study with the DFT(d) method because its van der Waals correction was originally not parameterised for iodine.[87] This parameterisation has now been carried out and the crystal structures of molecule IX have been optimised using the DFT(d) method.[99] The experimental structure is found to be the lowest energy crystal packing alternative among all the predicted structures submitted by the 2004 blind test participants.

The fourth target compound in the 2007 blind test, compound XV, is a co-crystal. Its structure was predicted correctly by two participants.[85] Because this compound has two independent molecules in the asymmetric unit which are interacting which each other through strong hydrogen bonds, some participants opted to reduce the computational burden of searching for all possible crystal structures by using the hydrogen bonded dimer as the search unit rather than treating the two molecules independently. Four possible planar dimer structures can be constructed.[85] It is interesting to note that both successful predictions for compound XV were achieved by treating the two molecules in the asymmetric unit fully independently during the search for crystal packing alternatives. Some of the participants who used the dimer approach also found the correct crystal structure, albeit not among their official three predictions. The strategy of using bimolecular units as the search unit in polymorph prediction of structures with more than one independent molecule in the asymmetric unit is attractive, but it is a prerequisite that the molecules in the unit interact strongly with each other, otherwise it is not possible to define a limited number of geometries. There is also the risk that the molecules interact in an unexpected manner, thus rendering the structure impossible to predict using this approach. This pitfall can be avoided by treating the molecules independently at considerable computational expense. In the case of compound XV, one of the successful predictions considered only about 30% of all possible structures because the computational effort involved in treating the molecules independently caused the simulation to run out of time.[86]

4.3.2 Co-crystals

Crystal structure prediction of co-crystals is more difficult than the prediction of crystal structures with a single molecule in the asymmetric unit because of the increased number of degrees of freedom. The difficulties in searching for co-crystal structures are illustrated in early work on urea–dihydroxy benzene co-crystals,[100] demonstrating the inefficiency of a systematic search approach even when the unit cell dimensions are known. Owing to the multitude of interaction patterns often available to the co-crystallising molecules, crystal structure prediction of co-crystals usually leads to more low-energy crystal packing alternatives than observed for crystals with a single molecule in the asymmetric unit. A few crystal structure prediction studies on co-crystals have been published and are summarised in this section.

4.3.2.1 Prediction of Co-crystal Formation

The question of whether crystal structure prediction may be used to predict the formation of co-crystals of pharmaceutical molecules was addressed in a pair of papers.[101,102] The first publication addressed the question as to whether lattice energy calculation methods are sufficiently accurate.[101] The authors calculated the lattice energies of 26 known co-crystals and compared their energies with

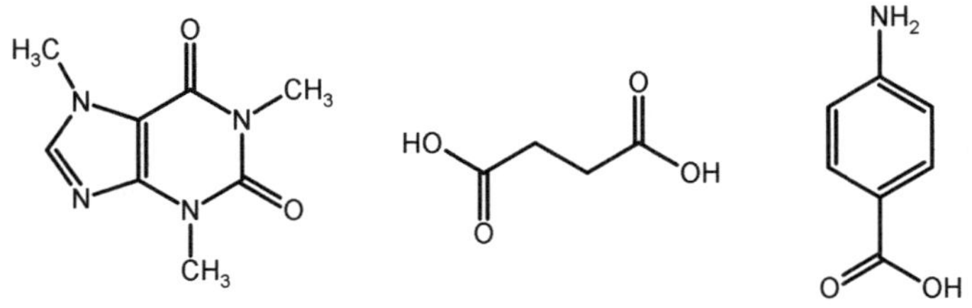

Figure 4.4 From left to right, molecular structures of caffeine, succinic acid and 4-aminobenzoic acid.

those of their component single molecule crystals. The co-crystals were chosen so that at least one component could be used as a co-former in a pharmaceutical co-crystal; succinic acid, 4-aminobenzoic acid and caffeine (see Figure 4.4).

Caffeine is particularly problematic as the experimental structure of the single component crystal is disordered. This issue was avoided as all calculations of caffeine co-crystal stability were reported relative to its most stable crystal structure, as predicted by crystal structure prediction.[101] The lattice energies were calculated using multipole expansions of the charge distributions of the molecules and a molecular quantum mechanical calculation of the energetic effect of packing on the flexible torsion angles of the molecules. Hartree–Fock calculations were used to calculate the increase in intramolecular energy due to packing and an MP2 charge density was used to calculate the atomic multipole moments. For caffeine and succinic acid the agreement with experimental observation was reasonable, with four out of six and six out of eight co-crystal structures respectively being lower in energy than the lowest single component lattice energies.[101] The co-crystal of benzamide (whose single component has two polymorphs) with succinic acid was found to be more stable than the single components when comparison was made with the energy of the metastable polymorph of benzamide, but less stable when compared with the energy of its stable polymorph. In general the lattice energy differences between the co-crystals and the single component crystals were of a similar size to those found between different polymorphs. This finding highlights the importance of calculating sufficiently accurate energies in order to make predictions about stability.

The results were not so encouraging for 4-aminobenzoic acid. If the heat of co-crystallisation was calculated relative to the less stable α polymorph then 11 of the 12 co-crystal structures were more stable than their single component crystals. But if the comparison was made with the more stable β polymorph, only 2 of 12 agreed with experiment.[101] This discrepancy was due to the large predicted lattice energy difference of 4.8 kcal mol^{-1} between the two polymorphs, compared with the experimental value of 1.3 kcal mol^{-1} for the enthalpy difference between the two forms. A recurring question in crystallisation is the role of kinetics in determining whether a crystal structure is observed or not. If kinetics is more important for co-crystals, then even the most accurate lattice energy calculations will not be able to predict whether a

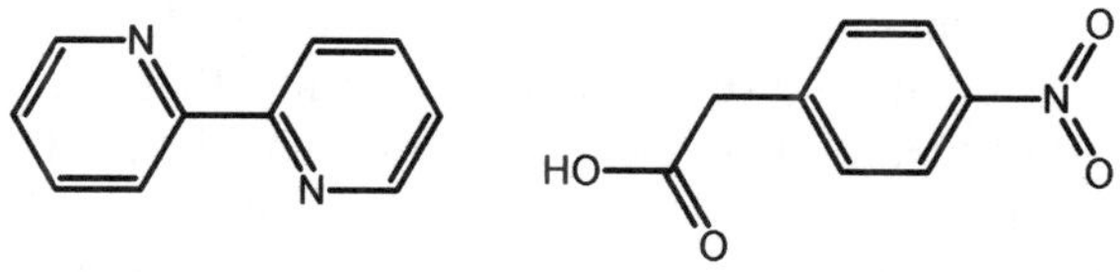

Figure 4.5 Molecular structures of 2,2′-bipyridine (left) and 4-nitrophenylacetic acid (right).

co-crystal crystallises or not. The results emphasize the importance of accurate lattice energy calculations. The force field for 4-aminobenzoic acid overestimates the stability of its stable polymorph, which leads to the prediction that the majority of its co-crystals are unstable.

In the second paper of the series,[102] a different question was posed. Given the ability to calculate accurate lattice energies, could the machinery of crystal structure prediction be used to generate the observed crystal packings? The systems chosen for examination were co-crystals of 4-aminobenzoic acid with 2,2′-bipyridine and 4-nitrophenylacetic acid (see Figure 4.5). The results of the first study had shown some of the limitations of the force field being used, especially for 4-aminobenzoic acid and as a result several improvements in the calculation of the lattice energies were employed in this second study.

The parameters defining the repulsion and dispersion non-bonded interactions for 4-aminobenzoic acid in the second paper were modified to use those of Williams,[28] with a specific modification of the Williams carboxylic proton–pyridine nitrogen repulsion potential for the 4-aminobenzoic acid–2,2′-bipyridine co-crystal. Calculations now predicted that the β polymorph of 4-aminobenzoic acid was 1.3 kcal mol^{-1} more stable than the α polymorph, which was in much better accord with experiment.[102] Crystal structure searching was performed for co-crystals and for the single component crystals using Crystal Predictor.[71,72] The search algorithm successfully found the experimental structures for the single component systems considered.[102]

The energy ranking was carried out in a three stage process. The initial rankings were performed with an isotropic non-bonded potential with a point charges model and torsion potentials derived from molecular quantum mechanical calculations. The second stage involved a more sophisticated electrostatic model using distributed multipoles derived from quantum mechanical calculations with inclusion of limited conformational dependence of the charge density by allowing analytical rotation of the multipole moments. The final and most accurate calculations were performed using DMAFlex[72] which couples the calculation of the distributed multipoles with an optimisation of the molecular geometries in the crystal.[102]

The reproduction of the experimental crystal structure geometries was good at all stages of the calculation. The root mean squared displacements between the six experimental structures and the corresponding predicted crystal structures were less than 0.5 Å using the COMPACK[73] algorithm. The correlations observed between the energies at the various stages of the calculations were interesting. The correlations were quite good for the single molecule crystals of 4-aminobenzoic acid and 2,2′-bipyridine, indicating that the global minima using the final stage method (the most accurate) would be captured. The correlations were poorer for the co-crystals and for 4-nitrophenylacetic acid and there was much less confidence in capturing the global minima. Of the two known co-crystals with 4-aminobenzoic acid, only that with 2,2′-bipyridine was predicted to be more stable than its single component crystals. Indeed this crystal was found to have the lowest lattice energy of any predicted with a 2:1 stoichiometry. Despite this, the crystal structure prediction for the 1:1 stoichiometry of the co-crystal indicated a more stable crystal than the experimentally observed 2:1 stoichiometry. Several low energy structures were predicted for the co-crystal with 4-nitrophenylacetic acid based on 4-aminobenzoic acid dimers, which were more stable than the experimental structures.[102] In addition, the structure most similar to the experimental co-crystal structure was higher in energy than the single component crystal lattice energies. Indeed no 1:1 co-crystals were found that were more stable than the most stable component crystals.

4.3.2.2 Stoichiometry Prediction

Another example of a computational study of co-crystal stability is given by the work on caffeine (see Figure 4.4) co-crystallising with monohydroxybenzoic acids (HBA) (see Figure 4.6).[103] Experimentally[104] it is found that 2-HBA and 3-HBA form 1:1 co-crystals with caffeine, while 4-HBA forms a concomitant mixture of 1:2 and 2:1 co-crystals.

To assess the relative stabilities of the various stoichiometries, global minima were identified by crystal structure prediction and compared to the energy of 2 moles of crystalline caffeine and 2 moles of crystalline HBA. As before for caffeine,[101] the lattice energy of the lowest energy crystal was used, ignoring the observed experimental disorder. 2-HBA has one known polymorph and 3-HBA and 4-HBA have two known polymorphs each. However, because of uncertainties in their structure determinations only one polymorph for each of 3-HBA and 4-HBA was considered in the analysis.[103]

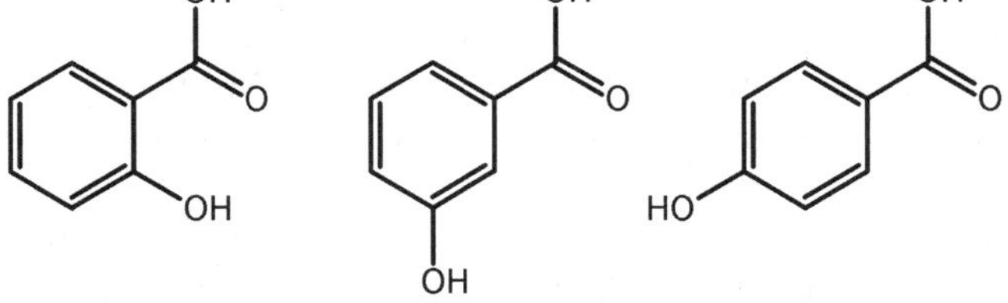

Figure 4.6 From left to right, molecular structures of 2-hydroxybenzoic acid (2-HBA), 3-hydroxybenzoic acid (3-HBA) and 4-hydroxybenzoic acid (4-HBA).

Crystal structure prediction was carried out using a multi-stage process. Crystal Predictor[71,72] was used to generate structures for the isolated components and the 1:1 co-crystals of caffeine and the HBAs in 15 of the most common space groups. An initial molecular quantum mechanical calculation was used to optimise possible structures for each conformer. The initial structure generation was performed using rigid molecules, a point charge model for the charge distributions and a Buckingham potential for the non-bonded interactions. Further refinement of the low energy structures was performed using a more accurate distributed multipole representation of the charge distribution, followed by a final refinement of the flexible internal degrees of freedom using a contribution to the intramolecular energy, calculated by molecular quantum mechanics.[103] Calculations of the lattice energies and the relaxed geometries of the known $Z'' = 3$ co-crystals (1:2 caffeine:3-HBA and 2:1 caffeine:4-HBA) were performed and the results were in reasonable agreement with experiment.

For the 1:1 co-crystal with 2-HBA, the global energy minimum was found to correspond to the experimental structure and was more stable than the isolated components by 0.1 kcal mol^{-1}. The structure corresponding to the experimental data for the 1:1 co-crystal with 3-HBA was found ranked third and was more stable than the isolated components by 0.6 kcal mol^{-1}.[103] Experimentally, the conformations of 3-HBA are different in the single component crystal in comparison to the co-crystal and the crystal structure prediction results are in accord with this. Thus the change in energy resulting from changing conformation in the isolated 3-HBA component crystal to the conformation in the co-crystal is more than compensated for by the additional packing energy gained. The results for the 4-HBA co-crystals were not as clear cut. The most stable structure was found to be a 1:1 co-crystal,[103] yet experimentally 1:2 and 2:1 co-crystals are found concomitantly. This discrepancy could be due to the lattice energy model being used or due to kinetic factors controlling the crystallisation process.

4.3.2.3 Pseudo-racemates of Amino Acid Co-crystals

Amino acids with hydrophobic side chains are known to prefer particular hydrogen bonding patterns when forming either racemic crystals or pseudo-racemates (sometimes called a quasi-racemate), depending on the branching in the side chain.[105] Racemic crystals contain equal numbers of *R* and *S* enantiomers of a single molecule, whilst quasi-racemates are crystals containing equal numbers of *R* and *S* enantiomers of different molecules. A crystal structure prediction study has been performed on three quasi-racemates (see Figure 4.7); (I) L-*allo*-isoleucine:D-leucine, (II) L-norvaline:D-methionine and (III) L-2-aminobutyric acid:D-norleucine.[105]

The subject of this study is particularly challenging for any crystal structure prediction methodology for several reasons. First of all the molecular entity in the solid state (a zwitterion) is different from that in the gas phase. This means that the transfer of molecular information from quantum mechanical

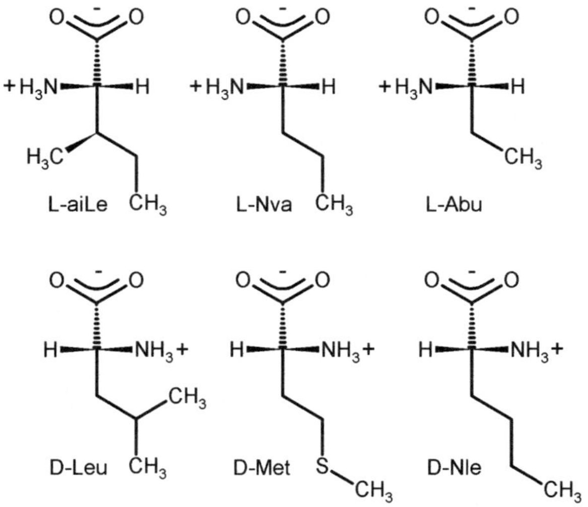

Figure 4.7 Chemical structures of the amino acids L-*allo*-isoleucine (L-aiLe), D-leucine (D-Leu), L-norvaline (L-Nva), D-methionine (D-Met), L-2-aminobutyric acid (L-Abu) and D-norleucine (D-Nle).

calculations in the gas phase to provide information on interactions in the solid state must be done extremely carefully. Secondly, there are two independent molecules in the asymmetric unit, which makes the search for packing alternatives challenging. Finally, these molecules are flexible, so the already difficult search problem becomes even more difficult.

After choosing the molecular conformations to be considered in the packing calculations and optimising them using molecular DFT calculations, co-crystal structures were generated in five of the most commonly observed space groups ($P1$, $P2_1$, $P2_12_12_1$, $P2_12_12$ and $C2$) using all combinations of molecular conformations.[105] A hierarchical approach was adopted to find the most stable structures for each co-crystal. Initial lattice energy minimisations were carried out using atomic centred point charges derived from the molecular electrostatic potential for each conformer, conformers having been carefully selected and optimised with a quantum mechanical calculation to reflect the range of conformations observed experimentally. A Dreiding force field with specific non-bonded van der Waals parameters for the solid state was used to allow torsion angles to vary, along with the molecular positions in the lattice and the lattice parameters. A final ranking was carried out with an intramolecular contribution to the lattice energy computed from a single-point quantum mechanical polarisable medium calculation for each molecule in the asymmetric unit and an intermolecular contribution computed from the van der Waals non-bonded energy, together with a charge–charge interaction based on distributed multipoles. The crystal structure prediction was carried out blind. That is, the calculations were performed without knowledge of the experimental results although the experimental work was published as part of the same paper.[105]

Co-crystal I (L-*allo*-isoleucine:D-leucine) consists of two amino acids with branched side chains and the structure with the lowest lattice energy was found to match the experimental crystal structure. Co-crystal II (L-norvaline:D-methionine), which has two amino acids with linear side chains, was predicted to have a global minimum energy structure 1.9 kcal mol^{-1} below the rank 6 structure, which was the most similar to experiment. The hydrogen bonding patterns in the rank 1 and 6 structures were found to be very similar. The final co-crystal considered (L-2-aminobutyric acid:D-norleucine) also has two amino acids with linear side chains, consequently the hydrogen bonding motif was expected to be similar to that for the L-norvaline:D-methionine co-crystal. Remarkably, both the experimental structure and its corresponding predicted structure, which was found to be the global minimum, had a hydrogen bonding motif similar to the L-*allo*-isoleucine:D-leucine co-crystal.[105]

4.3.3 Solvates

From a computational point of view, the prediction of the crystal structures of solvates poses the same issues as the prediction of co-crystal structures. Crystal structure prediction studies on solvates have only recently been attempted.

4.3.3.1 Prediction of Solvate Stoichiometry

Novel techniques for the creation of co-crystals and solvates such as neat and liquid assisted grinding have challenged the ability of crystal structure prediction to predict stoichiometry from first principles. Recent work has addressed the problem of predicting solvate stoichiometry of acetic acid (the solvent) with various molecules including carbamazepine (CBZ) and its 10,11-dihydro derivative (DHCBZ),[106] urea[107] and theobromine[108] (see Figure 4.8).

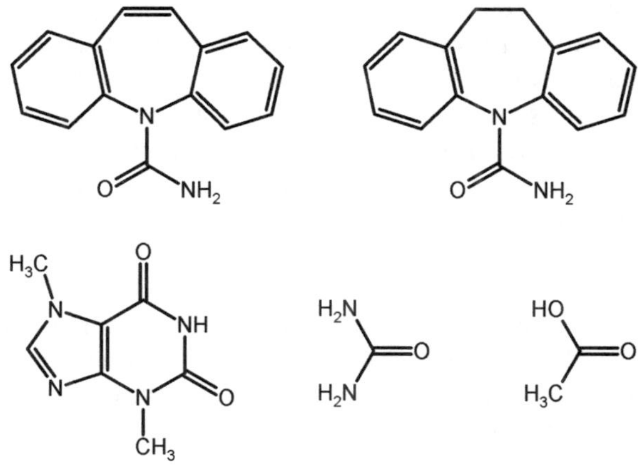

Figure 4.8 Molecular structures of carbamazepine (CBZ; top left), 10,11-dihydrocarbamazepine (DHCBZ; top right), theobromine (TB; bottom left), urea (bottom middle) and acetic acid (AcOH; bottom right).

For CBZ and DHCBZ solvates with acetic acid (AcOH), 1:1 crystals were generated in the most common space groups (six space groups for the CBZ solvate and two for the DHCBZ solvate).[106] Crystal structure searching was performed using the Polymorph Predictor from Accelrys.[21] Energy ranking was performed using a multipole moment of the charge density and the amide pyramidalisation was incorporated by considering the packing of structures with a pyramidal nitrogen and using an intramolecular contribution to the total energy from a molecular quantum mechanical calculation. At the time the calculations were performed, only the structure of a 1:1 solvate of CBZ was known. The lowest energy solvate structure for CBZ:AcOH was predicted to be more stable than its single components and agreed with the known experimental structure. For DHCBZ, no experimental solvate structure was known. The simulations on 1:1 DHCBZ:AcOh predicted a crystal structure which was isomorphic with the experimental CBZ solvate crystal and ranked third according to lattice energy.[106] This predicted structure (also stable with respect to the single components) was subsequently found to agree well with both the powder pattern and the single crystal X-ray analysis of the DHCBZ:AcOH crystal. This is an interesting finding given that it is known that the small difference in molecular structure between CBZ and DHCBZ is sufficient to result in different hydrogen bonding patterns in their low energy structures.[109]

The study on theobromine[108] explored the possible stoichiometries of its solvates with AcOH. The search for low energy structures with 1:1 and 1:2 TB:AcOH stoichiometries was performed using the Polymorph Predictor[21] in the space groups $P2_1/c$ and $P\bar{1}$, molecular geometries and potential derived charges having been determined by molecular quantum mechanical calculations. The final energy rankings and geometry optimisations were performed using distributed multipoles and rigid molecules. The 1:1 stoichiometry was found to have a structure that was 1.1 kcal mol^{-1} lower in energy than any 1:2 structures and 5.5 kcal mol^{-1} lower in energy than the component crystals. The computational work had been stimulated by the need to understand the liquid assisted grinding experiments that were also reported. Unable to grow single crystals and not having good enough powder diffraction data to determine the crystal structure, the comparison of the simulated powder diffraction pattern of the lowest energy 1:1 structure was close to the experimental powder pattern. This confirmed the 1:1 stoichiometry of the crystal and aided its structure determination.[108]

As part of the study of the solvate of urea with AcOH,[107] an experimental study had determined the crystal structure of the 1:2 urea:AcOH solvate. Crystal structure prediction studies were carried out on 1:1, 1:2 and 1:3 urea:AcOH solvates in order to investigate the ability of crystal structure prediction to predict stoichiometry. The methodology used was similar to the other investigations of AcOH solvates mentioned above. In addition, for the case of the 1:2 solvates ($Z'' = 3$) the free energies of the crystals were estimated by calculating the vibrational contribution to the entropy. Five, two and one space groups were searched for urea:AcOH solvates with stoichiometries 1:1, 1:2 and 1:3, respectively.[107] Limiting the search for the 1:3 solvate structure to $P\bar{1}$ only was especially restrictive, but understandable given the computational complexity of the calculations and the results provided some information on relative stabilities of the different stoichiometries. To simplify the search further, the most common urea–acetic acid dimer units found in the searches of 1:1 and 1:2 stoichiometries were used as rigid units, together with two acetic acid molecules, in the search of 1:3 stoichiometries. The most stable structure was found to have 1:2 stoichiometry, which is in agreement with experiment.[107] The predicted structure corresponding to the experimental observation was ranked second, 0.3 kcal mol^{-1} above the global minimum. To compare the relative energies of different stoichiometries at 0 K, the energies of all possible combinations of three moles of AcOH with one mole of urea were calculated for the crystalline solid state. The lowest energy structure was a 1:2 solvate of urea:AcOH, which is in accord with experiment. The effect of including an entropic contribution to the free energy for the 1:2 solvate stabilised the rank 2 structure more than the rank 1 structure so that at 300 K the rank 2 structure, which corresponds to the experimental one, was only 0.1 kcal mol^{-1} above the global minimum.[107]

Form II of CBZ which belongs to the $R\bar{3}$ space group was found by crystal structure prediction[110] to be relatively high in energy with more than 100 potential polymorphs lower in energy. Form II has a more open structure than the other three known polymorphs and this led to the speculation that solvent inclusion may be a contributing factor in its crystallisation. A computational study of the effect of filling voids in the crystal structure with toluene on the stability of the inclusion compound as a function of the packing fraction showed that no significant stabilisation in the energy of forming the inclusion compound from CBZ crystal and toluene liquid was gained beyond about 3% by weight of toluene.[111] This agreed well with the experimental results which showed that between 2.7 and 3.7% by weight of toluene was lost during thermogravimetric analysis of form II crystals of CBZ grown from toluene, strongly indicating that toluene inclusion is important in controlling the growth of form II crystals.

4.3.3.2 Hydrates

Given the potential utility of a hydrated form to modify the solubility and bioavailability of a drug molecule and the propensity of pharmaceutically active compounds to form hydrates,[112] it is not surprising that there is great interest in understanding the relative stabilities of anhydrous and hydrated forms. Whilst water is a small molecule, it has proved to be one of the most difficult molecules to model in either its liquid or its solid forms. This was illustrated in work which developed a new force field for ice and investigated the ability of this and other force fields to represent the structures of 22 hydrates.[27] The stabilities of seven of the hydrates relative to the anhydrous forms were calculated using a force field. Application of several standard, rigid body force fields to the calculation of the structures and relative lattice energies of four ice polymorphs showed that none of these standard force fields were capable of reproducing the ice structures well. Inclusion of a more accurate representation of the electrostatic potential around the water molecule through the use of distributed multipoles brought about a considerable improvement in the structural predictions.[27] Indeed, the best force field (giving the best structural predictions for all four polymorphs of ice) was shown to be one where the distributed multipole charge–charge interaction was combined with solid state dispersive and repulsive non-bonded interaction parameters for oxygen taken from studies of oxohydrocarbons[113] and for hydrogen taken from work on molecules with N–H groups hydrogen bonded to polar oxygen atoms.[114]

The majority of the non-hydrous components of the hydrates were chosen to be well represented by rigid molecule models and calculations of the hydrate structures using a distributed multipole charge–charge interaction model with a Buckingham potential for the other non-bonded interactions were promising; optimising the 22 hydrate structures, starting from the known experimental structures, led to all but five structures maintaining their hydrogen bonding pattern.[27] The stability of the hydrates for which an experimental anhydrous counterpart was known was found to be very sensitive to the assumed molecular geometry. Both experimental geometries and optimised geometries using molecular quantum mechanics were used. A crystal structure prediction study was performed on one of the hydrate systems, based on 5-azauracil monohydrate (see Figure 4.9). The force field had already been shown to find the experimental structure of the anhydrous form as the most stable in a crystal structure prediction study.[115] The search for packing alternatives of the monohydrate was performed using 66 different clusters of water and 5-azauracil, which were chosen to explore the hydrogen bonding possibilities between the two molecules. A final energy ranking was performed using a distributed multipole representation of the electrostatic interactions.[27]

The structure corresponding to the experimental structure of 5-azauracil monohydrate was found to lie within 1.1 kcal mol^{-1} of the global lattice energy minimum. When zero-point energies and thermal corrections for lattice vibrations were included, the same structure was ranked third and lay within 0.2 kcal mol^{-1} of the global minimum.[27]

4.3.4 Salts

Salts are regarded as more challenging than co-crystals in crystal structure prediction. Both have more than one molecule in the asymmetric unit, but salts have the added complexity of long range ionic interactions, charge transfer and strong polarisation effects. On the other hand, the salt bridge may be used to construct a neutral unit containing the anion and the cation provided that only a limited number of geometries are possible. This neutral unit can be used in the

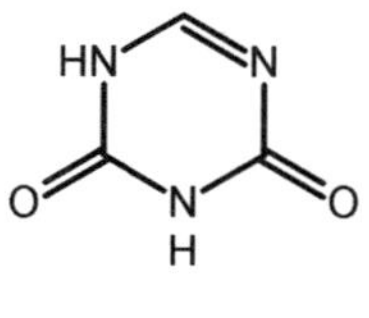

Figure 4.9 Molecular structure of 5-azauracil.

search for possible packing alternatives to reduce the number of degrees of freedom in the problem and thus reduce the computational effort. A few examples of crystal structure prediction of salts will be discussed in this section.

4.3.4.1 Terazosin Hydrochloride

Terazosin hydrochloride is an alpha-1-selective adrenoceptor blocking agent marketed by Abbott Labs under the name Hytrin. The molecular structure of this compound is shown in Figure 4.10. This pharmaceutical is used to treat benign prostatic hyperplasia and high blood pressure. Like many modern pharmaceuticals, the compound is found to exist in several polymorphic forms as well as solvates. Four solvent-free polymorphs have been isolated directly and one solvent-free form has been isolated indirectly by desolvation of a methanol solvate. In addition to the methanol solvate, a dihydrate has also been found.[116] These forms all exist at room temperature and their relative stabilities have been determined experimentally.

The crystal structures of two anhydrous polymorphs and the two solvates were determined experimentally by single crystal X-ray diffraction. Single crystals of sufficient size and quality could not be obtained to determine the crystal structures of the other two anhydrous forms in the same manner. However, low quality X-ray powder diffraction patterns were available and a crystal structure prediction study was carried out to generate possible crystal structures for these two anhydrous forms.[116]

Terazosin hydrochloride is a real challenge for crystal structure prediction because the cation is large and flexible and because its ionic nature complicates the calculation of the electrostatic contribution to the lattice energy. However, knowledge of the experimental crystal structures of four forms was used to deduce the most likely conformation of the cation as well as the protonation site and the most likely location of the chlorine anion. This neutral anion–cation complex was used as the search unit in the crystal structure prediction simulations. The known structures were also used to modify the standard Dreiding 2.21[36] force field in such a manner that it was able to reproduce the

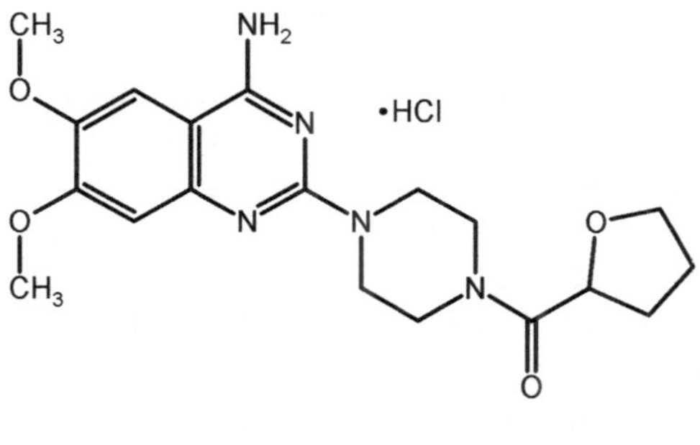

Figure 4.10 Molecular structure of terazosin hydrochloride.

structures reasonably well. It was found necessary to scale down the atomic charges to represent polarisation and charge transfer effects so that the anion had a charge of −0.8 and the cation had an overall charge of +0.8. The atomic radius of the chlorine anion was increased from the default 3.95 Å to 4.10 Å.[116] This modified force field was certainly not accurate enough to predict the crystal structures of terazosin hydrochloride from first principles, but it was good enough to generate crystal packing alternatives which could be used together with the experimental powder diffraction patterns to identify the crystal structures of the two anhydrous forms that had not yet been solved.

Since Hytrin is produced as a racemic mixture, only centro-symmetric space groups such as $P2_1/c$, $P\bar{1}$ and $C2/c$ were considered in crystal structure generation using the Polymorph Predictor[21] software. Furthermore, the X-ray powder diffraction pattern of one of the two elusive forms could be indexed to reveal space group $P\bar{1}$.[116] By simulating powder diffraction patterns for all the predicted structures and comparing them to the experimental patterns, both previously undetermined anhydrous polymorphs were identified in the $P\bar{1}$ space group. Although it was necessary to modify the conformation of the cation in one of the two predicted structures to optimise the fit in powder diffraction patterns, both structures were successfully Rietveld refined. The features of the solved structures, such as hydrogen bonding networks and voids, are consistent with the crystal structures of the other forms which were determined by single crystal X-ray diffraction studies.

4.3.4.2 Diastereomeric Salts

The relative stability of diastereomeric salts is relevant to the efficiency achieved in the separation of enantiomers through classical resolution. Most pharmaceuticals are chiral,[117] and the two enantiomers of a biologically active molecule interact differently with the chiral environment of the body leading to different biological effects, that is, one of the enantiomers has the desired biological effect, whereas the other enantiomer can be inactive, or it could be toxic. Racemic drugs can cause problems, not only because of the pharmacological differences between enantiomers, but also because the pharmacokinetics of enantiomers may differ significantly. Even if the unwanted enantiomer of a racemate is inactive, it should be considered as metabolic ballast, making the body work twice as hard as is necessary. Regulatory authorities nowadays do not allow racemic pharmaceuticals to be marketed, unless it can be shown that the racemate has a preferable biological effect. Many chiral compounds cannot be obtained in an optically pure form by asymmetric synthesis or isolation from natural products and a racemate is produced which needs to be separated into pure enantiomers. Racemate resolution can be achieved via a variety of methods including resolution by direct crystallisation, in which one enantiomer is selectively crystallised from a racemic solution, and selective crystallisation of diastereomeric salts (called 'classical resolution'), where a suitable resolving agent is added to a racemic solution and two diastereomeric complexes can be

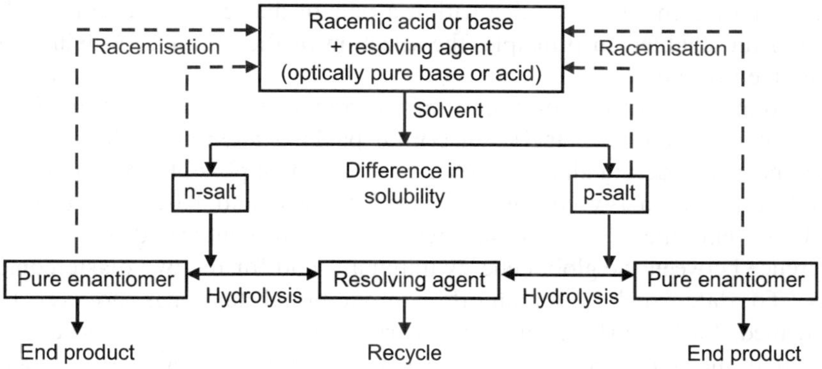

Figure 4.11 Schematic representation of racemate resolution through crystallisation of diastereomeric salts.

formed with a difference in solubility thus allowing the resolution of the racemate into its enantiomers (see Figure 4.11).

Direct resolutions and classical resolutions are both based on crystallisation and can be applied on the multi-ton scale required in industry. More than half of all chiral drugs currently on the market are produced by crystallisation of diastereomeric salts. Spontaneous (direct) resolution is the preferred route, however, only 10% of chiral compounds crystallise as enantiomorphs. For most chiral compounds it is energetically more favourable to crystallise as a racemate. If the stability difference between the racemic solid and the enantiomorphs is small, it may still be possible to force the compound to crystallise in an enantiomerically pure form by seeding a supersaturated solution of the racemate with an optically pure crystal, or through the use of a chiral solvent or a chiral additive. In all other cases, the racemate has to be resolved through the selective crystallisation of diastereomeric salts. Pioneering work has been carried out to evaluate the application of crystal structure prediction approaches to predict whether a given racemate will resolve spontaneously.[118–121] The results of these initial studies are very encouraging, but more resolutions need to be studied to establish the success rate of such predictions on spontaneous resolution. It is, however, presently not yet possible to predict which chiral solvent or additive could be used to force the crystallisation of a compound as an enantiomorph.

Currently, achieving an efficient separation of enantiomers through a classical resolution is a matter of trial and error; finding the optimal resolving agent and crystallisation conditions resulting in a large enough solubility difference between the two diastereomeric salts can be a tedious and frustrating process. The development of reliable predictive tools for classical resolution would therefore be a tremendous benefit to the pharmaceutical industry. The application of crystal structure prediction tools to predict racemate resolution is a recent scientific endeavour. The thermodynamics of classical resolutions has been worked out, pointing to a relationship between resolution efficiency and the lattice energy difference within a pair of diastereomeric salts.[122] This relationship enables, in principle, the selection of the optimal resolving agent for a given racemate.

To predict a classical resolution from first principles, one needs to predict, for each potential resolving agent, all crystal packing alternatives for both diastereomers considering all chiral space groups and all possible combinations of all relevant low-energy conformations of the acid and base molecules. The resolution efficiency of a candidate resolving agent is indicated by the energy difference between the global energy minima found for the two diastereomeric salts. To date, such a mammoth computational task has not yet been performed. Early work in this area clearly illustrates the lack of accuracy of molecular mechanics methods in calculating the lattice energies of diastereomeric salts.[122] A decade later, it was demonstrated for the first time that it is possible to predict the crystal structures of a diastereomeric salt pair.[123] The compounds consisted of a chlorine-substituted cyclic phosphoric acid and the two enantiomers of the alkaloid ephedrine, see Figure 4.12. The simulation used standard molecular mechanics, requiring tremendous computational effort and resulting in an error in the calculated lattice energies of about 3 kcal mol^{-1} – too large for a meaningful prediction.

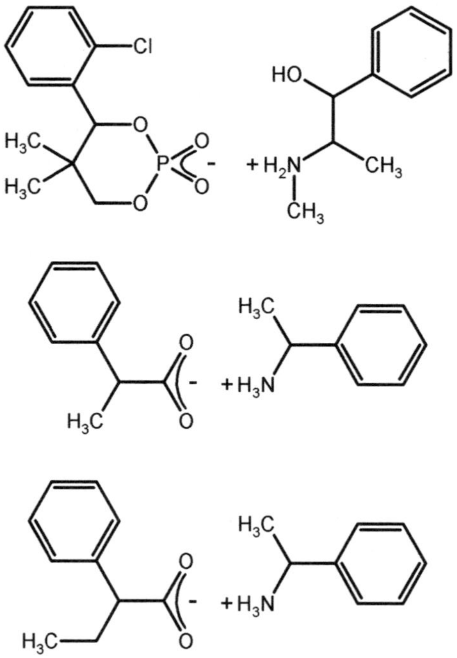

Figure 4.12 Molecular structures of diastereomeric salts; top: salt of ephedrine and a cyclic phosphoric acid; middle: salt of 2-phenylpropanoate and 2-phenylethylammonium; bottom: salt of 2-phenylbutyrate and 2-phenylethylammonium.

In a later study the two main problems in the crystal structure prediction of diastereomeric salts were addressed: the large number of degrees of freedom caused by having two independent molecules in the asymmetric unit and the error caused by the strong electrostatics associated with these ionic structures.[124,125] The model compounds in these studies were two salt pairs of 2-phenylethylammonium, as shown in Figure 4.12. By using experimental information on known crystal structures, the range of ion-pair geometries and ion conformations were limited to the most likely candidates. The number of degrees of freedom was reduced further by keeping the molecules rigid throughout the simulations,[124] or by restricting the flexibility of the ions to the most important torsional degrees of freedom.[125,126] The electrostatic contribution to the lattice energy was evaluated using a distributed multipole analysis.

Despite encouraging results, these molecular mechanics approaches are still not accurate enough to predict resolution efficiency from first principles. Also, the computational effort required to screen a number of potential resolving agents for a given racemate is prohibitive. The accuracy issue may be resolved by using solid state quantum mechanical calculations, but such calculations are even more demanding of computational resources. Therefore, although these studies have demonstrated that it is possible to predict the crystal structures of diastereomeric salts, the practical application of simulation tools to the prediction of an optimal resolving agent for a racemic mixture is still out of reach.

4.3.4.3 Pyridinium Chloride

A crystal structure prediction study was carried out on pyridinium chloride in order to investigate the transferability of the dispersion correction in the DFT(d) method,[58] which was parameterised against uncharged compounds, to ionic systems.[127] In previous studies, the DFT(d) method was shown to be very capable of predicting the crystal structures of small, neutral molecules.[86,120,128,129] The study considered $Z'=1$ and $Z'=2$ structures in all 230 space groups, using a tailor-made force field[42] for the search for crystal packing alternatives and the DFT(d) method for the stability ranking of the most promising structures. Experimentally, the pyridinium chloride salt (see Figure 4.13) has been reported in three polymorphic forms: a $Z'=\frac{1}{2}$ structure in space group $P2_1/m$, a $Z'=1$ structure in $P\bar{1}$ and a $Z'=2$ form in $P2_1/c$, although the $P2_1/m$ and $P\bar{1}$ structures are very similar and probably represent the same polymorph.[130] The $P2_1/c$ polymorph is believed to be meta-stable at room temperature, but there is no conclusive experimental evidence.

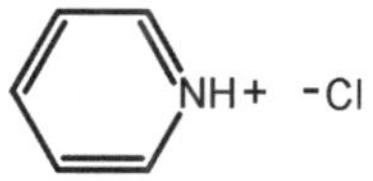

Figure 4.13 Molecular structure of pyridinium chloride.

Using the standard DFT(d) dispersion correction parameters derived from a small selection of accurately determined crystal structures of uncharged molecules,[58] the crystal structure prediction study with the GRACE software[69] ignored the effects of temperature and pressure. A structure corresponding to the experimentally observed $P2_1/c$ polymorph was found to be the most stable crystal packing alternative. The rank 2 structure, located 0.05 kcal mol^{-1} above the global minimum, did not correspond to the available experimental data. This *Pnma* structure has a considerably higher density than any of the other predicted low-energy structures, including the known polymorphs, and further simulations were carried out to show that the relative stability of this structure increases with pressure. It was therefore concluded that this structure represents an as yet undiscovered polymorph of pyridinium chloride that may be obtained in a high pressure crystallisation experiment.[127]

The energy of the third structure was extremely close to the lattice energy of the second structure and as the calculated energy difference between the second and third structures is similar to the numerical error of the DFT(d) method, it could be argued that the lattice energies of these two structures are essentially the same. This third structure corresponds to the $P\bar{1}$ polymorph. By performing simulations in $P1$ supercells it was shown that the experimentally observed $P2_1/m$ polymorph does not represent a true minimum on the lattice energy hypersurface. By removing superfluous space group symmetry, the $P2_1/m$ structure was optimised to its true minimum, the $P\bar{1}$ polymorph. The study also produced some interesting results on the as yet uncharacterised high temperature phase of pyridinium chloride, which may share structural features with the known high temperature phase of pyridinium iodide.

It is noteworthy that as part of the same study, a crystal structure prediction of phenylethylammonium lactate was also carried out.[127] The results were less impressive than the results obtained for pyridinium chloride: the experimentally observed structure of phenylethylammonium lactate was found as the fifth crystal packing alternative, 0.24 kcal mol^{-1} above the global lattice energy minimum structure. The reasons for this discrepancy are unclear. It is possible that the experimental structure of phenylethylammonium lactate is a kinetic polymorph and a thermodynamically more stable structure has not been found yet, or the known structure is stabilised by temperature effects that were not considered in the simulations, or the parameters of the dispersion correction do not transfer well from neutral systems to charged systems. Further validation work on molecular salts is required to pinpoint the reason for the inconclusive results obtained for these two compounds.

4.4 Outlook

Crystal structure prediction has come a long way in the last couple of decades. Using the most sophisticated computational tools currently available, it is now possible to predict, reliably and accurately, the crystal structures of small organic molecules from first principles. Despite the kinetic nature of the crystallisation process, it appears that for these compounds the calculated thermodynamic stability is a sufficient selection criterion to decide which crystal structures are likely to be observed experimentally. Crystal structure prediction is now progressing to tackle more complex systems such as large flexible molecules, co-crystals, solvates and salts. These structures have more degrees of freedom, making the search for all crystal packing alternatives more time consuming. Because of the multitude of potential polymorphs, the subtle balance between intra- and intermolecular energy contributions in flexible molecules and the strong electrostatic interactions in the case of salts, these complex systems demand a high level of accuracy in the lattice energy calculations, requiring substantial computational resources. With the ever increasing power and decreasing cost of modern computers, such simulations will become easily accessible in the future. A greater challenge is the inclusion of kinetic factors and solvent effects in the crystal structure prediction simulations to allow genuine prediction of relative polymorphic stability and co-crystal or salt formation.

References

1. J. Bernstein, *Polymorphism in Molecular Crystals*, Clarendon Press, Oxford, 2002.
2. J. Anwar and D. Zahn, *Angew. Chem. Int. Ed.*, 2011, **50**, 1996.
3. P. Bennema, H. Meekes, S. X. M. Boerrigter, H. M. Cuppen, M. A. Deij, J. van Eupen, P. Verwer and E. Vlieg, *Cryst. Growth Des.*, 2004, **4**, 905.
4. A. I. Kitaigorodskii, *Organic Chemical Crystallography*, Consultants Bureau, New York, 1961.
5. D. E. Williams, *Acta Crystallogr.*, 1966, **21**, 340.
6. D. E. Williams, *Acta Crystallogr., Sect. A*, 1969, **25**, 464.
7. P. Zugenmaier and A. Sarko, *Acta Crystallogr., Sect. B*, 1972, **28**, 3158.
8. D. E. Williams, *Acta Crystallogr., Sect. B*, 1973, **29**, 96.
9. W. R. Busing, *Acta Crystallogr., Sect. A*, 1972, **28**, S252.
10. C. P. Brock and J. A. Ibers, *Acta Crystallogr., Sect. B*, 1973, **29**, 2426.
11. Å. Kvick and J. H. Noordik, *Acta Crystallogr., Sect. B*, 1977, **33**, 2862.
12. J. Maddox, *Nature*, 1988, **335**, 201.
13. A. Gavezzotti, *J. Am. Chem. Soc.*, 1991, **113**, 4622.
14. D. W. M. Hofmann and T. Lengauer, *Acta Crystallogr., Sect. A*, 1997, **53**, 225.
15. W. T. M. Mooij, B. P. van Eijck, S. L. Price, P. Verwer and J. Kroon, *J. Comput. Chem.*, 1998, **19**, 459.
16. D. E. Williams, *Acta Crystallogr., Sect. A*, 1996, **52**, 326.
17. J. R. Holden, Z. Du and H. Ammon, *J. Comput. Chem.*, 1993, **14**, 422.
18. H. R. Karfunkel, F. J. J. Leusen and R. J. Gdanitz, *J. Comput. Aided Mater. Des.*, 1993, **1**, 177.
19. B. P. van Eijck, W. T. M. Mooij and J. Kroon, *Acta Crystallogr., Sect. B*, 1995, **51**, 99.

20. A. M. Chaka, R. Zaniewski, W. Youngs, C. Tessier and G. Klopman, *Acta Crystallogr., Sect. B*, 1996, **52**, 165.
21. F. J. J. Leusen, *J. Cryst. Growth*, 1996, **166**, 900.
22. P. Verwer and F. J. J. Leusen in *Reviews in Computational Chemistry*, ed. K. B. Lipkowitz and D. B. Boyd, John Wiley and Sons, New York, 1998, Volume 12, p. 327.
23. S. L. Price, *Adv. Drug Deliv. Rev.*, 2004, **56**, 301.
24. S. L. Price, *Int. Rev. Phys. Chem.*, 2008, **27**, 541.
25. G. M. Day, *Crystallogr. Rev.*, 2011, **17**, 3.
26. B. P. van Eijck and J. Kroon, *Acta Crystallogr., Sect. B*, 2000, **56**, 535.
27. A. T. Hulme and S. L. Price, *J. Chem. Theory Comput.*, 2007, **3**, 1597.
28. D. E. Williams, *J. Comput. Chem.*, 2001, **22**, 1154.
29. D. E. Williams, *J. Mol. Struct.*, 1999, **485**, 321.
30. J. Kendrick and M. Fox, *J. Mol. Graphics*, 1991, **9**, 182.
31. P. P. Ewald, *Ann. Physik*, 1921, **64**, 253.
32. A. J. Stone, *Chem. Phys. Lett.*, 1981, **83**, 233.
33. S. L. Price, M. Leslie, G. W. A. Welch, M. Habgood, L. S. Price, P. G. Karamertzanis and G. M. Day, *Phys. Chem. Chem. Phys.*, 2010, **12**, 8478.
34. G. M. Day and S. L. Price, *J. Am. Chem. Soc.*, 2003, **125**, 16434.
35. H. Sun, *J. Phys. Chem. B*, 1998, **102**, 7338.
36. S. L. Mayo, B. D. Olafson and W. A. Goddard, *J. Phys. Chem.*, 1990, **94**, 8897.
37. F. A. Momany and R. Rone, *J. Comput. Chem.*, 1992, **13**, 888.
38. J. M. Wang, R. M. Wolf, J. W. Caldwell, P. A. Kollman and D. A. Case, *J. Comput. Chem.*, 2004, **25**, 1157.
39. N. L. Allinger, F. B. Li and L. Q. Yan, *J. Comput. Chem.*, 1990, **11**, 848.
40. N. L. Allinger, M. Rahman and J. H. Lii, *J. Am. Chem. Soc.*, 1990, **112**, 8293.
41. N. L. Allinger, Y. H. Yuh and J. H. Lii, *J. Am. Chem. Soc.*, 1989, **111**, 8551.
42. M. A. Neumann, *J. Phys. Chem. B*, 2008, **112**, 9810.
43. M. F. Guest, I. J. Bush, H. J. J. Van Dam, P. Sherwood, J. M. H. Thomas, J. H. Van Lenthe, R. W. A. Havenith and J. Kendrick, *Mol. Phys.*, 2005, **103**, 719.
44. T. Helgaker, H. J. A. Jensen, P. Jørgensen, J. Olsen, K. Ruud, H. Ågren, A. A. Auer, K. L. Bak, V. Bakken, O. Christiansen, S. Coriani, P. Dahle, E. K. Dalskov, T. Enevoldsen, B. Fernandez, C. Hättig, K. Hald, A. Halkier, H. Heiberg, H. Hettema, D. Jonsson, S. Kirpekar, R. Kobayashi, H. Koch, K. V. Mikkelsen, P. Norman, M. J. Packer, T. B. Pedersen, T. A. Ruden, A. Sanchez, T. Saue, S. P. A. Sauer, B. Schimmelpfennig, K. O. Sylvester-Hvid, P. R. Taylor, O. Vahtras, *DALTON-an ab initio Electronic Structure Program,* http://www.kjemi.uio.no/software/dalton/dalton.html.
45. F. Neese, *ORCA – an ab initio, Density Functional and Semiempirical Program Package*, University of Bonn, 2008.

46. M. J. Frisch, G. W. Trucks, H. B. Schlegel, G. E. Scuseria, M. A. Robb, J. R. Cheeseman, G. Scalmani, V. Barone, B. Mennucci, G. A. Petersson, H. Nakatsuji, M. Caricato, X. Li, H. P. Hratchian, A. F. Izmaylov, J. Bloino, G. Zheng, J. L. Sonnenberg, M. Hada, M. Ehara, K. Toyota, R. Fukuda, J. Hasegawa, M. Ishida, T. Nakajima, Y. Honda, O. Kitao, H. Nakai, T. Vreven, J. A. Montgomery, Jr., J. E. Peralta, F. Ogliaro, M. Bearpark, J. J. Heyd, E. Brothers, K. N. Kudin, V. N. Staroverov, R. Kobayashi, J. Normand, K. Raghavachari, A. Rendell, J. C. Burant, S. S. Iyengar, J. Tomasi, M. Cossi, N. Rega, J. M. Millam, M. Klene, J. E. Knox, J. B. Cross, V. Bakken, C. Adamo, J. Jaramillo, R. Gomperts, R. E. Stratmann, O. Yazyev, A. J. Austin, R. Cammi, C. Pomelli, J. W. Ochterski, R. L. Martin, K. Morokuma, V. G. Zakrzewski, G. A. Voth, P. Salvador, J. J. Dannenberg, S. Dapprich, A. D. Daniels, Ö. Farkas, J. B. Foresman, J. V. Ortiz, J. Cioslowski and D. J. Fox, *Gaussian*, 2009, Gaussian, Inc., Wallingford CT, http://www.gaussian.com.
47. F. Jensen, *Introduction to Computational Chemistry*, John Wiley and Sons, New York, 2007.
48. W. Kohn and L. J. Sham, *Phys. Rev. A*, 1965, **140**, 1133.
49. A. D. Becke, *J. Chem. Phys.*, 1986, **84**, 4524.
50. T. G. Cooper, K. E. Hejczyk, W. Jones and G. M. Day, *J. Chem. Theory Comput.*, 2008, **4**, 1795.
51. G. Kresse and D. Joubert, *Phys. Rev. B*, 1999, **59**, 1758.
52. G. Kresse and J. Furthmüller, *Phys. Rev. B*, 1996, **54**, 11169.
53. S. J. Clark, M. D. Segall, C. J. Pickard, P. J. Hasnip, M. J. Probert, K. Refson and M. C. Payne, *Z. Kristallogr.*, 2005, **220**, 567.
54. R. Dovesi, R. Orlando, B. Civalleri, C. Roetti, V. R. Saunders and C. M. Zicovich-Wilson, *Z. Kristallogr.*, 2005, **220**, 571.
55. B. B. Laird, R. B. Ross and T. Ziegler, *Chemical Applications of Density-Functional Theory*, ACS Symposium Series, American Chemical Society, Washington DC, 1996.
56. R. M. Martin, *Electronic Structure: Basic Theory and Practical Methods*, Cambridge University Press, Cambridge, 2004.
57. S. Grimme, *J. Comput. Chem.*, 2004, **25**, 1463.
58. M. A. Neumann and M.-A. Perrin, *J. Phys. Chem. B*, 2005, **109**, 15531.
59. A. Gavezzotti, *Z. Kristallogr.*, 2005, **220**, 499.
60. A. Gavezzotti, *CrystEngComm*, 2003, **5**, 429.
61. A. Gavezzotti, *J. Phys. Chem. B*, 2002, **106**, 4145.
62. M. D. Gourlay, J. Kendrick and F. J. J. Leusen, *J. Mol. Struct.–Theochem*, 2007, **809**, 11.
63. F. H. Allen, *Acta Crystallogr., Sect. B*, 2002, **58**, 380.
64. A. J. Cruz-Cabeza, E. Pidcock, G. M. Day, W. D. S. Motherwell and W. Jones, *CrystEngComm*, 2007, **9**, 556.
65. H. L. Ammon, Z. Y. Du, J. R. Holden and L. A. Paquette, *Acta Crystallogr., Sect. B*, 1994, **50**, 216.
66. H. L. Ammon, Z. Y. Du, R. D. Gilardi, P. R. Dave, F. Forohar and T. Axenrod, *Acta Crystallogr., Sect. B*, 1996, **52**, 352.

67. B. P. van Eijck and J. Kroon, *J. Comput. Chem.*, 1999, **20**, 799.
68. J. M. Newsam, C. M. Freeman and F. J. J. Leusen, *Curr. Opin. Solid State Mater. Sci.*, 1999, **4**, 515.
69. *GRACE (the Generation, Ranking and Characterization Engine)* software package is a product of Avant-garde Materials Simulation, www.avmatsim.eu.
70. D. J. Earl and M. W. Deem, *Phys. Chem. Chem. Phys.*, 2005, **7**, 3910.
71. P. G. Karamertzanis and C. C. Pantelides, *J. Comput. Chem.*, 2005, **26**, 304.
72. P. G. Karamertzanis and C. C. Pantelides, *Mol. Phys.*, 2007, **105**, 273.
73. J. A. Chisholm and W. D. S. Motherwell, *J. Appl. Crystallogr.*, 2005, **38**, 228.
74. B. P. van Eijck and J. Kroon, *J. Comput. Chem.*, 1997, **18**, 1036.
75. P. G. Karamertzanis and S. L. Price, *J. Chem. Theory Comput.*, 2006, **2**, 1184.
76. G. M. Day, S. L. Price and M. Leslie, *J. Phys. Chem. B*, 2003, **107**, 10919.
77. S. M. Woodley and R. Catlow, *Nature Materials*, 2008, **7**, 937.
78. A. R. Oganov and C. W. Glass, *J. Chem. Phys.*, 2006, **124**, 244704.
79. A. R. Oganov and C. W. Glass, *J. Phys.: Condens. Matter*, 2008, **20**, 064210.
80. V. Buch, R. Martoňák and M. Parrinello, *J. Chem. Phys.*, 2005, **123**, 051108.
81. V. Buch, R. Martoňák and M. Parrinello, *J. Chem. Phys.*, 2006, **124**, 204705.
82. J. P. M. Lommerse, W. D. S. Motherwell, H. L. Ammon, A. Gavezzotti, D. W. M. Hofmann, F. J. J. Leusen, W. T. M. Mooij, S. L. Price, B. Schweizer, M. U. Schmidt, B. P. van Eijck, P. Verwer and D. E. Williams, *Acta Crystallogr., Sect. B*, 2000, **56**, 697.
83. W. D. S. Motherwell, H. L. Ammon, J. D. Dunitz, A. Dzyabchenko, P. Erk, A. Gavezzotti, D. W. M. Hofmann, F. J. J. Leusen, J. P. M. Lommerse, W. T. M. Mooij, S. L. Price, H. Scheraga, B. Schweizer, M. U. Schmidt, B. P. van Eijck, P. Verwer and D. E. Williams, *Acta Crystallogr., Sect. B*, 2002, **58**, 647.
84. G. M. Day, W. D. S. Motherwell, H. L. Ammon, S. X. M. Boerrigter, R. G. Della Valle, E. Venuti, A. Dzyabchenko, J. D. Dunitz, B. Schweizer, B. P. van Eijck, P. Erk, J. C. Facelli, V. E. Bazterra, M. B. Ferraro, D. W. M. Hofmann, F. J. J. Leusen, C. Liang, C. C. Pantelides, P. G. Karamertzanis, S. L. Price, T. C. Lewis, H. Nowell, A. Torrisi, H. A. Scheraga, Y. A. Arnautova, M. U. Schmidt and P. Verwer, *Acta Crystallogr., Sect. B*, 2005, **61**, 511.
85. G. M. Day, T. G. Cooper, A. J. Cruz-Cabeza, K. E. Hejczyk, H. L. Ammon, S. X. M. Boerrigter, J. S. Tan, R. G. Della Valle, E. Venuti, J. Jose, S. R. Gadre, G. R. Desiraju, T. S. Thakur, B. P. van Eijck, J. C. Facelli, V. E. Bazterra, M. B. Ferraro, D. W. M. Hofmann, M. A. Neumann, F. J. J. Leusen, J. Kendrick, S. L. Price, A. J. Misquitta, P. G. Karamertzanis, G. W. A. Welch, H. A. Scheraga, Y. A. Arnautova, M. U.

Schmidt, J. van de Streek, A. K. Wolf and B. Schweizer, *Acta Crystallogr., Sect. B*, 2009, **65**, 107.
86. M. A. Neumann, F. J.J. Leusen and J. Kendrick, *Angew. Chem. Int. Ed.*, 2008, **47**, 2427.
87. A. Asmadi, M. A. Neumann, J. Kendrick, P. Girard, M.-A. Perrin and F. J. J. Leusen, *J. Phys. Chem. B*, 2009, **113**, 16303.
88. W. T. M. Mooij and F. J. J. Leusen, *Phys. Chem. Chem. Phys.*, 2001, **3**, 5063.
89. G. M. Day, J. Chisholm, N. Shan, W. D. S. Motherwell and W. Jones, *Cryst. Growth Des.*, 2004, **4**, 1327.
90. A. T. Hulme, A. Johnston, A. J. Florence, P. Fernandes, K. Shankland, C. T. Bedford, G. W. A. Welch, G. Sadiq, D. A. Haynes, W. D. S. Motherwell, D. A. Tocher and S. L. Price, *J. Am. Chem. Soc.*, 2007, **129**, 3649.
91. R. K. R. Jetti, R. Boese, J. A. R. P. Sarma, L. S. Reddy, P. Vishweshwar and G. R. Desiraju, *Angew. Chem. Int. Ed.*, 2003, **42**, 1963.
92. S. Roy and A. J. Matzger, *Angew. Chem. Int. Ed.*, 2009, **48**, 8505.
93. M. T. Kirchner, L. S. Reddy, G. R. Desiraju, R. K. R. Jetti and R. Boese, *Cryst. Growth Des.*, 2004, **4**, 701.
94. C. H. Arnaud, *Chem. Eng. News*, 2009, **87** (nr. 40), 37.
95. T. Laird, *Org. Process Res. Dev.*, 2010, **14**, 1.
96. H. C. S. Chan, J. Kendrick and F. J. J. Leusen, *Angew. Chem. Int. Ed.*, 2011, **50**, 2979.
97. F.-L. Yu and C. H. Schwalbe, British Crystallographic Association Spring Meeting, Nottingham, 2002.
98. F.-L. Yu, C. H. Schwalbe and D. J. Watkin, *Acta Crystallogr., Sect. C*, 2004, **60**, o714.
99. J. Kendrick, M. A. Neumann and F. J. J. Leusen, in preparation.
100. R. B. Hammond, C. Ma, K. J. Roberts, P. Y. Ghi and R. K. Harris, *J. Phys. Chem. B*, 2003, **107**, 11820.
101. N. Issa, P. G. Karamertzanis, G. W. A. Welch and S. L. Price, *Cryst. Growth Des.*, 2009, **9**, 442.
102. P. G. Karamertzanis, A. V. Kazantsev, N. Issa, G. W. A. Welch, C. S. Adjiman, C. C. Pantelides and S. L. Price, *J. Chem. Theory Comput.*, 2009, **5**, 1432.
103. M. Habgood and S. L. Price, *Cryst. Growth Des.*, 2010, **10**, 3263.
104. D.-K. Bučar, R. F. Henry, X. Lou, R. W. Duerst, L. R. MacGillivray and G. G. Z. Zhang, *Cryst. Growth Des.*, 2009, **9**, 1932.
105. C. H. Görbitz, B. Dalhus and G. M. Day, *Phys. Chem. Chem. Phys.*, 2010, **12**, 8466.
106. A. J. Cruz-Cabeza, G. M. Day, W. D. S. Motherwell and W. Jones, *J. Am. Chem. Soc.*, 2006, **128**, 14466.
107. A. J. Cruz-Cabeza, G. M. Day and W. Jones, *Chem. Eur. J.*, 2008, **14**, 8830.
108. A. J. Cruz-Cabeza, S. Karki, L. Fábián, T. Friščić, G. M. Day and W. Jones, *Chem. Commun.*, 2010, **46**, 2224.
109. A. J. Cruz-Cabeza, G. M. Day, W. D. S. Motherwell and W. Jones, *Cryst. Growth Des.*, 2007, **7**, 100.

110. A. J. Cruz-Cabeza, G. M. Day, W. D. S. Motherwell and W. Jones, *Cryst. Growth Des.*, 2006, **6**, 1858.
111. A. J. Cruz Cabeza, G. M. Day, W. D. S. Motherwell and W. Jones, *Chem. Commun.*, 2007, 1600.
112. L. Infantes, J. Chisholm and W. D. S. Motherwell, *CrystEngComm*, 2003, **5**, 480.
113. S. R. Cox, L. Y. Hsu and D. E. Williams, *Acta Crystallogr., Sect. A*, 1981, **37**, 293.
114. D. S. Coombes, S. L. Price, D. J. Willock and M. Leslie, *J. Phys. Chem.*, 1996, **100**, 7352.
115. B. S. Potter, R. A. Palmer, R. Withnall, B. Z. Chowdhry and S. L. Price, *J. Mol. Struct.*, 1999, **485-486**, 349.
116. J. Bauer, J. Morley, S. Spanton, F. J. J. Leusen, R. Henry, S. Hollis, W. Heitmann, A. Mannino, J. Quick and W. Dziki, *J. Pharm. Sci.*, 2006, **95**, 917.
117. I. Agranat, H. Caner and J. Caldwell, *Nature Rev. Drug Disc.*, 2002, **1**, 753.
118. M. D. Gourlay, J. Kendrick and F. J. J. Leusen, *Cryst. Growth Des.*, 2007, **7**, 56.
119. M. D. Gourlay, J. Kendrick and F. J. J. Leusen, *Cryst. Growth Des.*, 2008, **8**, 2899.
120. J. Kendrick, M. D. Gourlay, M. A. Neumann and F. J. J. Leusen, *CrystEngComm*, 2009, **11**, 2391.
121. E. D'Oria, P. G. Karamertzanis and S. L. Price, *Cryst. Growth Des.*, 2010, **10**, 1749.
122. F. J. J. Leusen, J. H. Noordik and H. R. Karfunkel, *Tetrahedron*, 1993, **49**, 5377.
123. F. J. J. Leusen, *Cryst. Growth Des.*, 2003, **3**, 189.
124. P. G. Karamertzanis and S. L. Price, *J. Phys. Chem. B*, 2005, **109**, 17134.
125. P. G. Karamertzanis, P. R. Anandamanoharan, P. Fernandes, P. W. Cains, M. Vickers, D. A. Tocher, A. J. Florence and S. L. Price, *J. Phys. Chem. B*, 2007, **111**, 5326.
126. C. D. Antoniadis, E. D'Oria, P. G. Karamertzanis, D. A. Tocher, A. J. Florence, S. L. Price and A. G. Jones, *Chirality*, 2010, **22**, 447.
127. J. van de Streek, M. A. Neumann and M.-A. Perrin, *CrystEngComm*, 2010, **12**, 3827.
128. A. Asmadi, J. Kendrick and F. J. J. Leusen, *Phys. Chem. Chem. Phys.*, 2010, **12**, 8571.
129. A. Asmadi, J. Kendrick and F. J. J. Leusen, *Chem. Eur. J.*, 2010, **16**, 12701.
130. D. Mootz and J. Hocken, *Z. Naturforsch.*, 1989, **44b**, 1239.

CHAPTER 5

Shape and Polarity in Co-crystal Formation: Database Analysis and Experimental Validation

L. FÁBIÁN*[a] AND T. FRIŠČIĆ[b]

[a] School of Pharmacy, University of East Anglia, Earlham Road, Norwich, NR4 7TJ, UK; [b] University of Cambridge, Department of Chemistry, Lensfield Road, Cambridge, CB2 1EW, UK

5.1 Introduction

The synthesis of multi-component crystals, also known as co-crystals, has developed in the past 20 years from an academic exercise to one of the focal points of research in pharmaceutical materials science. In the latter context, pharmaceutical co-crystals provide a uniquely general method to obtain new solid forms of active pharmaceutical ingredients (APIs) by forming solid-state complexes with physiologically benevolent molecules known as pharmaceutical co-crystal formers (or co-formers). Consequently, a substantial amount of effort has been placed into methodologies for co-crystal screening and strategies for their design. The concepts of supramolecular synthons[1] and synthon competition[2–4] are fundamental to the design of co-crystals.

There are two types of supramolecular synthons (Formula 5.1): supramolecular homosynthons, which are formed by self-complementary functional groups (*e.g.*, carboxylic acid dimers) and supramolecular heterosynthons, which are formed by different, but complementary functional groups (*e.g.*, carboxylic acids and pyridine rings). Supramolecular heterosynthons that

RSC Drug Discovery Series No. 16
Pharmaceutical Salts and Co-crystals
Edited by Johan Wouters and Luc Quéré

Published by the Royal Society of Chemistry, www.rsc.org

connect a group of the API to a group of the co-former and that form in preference to all possible homosynthons are strong and easy-to-identify driving forces for co-crystallisation.[4] Despite the success of the synthon approach, it remains impossible to predict which molecules will form a co-crystal with a particular API and only about 20% of the co-crystal formers that could form appropriate heterosynthons with the API produce co-crystals in screening experiments.[5] The low success rate of screening experiments indicates the influence of other, yet unknown, factors for co-crystallisation.

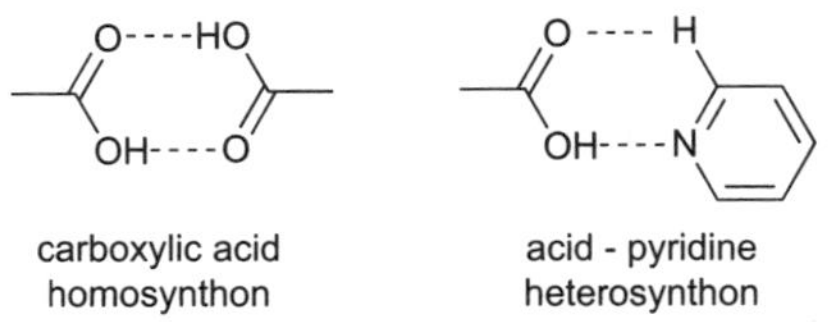

(5.1)

The purpose of this work was to find some of these additional factors and to exploit them in improving the efficiency of co-crystal screening. It was anticipated that known co-crystals, archived in the Cambridge Structural Database (CSD),[6] contain relevant information, which can be extracted by the use of appropriate data mining techniques. The results of the database analysis would then be used to prioritise screening experiments. Experiments selected with high priority should be more likely to succeed than those designed by considering only supramolecular synthons. Finally, the relationship between the design principles emerging from our database study and the established methodology based on supramolecular synthons was investigated.

The analysis focused on the influence of molecular properties, because isolated molecules can be characterised easily and reliably by molecular modelling methods. This eliminates the need for the full experimental characterisation of the API, which may not be feasible in the early stages of drug development. Rather than selecting a handful of properties that we anticipated to be important, we aimed to characterise molecules in as many ways as possible and to use statistical methods to identify the relevant characteristics.[7] Differences between experimental observations and predictions based on the statistical analysis of molecular descriptors could be used to point out specific and previously not recognised motifs of molecular recognition and assembly.

Several groups have utilised simple molecular descriptors to explain variations in the strength of hydrogen bond donors and acceptors, thereby providing a tool to predict the reliability of supramolecular synthons. Aakeröy and coworkers used pK_a values and the calculated maximum of positive electrostatic potential on the donor hydrogen site to describe the acidity of small molecule phenyloximes in co-crystallisation experiments with pyridine- and imidazole-based hydrogen bond acceptors (Scheme 5.1(a)).[8] Phenyloximes are particularly suitable for this purpose, as the variation of the substituent on the imine nitrogen atom enables direct manipulation of the pK_a value and Lewis acidity (Scheme 5.1(b)). Systematic screening for co-crystal formation between each of

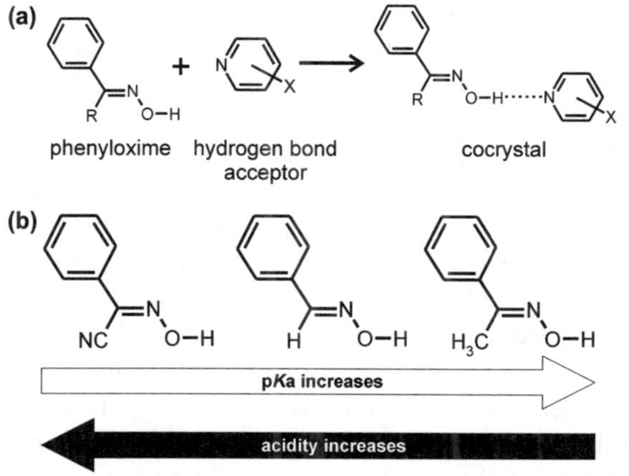

Scheme 5.1

the three phenyloxime model molecules and each of four different hydrogen bond acceptors revealed the strongest preference for co-crystal formation in cases where the difference in pK_a between the phenylcyanoxime and the acceptor was highest. Correspondingly, the most successful co-crystallisation agent in this study was phenylcyanoxime, which had the lowest pK_a value. These results also highlighted the OH group acidity as an important factor in co-crystal formation: AM1 calculations revealed that phenylcyanoxime, as the most successful co-crystallisation agent, also displayed the highest electrostatic potential centred on the hydrogen atom of the oxime hydroxyl moiety.

In a subsequent study involving substituted benzoic acids as hydrogen bond donors and substituted imidazoles as acceptors, Aakeröy and co-workers demonstrated that the likelihood of co-crystal formation is improved by increasing the pK_a, that is, the basicity of the hydrogen bond acceptor molecules.[9]

Chadwick *et al.* utilised the Hammett constants for different substituents as a means to rationalise the co-crystallisation of substituted carboxylic acids via the $R_2^2(8)$ acid–acid synthon. The study suggested that pairs of carboxylic acids bearing substituents with a large difference in Hammett constants are most likely to form a co-crystal.[10]

The results of Aakeröy *et al.* and of Chadwick *et al.* strongly suggest that the difference in acidities, whether expressed as pK_a, electrostatic potential, or electrostatic effects related to Hammett parameters, is an important factor in predicting co-crystal formation via hydrogen bonds. It is interesting to note that this conclusion is valid also in the context of halogen bond-driven co-crystal formation. The formation of an intermolecular halogen bond depends on the attraction of the electron pair on the acceptor atom with the positively charged σ-hole of an electron-deficient halogen bond donor. Consequently, it is clear that electron-withdrawing substituents on the halogen bond donor and electron-donating groups on the acceptor will enhance the probability of forming co-crystals. This is indeed observed with perfluorinated organic iodides or bromides which are significantly more efficient halogen bond donors than their hydrocarbon analogues. Correspondingly, anions are considerably more efficient as halogen bond acceptors than neutral molecules and are also more prone to form halogen-bonding interactions with multiple donors.

Fayos and co-workers have investigated the application of molecular descriptors to predict the crystal packing preferences of molecules.[11–13] They encoded the charge distribution,[11,12] shape and electrostatic potential[13] of molecules into vectors and grouped these vectors by similarity. An analogous procedure was applied to the corresponding crystal structures: interatomic distances in molecular clusters,[11] hydrogen bond motif types[12] and unit cell dimensions[13] were encoded and classified by similarity. In each case, diffuse but significant correlations were found between the classification of the molecules and the classification of their respective crystal structures. Although this work involved only single-component crystals and the results are hard to interpret qualitatively, they demonstrate that not only individual interactions, but also complex packing arrangements can be related to molecular properties.

5.2 Database Analysis of Molecular Properties in Co-crystals

5.2.1 Compilation of the Co-crystal Database

Statistical analysis requires a large number of observations, each of which is described by a set of well-defined attributes. The presence of a co-crystal structure in the CSD serves as an observation in our case. The structure is evidence that the constituent molecules can form a co-crystal with each other. Molecular descriptors, similar to those used in QSAR/QSPR studies,[14] were calculated for pairs of co-crystallising molecules. These paired molecular descriptors provided the numerical attributes for statistical analysis.

The first step in the generation of the database of molecular properties in co-crystals was the retrieval of co-crystal structures from the CSD. Co-crystals were defined as organic structures containing at least two charge neutral non-solvent molecules with different structural formulas. To avoid possible statistical bias by the frequent occurrence of some commonly used co-formers (*e.g.*, urea, 4,4′-bipyridine, see Table 1 in ref. 7), structures with these molecules were excluded. Initial survey of the data showed that inclusion compounds, which consist of a large and a small molecule, form a distinct group in the data set. Co-crystals of molecules with more than 30 and with less than six non-hydrogen atoms were thus removed and the analysis focused on co-crystals other than inclusion compounds. The final data set contained 710 structures.

For each molecule in these structures, 131 molecular descriptors were calculated. These included simple atom, bond and group counts, hydrogen bond donor and acceptor counts, topological indices, a variety of partitioned and charge weighted surface areas and descriptors of molecular size, shape, polarity and charge distribution.[7] Descriptors for molecules in the same structure were combined into pairs, so the observation that a pair of molecules co-crystallise with each other was described by 2 × 131 numbers that represent the properties of these molecules.

5.2.2 Statistical Analysis

If a molecular property has an influence on co-crystal formation then the corresponding descriptor values for pairs of co-crystallising molecules are correlated: favourable combinations of descriptor values occur more frequently than unfavourable ones. For descriptors that show no influence, there are no favourable and unfavourable combinations, so the values calculated for co-crystallising molecules are expected to be statistically independent of each other.

This means that important descriptors can be identified by simply calculating correlation coefficients. It is worth noting that the distribution of descriptor values was far from a normal distribution, so the use of the non-parametric Spearman's correlation coefficient (ρ) was preferred over the more commonly used Pearson's correlation coefficient (r).[15]

All possible descriptor pairs, that is, combinations of the same or two different descriptors for pairs of molecules that co-crystallise, were tested. The strongest correlations ($|\rho| = 0.3 - 0.4$) were found between identical descriptors for both molecules or between different descriptors related to the same molecular property.[7] Graphical representation of the data (Figure 5.1(a)) was then used for the qualitative interpretation of the correlations.

The CSD does not contain information about failed co-crystallisation experiments, so we did not have any negative data, which would show the contrast between molecule pairs that co-crystallise and those that do not. Instead, the results for co-crystallising molecules were compared with random pairs selected from the same data set (Figure 5.1(b)). The correlation coefficients for ideally random data should be zero and the actual values for each descriptor were close to this, between 0 and ± 0.03 and well below the correlations seen for molecules in co-crystals.

5.2.3 Results and their Interpretation

The strongest correlations found link descriptors that are associated with the polarities and the shapes of the molecules that form co-crystals with each other (Table 5.1):

- *FPV* and *FNO* quantify the fractional polarity of the molecules, taking values between 0 and 1. FPV is defined as the volume of N, O, S and polar H atoms in the molecule divided by the total molecular volume. FNO is the total number of N and O atoms in the molecule divided by the total number of non-hydrogen atoms. FNO can be used as an easy-to-calculate substitute for FPV.

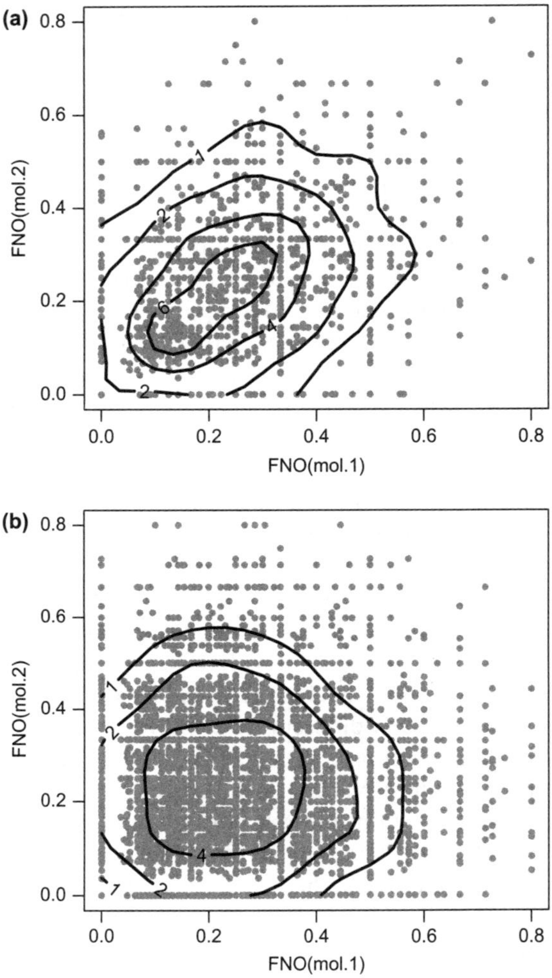

Figure 5.1 Scatter plot (grey dots) and density plot (contour lines) showing the distribution of FNO values for (a) pairs of molecules in co-crystals and (b) random pairs of molecules. A uniform random distribution would correspond to a constant contour value of 1 over the whole plot area.

- *Dipole moment* is the molecular dipole moment in Debyes. Our original analysis was based on dipole moments calculated from Gasteiger–Hückel point charges and the experimentally observed molecular geometries.[7] Later tests showed that dipole moments from geometry optimised AM1 wave functions give virtually the same statistical results.
- *S*/*L*, *M*/*L*, *S*/*M*: Descriptors *S*, *L* and *M* were taken from the box model of crystal packing.[16] They are obtained by aligning the molecule so that its principal axes of inertia are parallel to the *x*, *y* and *z* axes of the Cartesian coordinate system. The van der Waals volume of the molecule is then enclosed by a rectangular box with axes along the *x*, *y* and *z* directions.

Table 5.1 Highest descriptor correlations in the co-crystal data set. Only correlations calculated for the same descriptor of both molecules in a co-crystal are given. Descriptors are defined in the text.

Descriptor	*ρ*	*r*	*Median Difference*	*Difference 9th decile*	*Median Diff. in Random Pairs*	*9th Decile of Diff. in Random Pairs*
FPV	0.41	0.37	0.142	0.406	0.229	0.550
FNO	0.31	0.30	0.100	0.294	0.143	0.357
Dipole (D)	0.39	0.28	1.65	5.94	2.36	10.74
S/L	0.40	0.38	0.091	0.275	0.134	0.336
S/M	0.38	0.38	0.117	0.354	0.183	0.417
M/L	0.38	0.41	0.114	0.310	0.164	0.398

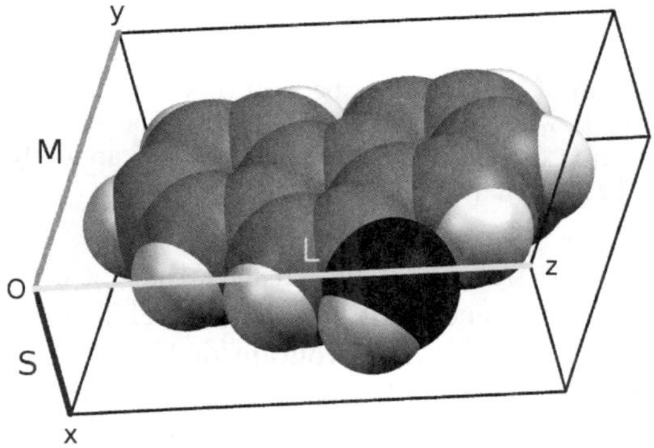

Figure 5.2 Illustration of the box model for 9-phenanthrol. The principal axes of the molecule are parallel to the x, y and z directions. L, M and S denote the dimensions of the box.

The short, medium and long axis of the enclosing box gives S, M and L, respectively (Figure 5.2). The ratios of these axes are representative of the shape of the molecule. A low S/L ratio corresponds to a flat molecule, a low M/L ratio to a rod-shaped molecule. These axis ratios showed much stronger correlation than the absolute values of S, M and L. Analogous descriptors are referred to as 'shadow descriptors' in the QSAR literature.[17]

Graphical representation of the data showed a similar picture for each of these descriptors, so only FNO graphs will be presented. Scatter plots (grey dots in Figure 5.1) are of limited use, because of the large number and overlap of data points. Density plots show the number of data points in each area of the graph. The density of data points in Figure 5.1 is shown by contour lines and clearly reveals trends that can hardly be identified from the scatter plots. A concentration of data points near the graph diagonal is apparent for co-crystals (Figure 5.1(a)), while it is not seen for random pairs (Figure 5.1(b)). This indicates that co-crystals are preferentially formed by molecules that have similar FNO values.

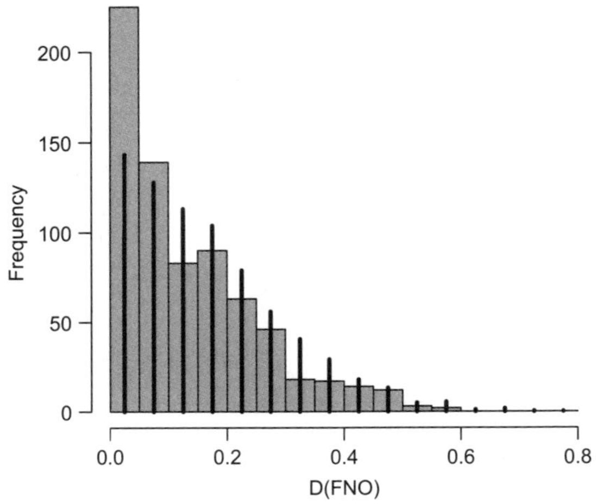

Figure 5.3 Histogram of the difference between FNO values of molecules that form co-crystals with each other (grey bars). The thin black lines show the distribution of FNO differences for pairs of randomly selected molecules from the same data set.

Data points relatively far from the diagonal show that the observed trend can be interpreted only as a preference, not as a strict rule. The preference for small differences can also be seen in the histogram of Figure 5.3. Small differences between the FNO values are more frequent for co-crystallising molecules, while larger differences are more frequent for random pairs of molecules.

It is interesting that neither hydrogen bond donor/acceptor counts nor the difference between the number of donors and acceptors in both molecules showed any discernible trend ($|\rho| < 0.1$). This result may seem to contradict the well-known utility of hydrogen bonds in co-crystal design, but it is the strength of the heteromolecular hydrogen bonds that determines whether a co-crystal is formed, not an imbalance in the number of donors and acceptors in the co-formers. Indeed, most organic molecules can accept more hydrogen bonds than they can donate, so unsatisfied donors are unusual, while unsatisfied acceptors are common in organic crystal structures.[18]

Statistical results in themselves cannot explain why molecules prefer to co-crystallise with partners that have similar dipole moments, fractional polarities and shapes to them. A definitive answer would require systematic structure analyses and is beyond the scope of the current work. Yet, it is important to discuss the possible background of these observations and to show that they are consistent with plausible chemical expectations.

Similar dipole moments are presumably preferred because of competing homo- and hetero-molecular interactions. A strongly dipolar molecule, for example a zwitterion, is likely to interact strongly with other molecules of the same kind. Intermolecular interactions with a less polar molecule (one with a small dipole moment) are unlikely to compete successfully with the strong homomolecular association. This explanation suggests that molecules with large local, but small overall dipoles (*e.g.*, oxalic acid) may form co-crystals, despite a substantial difference between the dipoles of the co-crystal formers, and hence be outliers from the observed trend.

Supramolecular heterosynthons constrain the mutual orientation of co-crystal former molecules in the crystal. If the molecules have similar shapes, it is easier to find close packed arrangements and it is less likely that the spatial requirements of supramolecular synthons and close packing will be in conflict. Similarly, the effect of FNO can be related to minimising the conflict between supramolecular synthons and the tendency to form separate hydrophobic and hydrophilic regions in the crystal. The relative volume of polar and apolar fragments is known to influence the topology of hydrophobic–hydrophilic segregation in crystals,[19] so it is reasonable to expect that molecules with this ratio similar will pack together more easily.

The preference for similar polarities could also be attributed to similar solubilities of co-crystal formers. Most co-crystals in the CSD were prepared as single crystals grown from solution, and similar co-former solubilities make finding the right conditions for single crystal growth easier.[20] However, the lack of correlation in descriptors usually associated with aqueous solubility (Log *P*, polar surface area) suggests that solubility is not the decisive factor behind the preferred similarity of polarity descriptors.

5.3 Experimental Testing of Shape and Polarity Preferences

The above database results can be translated into simple qualitative rules for selecting co-crystal formers. Choosing co-formers with similar shapes, similar fractional polarities and similar dipole moments to the API should increase the chances of successful co-crystal formation. Because of the relative weakness of the underlying trends, a large number of experiments is required to demonstrate convincingly that consideration of these factors improves the success rate of co-crystal screening experiments.

We envisaged two complementary approaches to testing the usefulness of shape and polarity matching. First, a large set of arbitrary compound pairs was considered. These 218 compound pairs were selected with supramolecular synthons in mind and were tested thoroughly for co-crystal formation in the group of Prof. William Jones (University of Cambridge, UK). It was then checked whether prioritising the experiments on the basis of shape and polarity relationships could increase the success rate. The second set of experiments focused on the relative influence of supramolecular synthons and shape matching. Weak and strong synthons were tested between molecules of similar and dissimilar shapes.

The toxicity of the co-crystal formers was not considered, since the purpose of the experiments was to confirm the physical significance of the results from the database study. Of course, only non-toxic substances can be used when developing a pharmaceutical co-crystal, but this difference is expected to have no significance here. The same physical forces drive the formation of co-crystals with both toxic and non-toxic compounds.

5.3.1 Co-crystal Screening Methods

This section is dedicated to a brief description of an experimental co-crystal screening and characterisation approaches designed to aid the statistical study of molecular parameters that determine the success of co-crystallisation. Although a great deal of information on co-crystal structure and design can be readily extracted through the analysis of data already deposited within the CSD, the usefulness of the database for validation purposes is limited. The limitation principally stems from the fact that the contents of the database are inherently limited to successful co-crystallisation experiments. In that way important information on the contrast between successful and unsuccessful experiments can be lost, requiring the use of in-house generated data which includes both successful and unsuccessful experiments.

We have recognised mechanochemistry as the most suitable experimental approach for screening for co-crystal formation or its absence, particularly liquid-assisted grinding (LAG) methodology. The conventional approach to co-crystal synthesis in solution is limited by effects related to solubility equilibria. These effects are expected to be particularly relevant in the case of incongruently soluble co-crystals, that is, co-crystals whose individual components exhibit considerable differences in solubility. In such cases, the solution is more likely to become supersaturated with respect to the low solubility co-crystal component than with respect to the desired co-crystal. As a result, the co-crystallisation experiment will be unsuccessful and lead to the separate crystallisation of individual co-crystal components, although these could readily form a co-crystal under a different set of conditions. As demonstrated by Rodríguez-Hornedo and co-workers,[21] the difficulties arising from incongruent solubility of co-crystal components can be resolved by ensuring that the solution is always saturated with respect to all potential co-crystal components. However, screening through such reaction co-crystallisation conditions can be complicated by the need to use reagents in a large excess and still presents difficulties in nucleating the co-crystal rather than individual components. An alternative to simple solution co-crystallisation is the use of solvent-free approaches, such as mechanochemical grinding or melting together of individual co-crystal components. Although both approaches avoid difficulties inherent in solution chemistry, each presents limitations of a different kind. In the case of neat grinding, co-crystallisation is affected by kinetic effects related to the reactivity of solid reagents or the stability of the amorphous phase. Screening by crystallisation from the melt is limited only to thermally stable or low-melting organic molecules.

However, grinding in the presence of a small amount of a liquid phase (liquid-assisted grinding or LAG) was found to be highly efficient for the synthesis of co-crystals, resulting in rapid and quantitative formation of co-crystals, typically within 20 minutes. A suitable illustration of the greater efficiency of LAG for co-crystal screening is the proof-of-principle study of the formation of three-component co-crystals based on a hydrogen-bonded inclusion host framework of caffeine and succinic acid. In particular, a set of screening experiments involving caffeine, succinic acid and 25 different guest molecules resulted in the formation of a three-component solid in four (16%) cases when using solution co-crystallisation, 15 cases (60%) when using neat grinding co-crystallisation and in 18 cases (72%) when using LAG. Similar results were obtained in comparative screening experiments by Karki *et al.*,[22] who investigated the formation of co-crystal stoichiometric variations involving the model API nicotinamide and dicarboxylic acids using solution crystallisation, crystallisation from the melt, neat and liquid-assisted grinding, and by Childs *et al.*[23] who investigated the co-crystallisation of the API carbamazepine with carboxylic acids using slurry sonication, high-throughput evaporation, liquid-assisted grinding and reaction crystallisation. Further details and examples of applications of mechanochemistry in the construction of pharmaceutical solid forms are given in Chapter 9.

In addition to efficiency and speed, LAG also provides a highly crystalline product. As a result, co-crystal formation is readily observed through X-ray powder diffraction, which can also be used to characterise the new materials structurally.[24] Indeed, Karki and co-workers have recently demonstrated that the reaction speed, high crystallinity and quantitative yield associated with LAG co-crystallisation enable the rapid structural characterisation of new co-crystals involving simple components, such as the model API theobromine, within 24 hours.[25] More recently, it was demonstrated how products of mechanochemical co-crystal formation can also be structurally characterised through crystal structure prediction methods, followed by comparison of measured and calculated X-ray powder diffraction patterns.[26]

5.3.2 Prioritising Screening Experiment by Molecular Descriptors

The co-crystal database does not contain information about pairs of molecules that do not form co-crystals with each other, so common machine learning techniques cannot be applied to derive a predictive model, for example, by fitting a scoring function. Instead, we based our selection criteria on descriptor differences between co-crystal formers in the CSD. Four of the descriptors, S/L, M/L, FNO and dipole, were chosen for further use. The other descriptors in Table 5.1 contain redundant information and show wider distributions than those selected.

First, a simple classification scheme was tested. A pair of molecules was classified as likely to form a co-crystal if the difference between the molecules in all four descriptors is smaller than a predefined cut-off value. The 9th deciles of the differences in the CSD (Table 5.1) were used as cut-offs, that is, 90% of known co-crystals show a difference smaller than the cut-off values. The scheme was applied to the 218 screening experiments that had been performed previously (see above). Effectively, this test probes whether selecting co-formers using shape and polarity *in addition* to supramolecular synthons gives a benefit.

Table 5.2 Classification of 218 screening experiments based on molecular descriptors.

	Experiment	
Prediction	*Co-crystal Formed*	*No co-crystal*
Likely to form	40	67
Unlikely to form	8	103

The molecular models for the calculation of the descriptors were not taken from the co-crystals, since these would not be known prior to performing the experiments. They were extracted from the single-component crystals of the test compounds when available and they were calculated using molecular mechanics (MMFF94 force field) otherwise.

Of the 218 compound pairs, 48 (= 40 + 8) formed a co-crystal, which corresponds to a success rate of 22% (Table 5.2). Only about half of the 218 experiments were classified as likely to give a co-crystal (40 + 67 = 107), which included 83% (40/48) of the successful screens. This means that performing only the experiments labelled as 'likely' would reduce the number of experiments by 50% and increase the success rate from 22% to 37% (=40/107).

The same scheme was further used to select likely co-crystal formers for artemisinin.[27] Only two of the 75 compounds tested experimentally gave a co-crystal, orcinol and resorcinol (Formula 5.2). The difficulty of rationalising co-crystal formation with artemisinin is illustrated by the fact that the structures and supramolecular motifs in the two resulting co-crystals show almost no structural similarity, except the existence of O–H···O hydrogen bonds. Yet, both orcinol and resorcinol were among the 41 likely co-formers (56%).

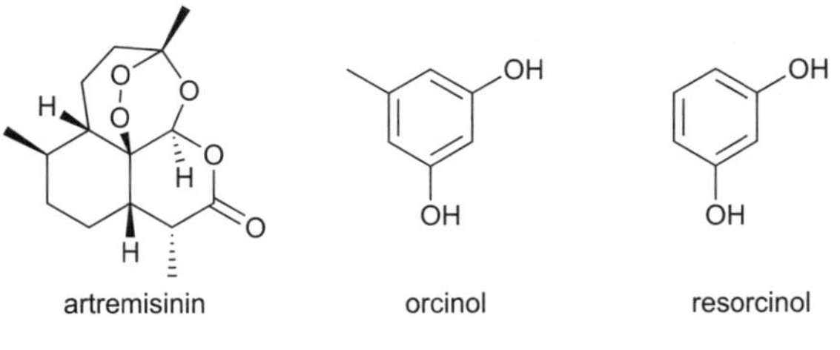

(5.2)

When several co-crystal formers are tested against a single target molecule, it would be better to rank potential co-formers, so that the most promising experiments can be performed first. A simple dissimilarity score was devised using the above four descriptors and the corresponding cut-off values. It is obtained by dividing the difference between the target molecule and the co-crystal former (Δ_D) by the cut-off value (δ_D) and summing the results for each descriptor (D):

$$score = \sum_D (\Delta_D/\delta_D) = |S_t/L_t - S_c/L_c|/0.275 + |M_t/L_t - M_c/L_c|/0.310 + |FNO_t - FNO_c|/0.294 + |dipole_t - dipole_c|/5.94D \qquad (5.1)$$

where the subscripts t and c refer to the descriptor value calculated for the target molecule (*e.g.*, an active pharmaceutical ingredient) and the co-former, respectively. When ranking the co-crystal formers tested with artemisinin using this score, orcinol (score = 1.22) and resorcinol (score = 1.72) are ranked third and 11th, respectively. This result suggests that testing co-crystal formers with lower score values first could significantly reduce the number of experiments required to find co-crystals.

If the 218 test experiments were performed in the order of calculated dissimilarity scores, then most co-crystals would be recovered earlier than by performing the experiments in a random sequence (Figure 5.4). The gain in efficiency is moderate in the first 10–15 experiments (scores below 1.0), but from then it increases up to about 125 experiments (score values of ~2.3). This trend suggests that even very low score values do not guarantee co-crystal formation, but with increasing values it becomes gradually less likely that a co-crystal can be prepared (Figure 5.5).

This point is further illustrated by comparing the screening results of theophylline and theobromine with carboxylic acids (Formula 5.3; Table 5.3). Both xanthines are expected to associate with carboxylic acids via COOH ··· N(imidazole) hydrogen bonds. Theophylline formed co-crystals with five of the nine acids, including those with the four lowest dissimilarity scores. Theobromine, on the other hand, co-crystallised with only three acids and score values would provide little advantage in finding them (the three acids are ranked 3rd, 5th and 7th). Comparing data in Table 5.3 and Figure 5.5, it is apparent that each of the score values is relatively low except those for ibuprofen and sebacic

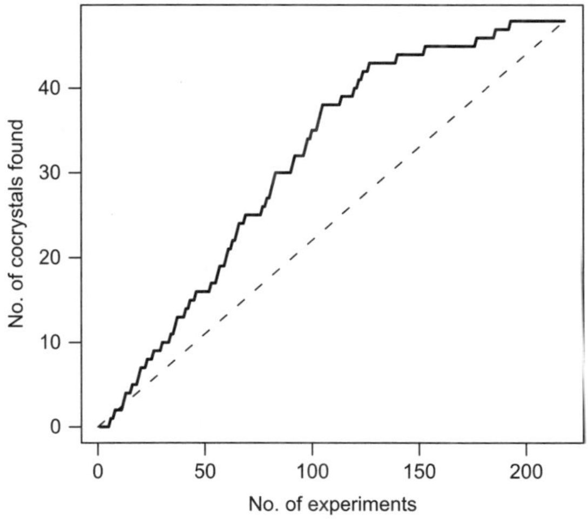

Figure 5.4 Results of performing 218 screening experiments in the order of increasing score values. The continuous line shows how many of the 48 co-crystals can be found by performing a given number of the lowest scoring experiments, while the dashed line shows how many co-crystals would be found by performing the same number of experiments without score ranking (*i.e.*, randomly).

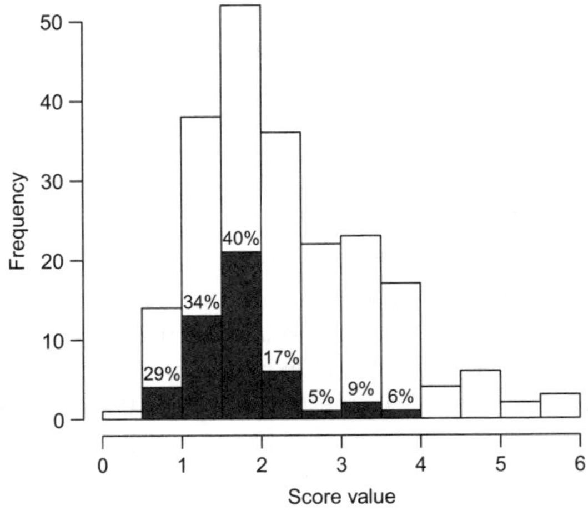

Figure 5.5 Distribution of score values calculated for 218 screening experiments. The shading and the percentage values show how many of the experiments gave co-crystals within each histogram bin.

acid. (Relative to the two xanthenes, ibuprofen has a low fractional polarity, while sebacic acid has a more elongated shape.) Theobromine shows a low overall propensity for co-crystal formation: of 28 compounds tested, only the three acids listed in Table 5.3 produced a new phase. The low success rate among low scoring co-formers with theobromine is the consequence of this low propensity, which, unfortunately, cannot be explained by the descriptors used here.

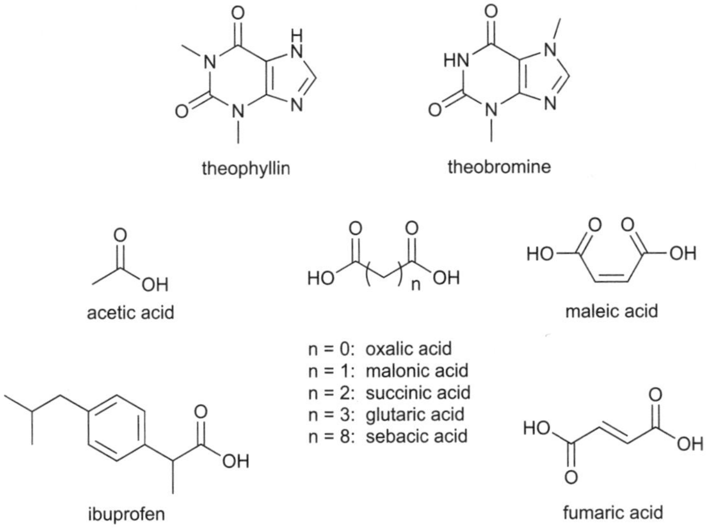

(5.3)

5.3.3 Combined Effect of Synthons and Shape

Development of an optimal design strategy that involves both supramolecular synthons and molecular properties requires an understanding of the relative

Table 5.3 Screening results of theophylline (tp) and theobromine (tb) with carboxylic acids.

	Experimental result[a]		*Score value*	
Coformer	*tp*	*tb*	*tp*	*tb*
Maleic acid	x		0.96	0.78
Acetic acid	x	x	1.09	1.47
Oxalic acid	x	x	1.57	1.54
Glutaric acid	x		1.58	1.30
Fumaric acid			1.61	1.52
Succinic acid			1.63	1.56
Malonic acid	x	x	1.80	1.57
Ibuprofen			2.52	2.45
Sebacic acid			3.58	3.49

[a]An 'x' indicates that a co-crystal was obtained.

influence and combined effect of these two principles. The relative importance of molecular shapes and supramolecular synthons was tested by attempting to co-crystallise molecules with similar shapes and strong or weak synthons and molecules with strong synthons, but with similar or dissimilar shapes.

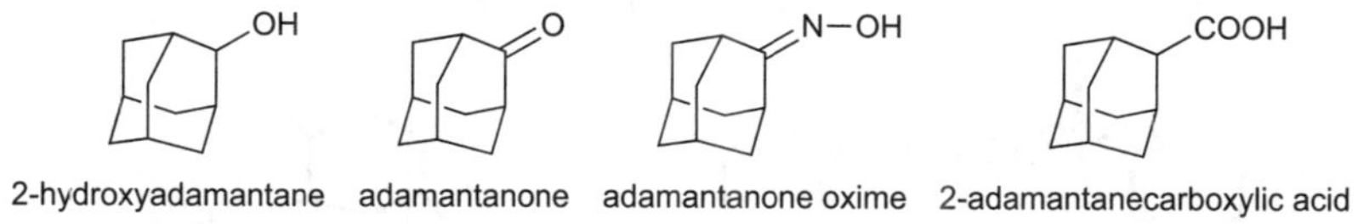

(5.4)

The four adamantane derivatives, 2-hydroxyadamantane, 2-adamantanone, 2-adamantanone oxime and 2-adamantanecarboxylic acid (Formula 5.4) all have a nearly spherical shape. Screening experiments were performed with each possible pair of the four compounds, but only two co-crystals were found: hydroxyadamantane– adamantanone and adamantanone oxime–adamantane-carboxylic acid. The alcohol–ketone and acid–oxime hydrogen bonds provide strong supramolecular synthons for the formation of these co-crystals, while for the other combinations possible heteromolecular synthons are not preferred to homomolecular ones. The results thus show that shape similarity cannot overcome the effect of good homomolecular synthons.

The next set of experiments utilised a robust synthon, formed by oximes and carboxylic acids (Figure 5.6). The shape of the molecules was varied between spherical and flat molecules. Co-crystals formed if either both molecules were flat or both were spherical, suggesting that even the relatively strong oxime–acid synthon cannot compensate for the negative effect of shape mismatch. The strongest acid in Figure 5.6, 3,5-difluorobenzoic acid was an exception, since it formed co-crystals with the flat and spherical oximes alike. The electrostatic interaction between the cations and anions of salts can be viewed as an extremely strong heteromolecular synthon. Salts are known to form with ions of different shapes (*e.g.*, pharmaceutical hydrochlorides), so it can be expected that the stronger the heterosynthon, the larger is the shape difference it can overcome.

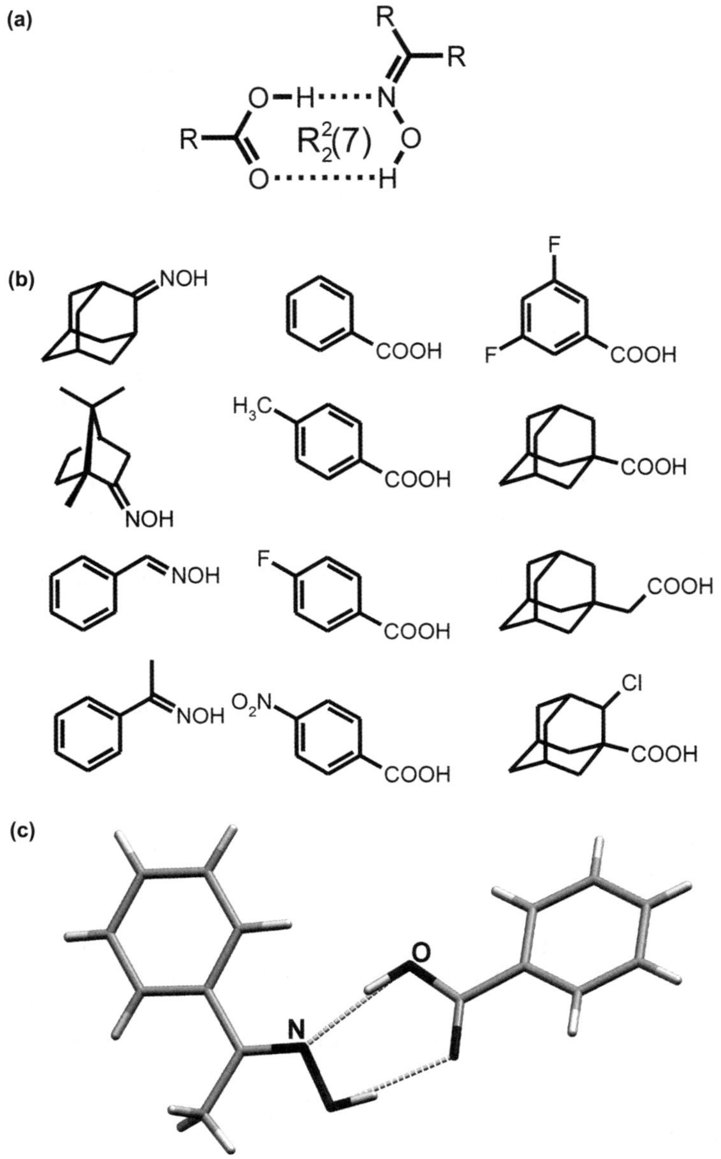

Figure 5.6 (a) Diagram of the hydrogen-bonded $R_2^2(7)$ synthon expected to form between carboxylic acids and oximes; (b) model oxime and carboxylic acid molecules with different shapes and (c) a hydrogen-bonded dimer based on the $R_2^2(7)$ oxime–carboxylic acid synthon in the crystal structure of the co-crystal of acetophenoneoxime with benzoic acid.[28]

5.3.4 Identification of New Intermolecular Interactions

The results described above showed that the use of shape- and polarity-matching generally improves the success rate of co-crystallisation experiments.

However, for some specific compounds tested this was not the case. Specifically, liquid-assisted grinding experiments involving four steroids as target molecules and 24 small planar molecules as co-crystal formers showed surprising results (Figure 5.7).[29] The shapes and polarities of the steroids

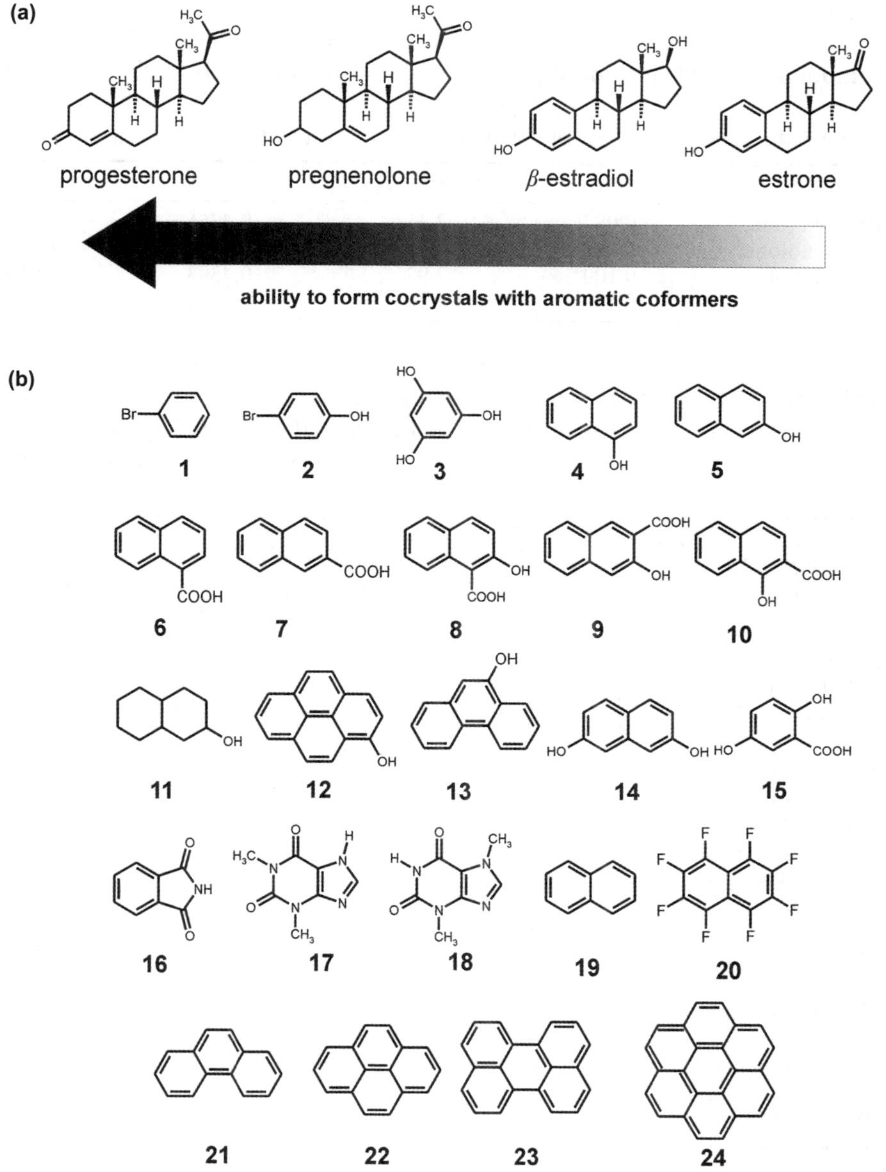

Figure 5.7 (a) Molecular diagrams of the four model steroids used in the co-crystallisation screen and (b) molecular diagrams of explored co-crystal formers. Reproduced from ref. 29.

progesterone, pregnenolone, β-estradiol and estrone are similar (Table 5.4). Based on descriptor cut-offs, all four steroids would be expected to co-crystallise with at least 16 of the 24 co-formers. Furthermore, they all could form hydrogen-bonded associates with phenols or carboxylic acids and both groups were well represented among the 24 co-formers. Consequently, it was not expected that the screening results for the four steroids would be very different.

As shown in Table 5.5, progesterone forms co-crystals with 19 of the 24 compounds tested, including phenols (*e.g.*, 4-bromophenol), carboxylic acids

Table 5.4 Descriptor values calculated for the four steroid molecules.

Steroid	*FNO*	*Dipole(D)*	*S/L*	*M/L*
Progesterone	0.087	2.35	0.4366	0.5412
Pregnenolone	0.087	4.41	0.4154	0.5198
β-Estradiol	0.100	2.92	0.4383	0.5766
Estrone	0.100	2.20	0.4494	0.5990

Table 5.5 Results of co-crystal screening involving the four model steroid molecules. The formation of a co-crystal is denoted by '+', and was determined by X-ray powder diffraction and solid-state infrared spectroscopy. Reproduced from ref. 29.

Co-former \ *Steroid*	*Progesterone*	*Pregnenolone*	*β-Estradiol*	*Estrone*
1, Bromobenzene	–	–	–	–
2, 4-Bromophenol	+	+	–	–
3, Phloroglucinol	+	–	–	
4, 1-Naphthol	+	+	–	–
5, 2-Naphthol	+	+	–	–
6, 1-Naphthoic acid	+	–	–	–
7, 2-Naphthoic acid	+	–	–	–
8, 2-Hydroxy-1-naphthoic acid	+	–	–	–
9, 3-Hydroxy-2-naphthoic acid	+	–	–	–
10, Xinafoic acid	+	–	–	–
11, Decahydro-2-naphthol	–	–	–	–
12, Pyrenol	+	–	–	–
13, 9-Phenanthrol	+	–	–	–
14, 2,7-Dihydroxynaphthalene	+	+	–	–
15, Gentisic acid	+	–	–	–
16, Phthalimide	+	–	–	–
17, Theophylline	–	–	–	–
18, Theobromine	–	–	–	–
19, Naphthalene	+	–	–	–
20, Octafluoronaphthalene	–	–	+	+
21, Phenanthrene	+	–	+	–
22, Pyrene	+	–	+	–
23, Perylene	+	–	–	–
24, Benzocoronene	+	–	–	–

(*e.g.*, 2-naphthoic acid) and even compounds that cannot partake in hydrogen bonding (*e.g.*, naphthalene). Pregnenolone co-crystallised with four phenols and β-estradiol with three compounds, none of which has any hydrogen-bond donors or acceptors. The only co-crystal of estrone was obtained with octafluoronaphthalene. Interestingly, octafluoronaphthalene is one of the five compounds that did not co-crystallise with progesterone.

While puzzling, these differences indicate that specific intermolecular interactions are at work.[29] The hydrogen atoms pointing outwards from the α-face of progesterone (Figure 5.8(a), below the molecule) generate a flat, positively polarised surface. This surface interacts favourably with the negatively polarised π-electron cloud of aromatic compounds, leading to the formation of co-crystals involving singly or doubly stacked steroid and co-former molecules (Figure 5.8(b) and (c)). The stabilising effect of such α···π stacking was confirmed through crystal structure prediction experiments, which revealed the persistence of the stacking motif in almost all low-energy predicted structures. Calculated molecular electrostatic surface potentials reveal that the area of the positive region at the α-face decreases gradually in the sequence progesterone > pregenenolone > β-estradiol ≈ estrone (Figure 5.8(d)), in this way gradually hindering the formation of co-crystals through

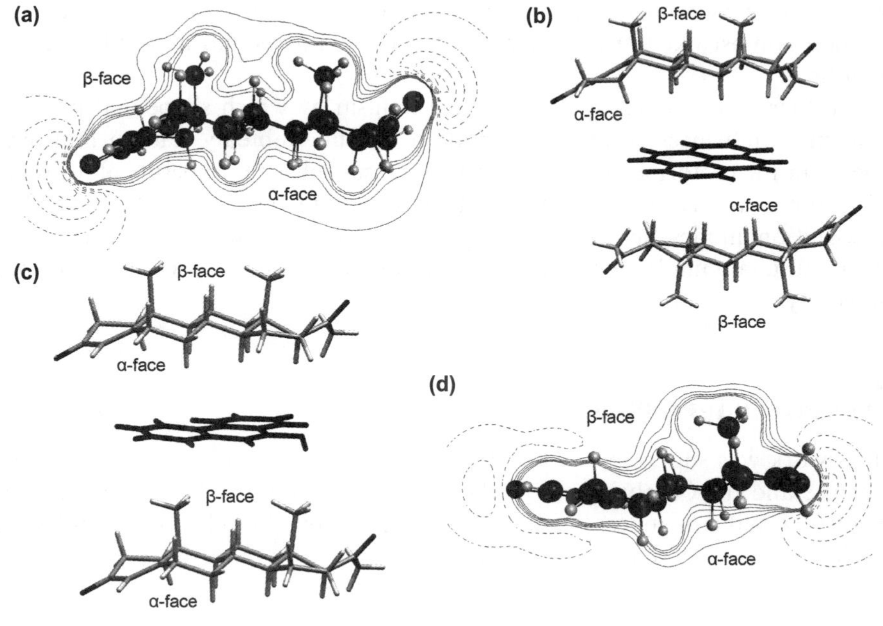

Figure 5.8 (a) Electrostatic potential around progesterone; (b) a doubly stacked complex of progesterone and pyrene (shown in black) in their co-crystal; (c) a singly stacked complex of progesterone and 9-phenanthrol (shown in black) in the solid state and (d) electrostatic potential around estrone. Continuous contour lines indicate positive, dashed lines negative potential. The interval between subsequent contour lines is 0.02 a.u. Parts (b) and (c) are reproduced from ref. 29.

$\alpha \cdots \pi$ stacking. The formation of co-crystals between octafluoronaphthalene and the aromatic steroids β-estradiol and estrone was interpreted as the result of positive electrostatic potential above the octafluoronaphthalene rings. This charge inversion, induced by the electron-withdrawing effect of the fluorine atoms, enables the sufficiently negative β-estradiol and estrone to form co-crystals with octafluoronaphthalene via conventional $\pi \cdots \pi$ stacking interactions.

It can thus be concluded that the differences in the co-crystallisation behaviour of the four steroids are caused by specific intermolecular interactions between the steroid α-face and aromatic rings. The discrepancy between the observed reactivity and expectations based on molecular descriptors led to the identification of this new mode of molecular recognition.

5.4 Conclusions

Analysis of molecular properties in co-crystals revealed a preference for co-crystal formation between molecules of similar shapes and polarities. Statistical parameters, derived from co-crystals in the Cambridge Structural Database, were used to derive simple methods to select and rank likely co-crystal formers. Application of these methods has been demonstrated to improve the efficiency of screening experiments significantly. It has been shown that co-crystal formation is most likely if both supramolecular synthons and molecular shapes favour it.

Further work to clarify the exact mechanism by which shape and polarity matching influences co-crystal formation will undoubtedly lead to improvements in the co-former selection strategy presented here. Experimental results that apparently contradict these principles may uncover new supramolecular recognition modes, as illustrated by the co-crystallisation of steroids with aromatic compounds, or help identify further factors that need to be considered in co-crystal design.

Acknowledgements

This work was supported by the Pfizer Institute for Pharmaceutical Materials Science and the Cambridge Crystallographic Data Centre. We are grateful to Professor William Jones, Shyam Karki and Tiziana Chiodo for providing us the results of their screening experiments.

References

1. G. R. Desiraju, *Angew. Chem. Int. Ed. Engl.*, 1995, **34**, 2311.
2. M. C. Etter, *Acc. Chem. Res.*, 1990, **23**, 120.
3. C. B. Aakeröy, A. M. Beatty and B. A. Helfrich, *Angew. Chem. Int. Ed.*, 2001, **40**, 3240.

4. T. R. Shattock, K. K. Arora, P. Vishweshwar and M. J. Zaworotko, *Cryst. Growth Des.*, 2008, **8**, 4533.
5. N. Blagden, D. J. Berry, A. Parkin, H. Javed, A. Ibrahim, P. T. Gavan, L. L. De Matos and C. C. Seaton, *New J. Chem.*, 2008, **32**, 1659.
6. F. H. Allen, *Acta Crystallogr., Sect. B: Struct. Sci.*, 2002, **58**, 380.
7. L. Fábián, *Cryst. Growth Des.*, 2009, **9**, 1436.
8. C. B. Aakeröy, D. J. Salmon, M. M. Smith and J. Desper, *Cryst. Growth Des.*, 2006, **6**, 1033.
9. C. B. Aakeröy, J. Desper and M. E. Fasulo, *CrystEngComm*, 2006, **8**, 586.
10. K. Chadwick, G. Sadiq, R. J. Davey, C. C. Seaton, R. G. Pritchard and A. Parkin, *Cryst. Growth Des.*, 2009, **9**, 1278.
11. J. Fayos and F. H. Cano, *Cryst. Growth Des.*, 2002, **2**, 591.
12. J. Fayos, L. Infantes and F. H. Cano, *Cryst. Growth Des.*, 2005, **5**, 191.
13. J. Fayos, *Cryst. Growth Des.*, 2009, **9**, 3142.
14. D. J. Livingstone, *J. Chem. Inf. Comput. Sci.*, 2000, **40**, 195.
15. A. Field, *Discovering Statistics Using SPSS*, SAGE Publications, London, 2nd edn, 2005, ch. 4, pp. 107–134.
16. E. Pidcock and W. D. S. Motherwell, *Chem. Commun.*, 2003, 3028.
17. F. Zheng, C. Li, Q. Yuan and F. Vriesekoop, *J. Membr. Sci.*, 2008, **318**, 114.
18. P. A. Wood and P. T. A. Galek, *CrystEngComm*, 2010, **12**, 2485.
19. M. D. Ward and M. J. Horner, *CrystEngComm*, 2004, **6**, 401.
20. R. A. Chiarella, R. J. Davey and M. L. Peterson, *Cryst. Growth Des.*, 2007, **7**, 1223.
21. N. Rodríguez-Hornedo, S. J. Nehru, K. F. Seefeldt, Y. Pagan-Torres and C. J. Falkiewicz, *Mol. Pharmaceutics*, 2006, **3**, 362.
22. S. Karki, T. Friščić and W. Jones, *CrystEngComm*, 2009, **11**, 470.
23. S. L. Childs, N. Rodríguez-Horinedo, L. S. Reddy, A. Jayasankar, C. Maheshwari, L. McCausland, R. Shipplett and B. C. Stahly, *CrystEngComm*, 2008, **10**, 856.
24. S. H. Lapidus, P. W. Stephens, K. K. Arora, T. R. Shattock and M. J. Zaworotko, *Cryst. Growth Des.*, 2010, **10**, 4630.
25. S. Karki, L. Fábián, T. Friščić and W. Jones, *Org. Lett.*, 2007, **9**, 3133.
26. A. J. Cruz-Cabeza, S. Karki, L. Fábián, T. Friščić, G. M. Day and W. Jones, *Chem. Commun.*, 2010, **46**, 2224.
27. S. Karki, T. Friščić, L. Fábián and W. Jones, *CrystEngComm*, 2010, **12**, 4038.
28. J. K. Maurin, M. Winnicka-Maurin, I. C. Paul and D. Y. Curtin, *Acta Crystallogr. Sect. B*, 1993, **49**, 90.
29. T. Friščić, R. W. Lancaster, L. Fábián and P. G. Karamertzanis, *Proc. Natl. Acad. Sci. U.S.A.*, 2010, **107**, 13216.

CHAPTER 6

Role of Co-crystals in the Pharmaceutical Development Continuum

NATE SCHULTHEISS† AND JAN-OLAV HENCK*

SSCI, a division of Aptuit, 3065 Kent Avenue, West Lafayette, IN, 47906, USA

6.1 Introduction

The field of solid-state chemistry of drugs encompasses many scientific disciplines and is critical for drug development.[1,2] The aim of this chapter is to show how knowledge of solid-state properties can reduce time to preclinical trials, balance risks, and ultimately reduce costs. The general approach will be to demonstrate how an understanding of the molecular structure in the solid-state of an active pharmaceutical ingredient (API) and the use of co-crystals can lead to better design and control of drug performance. It is nowadays well understood that early engagement with pharmaceutical development scientists on the identification of an optimal solid form and formulation during the drug discovery stage can lead to significant benefits throughout the drug development process. Solid forms such as different polymorphs of an API, hydrates or solvates, as well as salts, co-crystals, and amorphous materials, including amorphous dispersions, need to be considered in this context.

† Present address – Halliburton, 2600 South 2nd Street, Duncan, OK 73536, USA

RSC Drug Discovery Series No. 16
Pharmaceutical Salts and Co-crystals
Edited by Johan Wouters and Luc Quéré

Published by the Royal Society of Chemistry, www.rsc.org

Co-crystals are increasingly being considered in order to address solubility, bioavailability and/or other physical or chemical deficits of a given API. One of the most important benefits of applying efforts toward early co-crystal identification is the potential for reducing preclinical candidate attrition and reaching *in vivo* studies more quickly. These activities can, for example, demonstrate dose limiting toxicity in order to establish acceptable and reproducible safety margins in preclinical toxicity studies. Early identification of an optimal solid form such as a co-crystal also has the potential to minimize the need for multiple changes of the solid form of an API during drug development, which in turn reduces costs directed towards *in vitro* physicochemical studies and *in vivo* comparison studies. Obviously, early solid form optimization is just one aspect of a targeted drug development strategy; many other factors play an important role in whether or not a given API can be brought to market successfully.

6.2 Common Solid-state Strategies for API Property Modification

It is well understood that crystalline materials obtain their fundamental physical properties from the molecular arrangement of the molecules within the solid, and altering the placement and/or interactions between these molecules can, and usually does, have a direct impact on the properties of the particular solid.[3] Currently, solid-state chemists call upon a variety of different strategies when attempting to alter the chemical and physical solid-state properties of APIs, namely the formation of salts, polymorphs, amorphous materials, hydrates, solvates, and co-crystals. Amorphous dispersions also provide a viable solid-state route, where an amorphous material is embedded within a polymer matrix, sometimes leading to improved stability and solubility in comparison to the amorphous-free form. However, in comparison to a crystalline form, an amorphous material is thermodynamically less stable and thus, typically is not the first solid form selected. Salt formation is one of the primary solid-state approaches used to modify the physical properties of APIs and it is estimated that over half of the medicines on the market are administered as salts.[4] However, a necessary criterion within this approach is that the API must possess a suitable (basic or acidic) ionizable site. In comparison, co-crystals offer a different pathway, where any API, regardless of acidic, basic, or ionizable groups, could potentially be co-crystallized. This aspect helps compliment existing methods by reintroducing molecules that had limited pharmaceutical profiles based on their non-ionizable functional groups. In addition, the number of potential non-toxic co-crystal formers (or co-formers) that can be co-crystallized with the API is numerous.[5] Of course, all molecules are toxic at some critical level, so if the drug loading is known, consideration should be given when selecting suitable co-formers.

The efforts of the pharmaceutical industry over last 10 to 15 years to invent new chemical entities to be used as APIs have led to more and more compounds with limited water solubility and bioavailability. It is estimated that approximately 70–80% of the drugs currently under development belong to Class 2 or Class 4 of the biopharmaceutical classification system.[6] Class 2 compounds show low solubility and high permeability, whereas Class 4 compounds exhibit low solubility and low permeability. Therefore Class 2 compounds are of particular interest in the context of co-crystals because improving the apparent solubility (combination of dissolution and solubility) of a given material through a co-crystal may be a useful and successful strategy.

Other properties of an API that could potentially be altered by forming a co-crystal are the melting point, hygroscopicity, physical and chemical stability, filterability, color, as well as crystallinity. It should be made clear that no one particular strategy offers a solution for the property improvement of all APIs. Each API must be examined and evaluated on a case-by-case basis in terms of molecular structure and desired final properties. In addition, there is no guarantee that a particular property of interest will be altered, as physical property prediction and/or trending are not yet available tools.

6.3 Co-crystals and their Characterization

To date, a universal and agreeable definition of what constitutes a co-crystal is still not available. Other chapters of this monograph will deal with this issue in more detail. However, in this chapter, co-crystalline materials will possess the following descriptors:

- an API, neutral or zwitterionic, along with a neutral co-former, held together through non-covalent, freely reversible interactions
- and a co-former (liquid or solid), which may or may not be pharmaceutically acceptable.

Single crystal X-ray diffraction is the preferred characterization technique in determining if a co-crystalline material has been generated; however, suitable X-ray quality crystals of appropriate size cannot always be produced. Additionally, even if single crystals can be grown to a sufficient size and quality, the exact location of the hydrogen atom (determination if proton-transfer has occurred from the acid to the base or not) may be ambiguous.[7] Thus, if single crystals are unobtainable, utilizing other solid-state, spectroscopic techniques, for instance Raman, infrared, solid-state nuclear magnetic resonance (SSNMR), and X-ray photoelectron would be advantageous in attempting to characterize potentially new co-crystalline materials.[8]

In particular, infrared spectroscopy can be a very powerful tool in detecting co-crystal formation, especially when a carboxylic acid is used as a co-former and/or when a neutral O–H···N hydrogen bond is formed between the components. Distinct differences, within the IR spectra, can be observed between a neutral carboxylic acid moiety and a carboxylate anion. A neutral carboxylic acid (–COOH) displays a strong C=O stretching band around

1700 cm^{-1} and a weaker C–O stretch around 1200 cm^{-1}, while a carboxylate anion (–COO^-), due to resonance, displays a single C–O stretch in the fingerprint region of 1000–1400 cm^{-1}. Additionally, if a neutral intermolecular O–H···N hydrogen bond has formed between the components, then two broad stretches around 2450 cm^{-1} and 1950 cm^{-1} will be observed.[9]

Recently, it was reported that by combining SSNMR and X-ray photoelectron spectroscopy (XPS) one could unambiguously determine the degree of protonation between a proton donor and nitrogen acceptor.[10] Although this combination of techniques might be limiting in scope, it still provides clarity to assignment of proton location and whether a salt or co-crystal has been formed. Also, the use of molecular electrostatic potential (MEP) surface calculations has been recently reported in order to assign relative hydrogen-bond donor/acceptor strengths to various chemical functional groups.[11] The difference in MEP values calculated from a series of Brönsted–Lowry acids and bases can then be used to 'anticipate' whether proton transfer occurs or not.

A number of different characterization techniques mentioned above are used to help distinguish between co-crystals and salts and it should be noted that in some cases differentiation between the two may still be difficult. It should also be pointed out that although salts and co-crystals often posses different properties (*e.g.*, stability or solubility), these issues from a pharmaceutical development standpoint may not be influential as long as the process can be monitored and closely controlled. However, for intellectual property (IP) rights and regulatory issues, differentiating between a co-crystal and a salt is important.

6.4 Co-crystals and their Role in the Pharmaceutical Development Process

In attempts to generate co-crystalline materials, the ability to predict whether or not two or more individual components will come together in the solid state is not yet possible; however, one can often rely upon chemical intuition to begin the selection process. A thorough evaluation of the API should be considered, including but not limited to, the number and arrangement of hydrogen bond donors and acceptors, molecular and conformational flexibility, salt forming propensity, and kinetic solubilities in a variety of solvent and solvent mixtures. Furthermore, the ability to conduct experiments with all possible co-formers is unrealistic and thus selection is often based on hydrogen bonding rules, probable molecular recognition events,[12] and toxicological profiles of the co-formers;[4a,13] however, one should not limit themselves to just these criteria, because co-crystals could easily be missed.

After an initial molecular evaluation, a variety of crystal growth strategies are available, but not limited to evaporation and cooling (at varying rates and temperatures),[14] milling (liquid-assisted or neat),[15] sonication,[16] melting,[17] and reaction crystallization.[18] More 'non-traditional' techniques include spray drying,[19] twin-screw extrusion,[20] and supercritical fluids[21] which have all been demonstrated to be scalable processes.

So, where or when in the pharmaceutical development process should co-crystals be considered?

The physical and chemical properties of a co-crystal need to be investigated in the same manner as any other solid form in order to determine whether they can be developed into a marketed dosage form.[22] However, co-crystals may also play a role in the earlier stages of drug development. Often the dissolution, solubility, and bioavailability behavior of a drug at the interface between discovery and preclinical testing are found too challenging to achieve the levels of exposure necessary to determine the toxicity profile of a drug. One potential way to increase dissolution, solubility, and bioavailability could be by generating a co-crystal with the given API.[23] This co-crystal than could be filled into capsules without further drug development activities. Obviously in many cases this is not going to be the final drug formulation for this drug candidate, but it allows a method of generating very important data points in order to obtain decision points and milestones in the drug development process. Physicochemical properties, such as crystallinity, melting point, solubility, dissolution, and stability, are important when moving a new compound, such as a co-crystal, through early development. The information from these studies can be used to prioritize the available forms and a flow chart can help to organize the process of solid form selection. An example of a flow chart that could be used to choose the best co-crystal candidate is given in Figure 6.1. It combines a number of properties inherent in drug development; however, it should be used

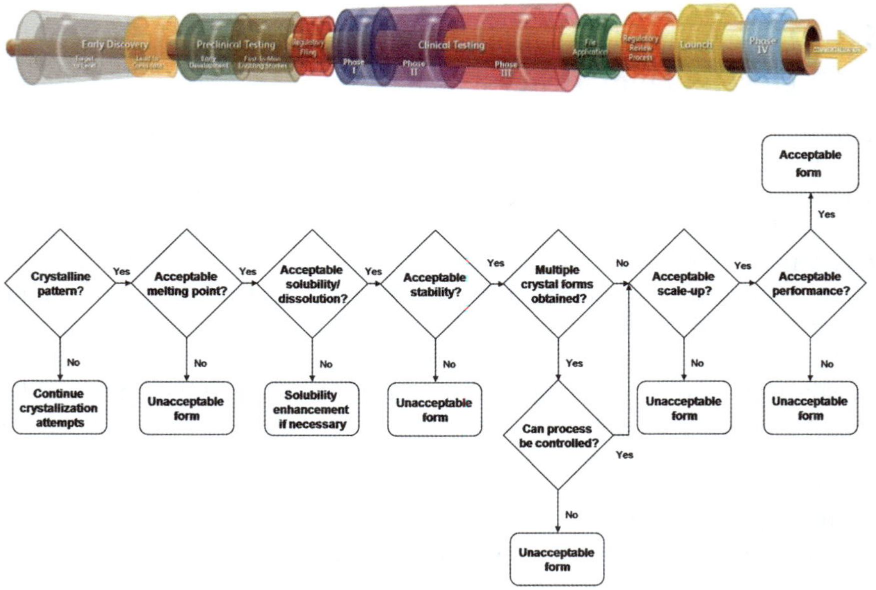

Figure 6.1 Plausible flow chart for choosing a co-crystal candidate. Reproduced with permission from Schultheiss and Newman, ref. 23, Copyright 2009 American Chemical Society.

as a guide only, since every compound will have its own challenges, and the properties in the flow chart will be unique to the situation at hand.

6.5 Using Co-crystals to Alter Physicochemical Properties of APIs

6.5.1 Stability

Stability is a heavily studied parameter during the development of a new chemical entity. Different types of stability need to be considered, depending on the characteristics of the molecule, including but not limited to, chemical structure, functional groups, acid/base tolerance, moisture, and air sensitivity. Chemical and physical stability data are commonly obtained at accelerated stability conditions to determine developability and shelf life.[24] Water uptake is included from a handling and packaging point of view. The amount of water present can also lead to form changes, degradation, and worse if the effect of the water uptake is not investigated early in the development process.[25] Thermal stress studies are also incorporated and extra work may be warranted for hydrates or thermally labile materials. In the case of co-crystals and salts, solution stability may be a factor due to dissociation of the material resulting in precipitation of the less soluble parent compound or a less soluble form (such as a hydrate in an aqueous medium).

A variety of reports are available on co-crystals that have been subjected to different stability conditions including, relative humidity,[26] thermal,[27] chemical,[27b] solution and thermodynamic.[28] For potential co-crystal candidates, each of these conditions should be tested and the information gained should be used to evaluate and design processing conditions, scale-up, formulations, packaging, and so on.

Solution stability is a very important parameter to assess during development and for this discussion will be defined as the ability of the co-crystal components to stay in solution and not readily crystallize or precipitate. A variety of media can be introduced, including water, simulated gastric fluid (SGF), simulated intestinal fluid (SIF), formulation vehicles, and buffered solutions. In many instances, these experiments can be coupled with solubility or dissolution experiments to obtain a more complete picture of the behavior and the solid form remaining at the end of the experiment. Because dissociation of a co-crystal often occurs and precipitation of a less soluble form (either one of the components or another co-crystal) can result, solution stability studies should be considered early in development of a potential drug product. Attention has been focused on formulation approaches to stabilize the co-crystal (or co-crystal components) in solution by addition of surfactants to avoid the problem of recrystallization. One of the first reports on this approach was published by Remenar *et al.* on a celecoxib:nicotinamide co-crystal in various formulation vehicles (1–10% sodium dodecylsulfate (SDS) and polyvinylpyrrolidone (PVP)) in order to understand the solid form that may result upon contact with simulated gastrointestinal (GI) fluids.[29] It was determined that the dissolution and solubility of the celecoxib:nicotinamide co-crystals were dependent on the medium and have been attributed to the dissociation of the co-crystal and recrystallization of celecoxib Form I and III. Ultimately, it was found that the addition of surfactants affected this conversion and its rate.

Another, more recent example by Rodríguez-Hornedo and co-workers elegantly displayed how micellar additives, such as surfactants, can bring thermodynamic stability to unstable co-crystal phases.[30] An incongruently dissolving carbamazepine:salicyclic acid co-crystal was tested and it was found that by introducing various concentrations of aqueous solutions of sodium lauryl sulfate (surfactant), a congruently saturated system resulted.

Solution stability is important not only for solution or suspension products, but can also impact other properties such as solubility and dissolution data for other types of dosage forms. The effects of dissociation, recrystallization, ionization, complexation, and solution chemistry are all areas that will need further work to understand better how co-crystals will behave in various media.

6.5.2 Solubility

One of the main reasons to investigate co-crystals is to increase the solubility of a poorly soluble compound (BCS Class 2 and 4). For neutral molecules, co-crystals can certainly expand the solid forms possible for development, while for a free acid or free base, both salts and co-crystals can be used to improve the solubility profile. There are a number of considerations when discussing solubility data. The first is equilibrium versus kinetic (or apparent) solubility measurements. Kinetic solubility values are approximate values usually based on one measurement at one time point. Unless preliminary experiments have been performed, it is not known if equilibrium has been reached in the time frame used. For equilibrium solubility, a number of time points and measurements are taken to ensure that the solution has reached equilibrium as evidenced by a plateau in the concentration data. This is sometimes called powder dissolution. The time required to reach the equilibrium solubility can also be a factor in development based on the residence time in the stomach and intestines. It is desirable for the drug to dissolve while it is in the gastrointestinal tract and very long dissolution times may result in less absorption of the drug. Powder dissolution rates can also be dependent on particle size; therefore the intrinsic dissolution rate may be a better assessment of this parameter.

A second consideration is form change during the experiment. When form changes occur, the solubility data obtained may not be relevant to the starting compound in the experiment. Form changes can be suggested by solubility data collected at various time points by a precipitous drop in concentration indicating crystallization of a less soluble form. In these cases, the maximum solubility observed over the time profile (S_{max}) may be reported along with the time it occurred (it should be noted that the S_{max} value is likely not an equilibrium solubility value). Any suggested form changes can then be confirmed by analysis of the solid form remaining at the end of the experiment. In all, when a co-crystal dissolves, the concentration in solution may be supersaturated, saturated, or under-saturated with respect to the individual components. Co-crystal dissolution that results in under-saturated concentrations of individual components leads to solution stability (Figure 6.2, curve A). Co-crystal dissolution that leads to supersaturated concentrations of one or both components may result in precipitation of the component(s) and therefore the co-crystal is not stable in solution, Figure 6.2, curve C. Another possibility, although considerably rare, is that supersaturated concentrations are achieved, but recrystallization does not occur, at least for a significant period of time, Figure 6.2, curve D.

A recent study conducted at SSCI investigated the propensity of the nutraceutical[31] compound pterostilbene to form co-crystals and ultimately attempt to improve its poor aqueous solubility profile through this type of solid form approach. Pterostilbene is a non-ionizable, methylated analogue of resveratrol and has been found in a number of edible berries, although its pharmacological benefits are negligible in its free form due to its limited aqueous solubility, $\sim 21 \mu g\ mL^{-1}$.[32] Five co-crystals of pterostilbene were identified and fully characterized: two 1:1 pterostilbene:caffeine polymorphs (Form I and Form II), a 2:1 pterostilbene:piperazine co-crystal, a 1:1 pterostilbene:carbamazepine co-crystal, and a 1:1 pterostilbene:glutaric acid co-crystal.[33] The aqueous solubility of each co-crystal was measured at ambient conditions (Table 6.1) over a variety of durations depending on the behavior of the co-crystalline material throughout the experiment. Surprisingly, at five hours the concentration of the pterostilbene:caffeine co-crystal was 27 times higher, while the

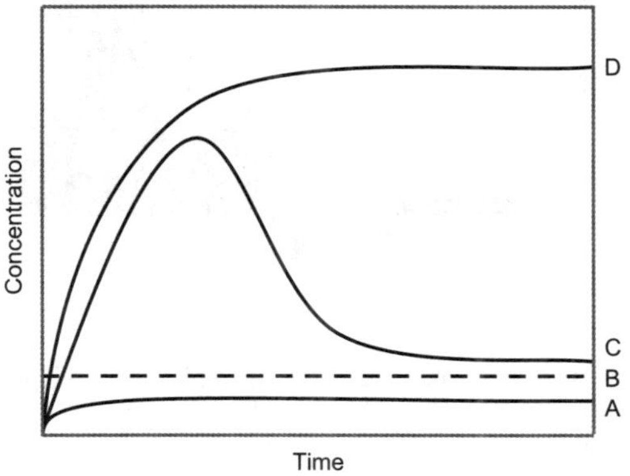

Figure 6.2 Schematic of a solubility curve due to a form change and precipitation of a less stable form (curve C); at equilibrium, the curve will level out to the solubility of the less soluble form. Line B represents the less soluble form. Curve A represents a stable form which is under-saturated with respect to the less soluble form, while curve D represents a stable form that is supersaturated with respect to the less soluble form. The phenomenon observed for curve D is rare, due to the large driving force towards recrystallization. Reproduced with permission from Schultheiss and Newman, ref. 23, Copyright 2009 American Chemical Society.

Table 6.1 Aqueous solubility measurements of individual components and co-crystals at ambient temperature.

Compounds	*Solubility* (± *standard deviation*) (*mg* mL^{-1})
Pterostilbene	21[a]
Caffeine	18.5 (±0.5) × 103
Carbamazepine	56 ± 4
Piperazine	150 × 103[b]
Glutaric acid	850 ×103[b]
Caffeine co-crystal	560 ± 9[c]
Piperazine co-crystal	123 ± 6[c]
Carbamazepine co-crystal	8.5 ± 0.7[d]
Glutaric acid co-crystal	–

[a]Form I of pterostilbene.
[b]Kinetic solubility is estimated visually and is approximate.
[c]Concentration measurements at ~5 hours.
[d]Concentration measurements at ~72 hours.

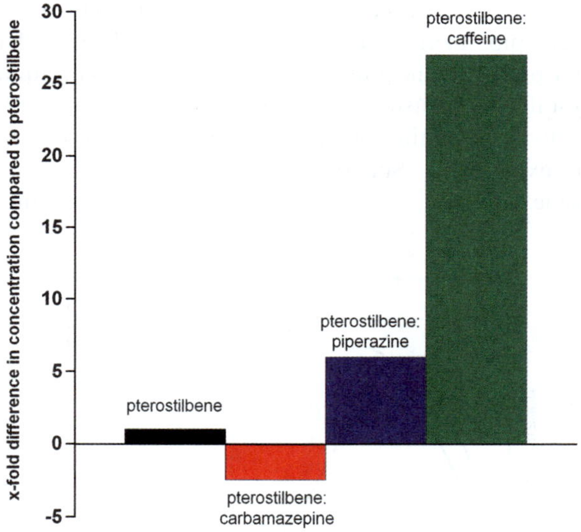

Figure 6.3 Comparison between the aqueous solubility of pterostilbene and the solution concentration of pterostilbene achieved with three co-crystals. The x-fold increase or decrease is calculated from equilibrium or kinetic solubility measured for the respective co-crystal. Pterostilbene solubility (black) is represented as one.

pterostilbene:piperazine co-crystal was six times higher in comparison with the solubility of pterostilbene, Figure 6.3. Both co-crystals possess solubility profiles represented as curve D outlined in Figure 6.2. The pterostilbene:carbamazepine co-crystal solubility was 2.5 times lower than pterostilbene measured for 72 hours (represented by curve A), while in extreme contrast the pterostilbene:glutaric acid dissociated within five minutes of slurring (represented by curve C) and thus no dissolution data was obtained. However, it should be noted that accurate dissolution data is obtainable for co-crystals where dissociation occurs.

In another example from SSCI where the solubility of an API was altered through formation of a co-crystal is the case study on fluoxetine HCl. For co-crystals of fluoxetine HCl, the aqueous solubility (called powder dissolution in the paper) was measured at various time points up to 120 minutes and the solutions were analyzed by UV.[34] Samples were sieved to obtain a particle size range of 53–150 μm. The fluoxetine hydrochloride solubility was 11.4 mg mL^{-1}. The benzoic acid co-crystal solubility was lower at 5.6 mg mL^{-1} and the fumaric co-crystal solubility was higher at 14.8 mg mL^{-1}. The succinic acid had a peak solubility of 20.2 mg mL^{-1} after about one minute, but then decreased to that of fluoxetine HCl based on the dissociation of the co-crystal to crystalline fluoxetine hydrochloride. This system exhibited higher and lower solubilities along with dissociation, making it a very interesting and complex set of co-crystal solubilities.

Solubility is a very important parameter to obtain during development of a new compound. As expected, the limited examples in this section show that solubility may be improved using co-crystals, but not in all cases. For a poorly soluble compound, especially if it does not have ionizable groups, it is worth trying co-crystals to see if an improvement can be obtained. In the case of co-crystals versus salts, there appears to be a trend that salts may offer a greater solubility advantage, but more work needs to be performed to determine if this is a general trend. Solubility measurements of co-crystals will continue to present interesting challenges for pharmaceutical scientists.

6.5.3 Bioavailability

So far we have discussed that the solid form of an API can have a significant impact on its physical and chemical properties such as stability, dissolution rate, and solubility. A co-crystal is a particular solid form of an API that can be considered in this context. Others could be salts, different polymorphic forms, and amorphous materials including amorphous dispersions of a given API. The dissolution rate (kinetic factor) and the solubility (thermodynamic factor) can play an important role with regard to the bioavailability of a compound. It is being estimated that somewhere between 70% and 80% of the drug substances currently being focused on in the interface between discovery and development by the pharmaceutical industry are poorly soluble. As discussed previously, a co-crystal offers the potential of improving both the dissolution rate as well as the solubility of a poorly soluble API and therefore can have a positive impact on improving the bioavailability of a BCS Class 2 or 4 compound.

A couple of examples have been discussed in the literature. A study that has often been referred to in this context was conducted by McNamara *et al.* on the drug 2-[4-(4-chloro-2-fluorophenoxy)phenyl]pyrimidine-4-carboxamide.[35] The compound showed very low solubility and bioavailability in dogs. A 1:1

co-crystal with glutaric acid of this compound was discovered and showed significantly improved dissolution behavior. Therefore this co-crystal was tested in bioavailability studies in dogs and the results showed an approximately four times higher area under curve (AUC) compared to the original drug. Since then the interest of companies developing drugs in using co-crystals has increased significantly and more and more studies have been published describing the dissolution rate, solubility, and bioavailability improvements of such compounds. In the same year a study was published by Variankaval *et al.* describing a bioavailability study of the 1:2 co-crystal of L-883555 with L-tartaric acid in Rhesus monkeys.[27b] This co-crystal showed approximately 20 times higher AUC compared to the original API. Later Hickey *et al.* studied a carbamazepine:saccharin co-crystal in terms of a series of attributes, including suitability for multi-gram scale-up, propensity for crystal polymorphism, physical stability, *in vitro* dissolution and oral bioavailability, with the goal of comparing it with the marketed form of carbamazepine.[36] The authors concluded that the benefits of using the carbamazepine:saccharin co-crystal include relative lack of polymorphism and equivalent chemical stability to the anhydrous polymorph, favorable dissolution properties and suspension stability, and comparable oral absorption profile in dogs compared with the commercial immediate release product. In 2008, Bak *et al.* published a co-crystal study on the API, AMG517.[37] The authors compared the properties of a sorbic acid:-AMG517 co-crystal to the original API and found that the co-crystal showed increased exposure especially in higher doses after oral application in rats. Cheney *et al.* worked on the drug substance lamotrigine and found several different co-crystals and polymorphs thereof.[38] A subset of these crystal forms were tested with regard to their dissolution, solubility, and pharmacokinetics properties.

6.5.3.1 Solid Form Research at the Interface between Discovery and Development

A key element of the strategy to improve the time required for preclinical development of a drug is to link solid form discovery with formulation, formulation design, and manufacturing. This strategy involves:

- Focusing early on discovering a suitable crystal form.
- Determining the solubility of the crystal form.
- Setting the first in human (FIH) formulation development strategy based on this solubility.
- Merging solid form discovery, with the formulation design and development, and with FIH manufacturing into the same cohesive team.

This strategy involves performing a solid form screen as soon as possible after the candidate is selected. This solid form screen should include co-crystals. Once the screen is completed the most beneficial solid form is identified and a decision on the formulation is made. Thus, the FIH formulation and development strategy could involve any of the following:

- solid formulation of the most stable form
- suspension formulation
- liquid formulation based on toxicological formulation
- salt formulation
- co-crystal formulation
- amorphous formulation or liquid crystal formulation.

The solid formulation would preferably be a chemical in bottle (CIB), chemical in capsule (CIC), or chemical in capsule in bottle (CICIB). The CIB formulation would be suspended in solution and administered. However, such suspensions must be stable and are inconvenient for a large clinical trial. An alternative is to weigh the chemical into capsules and administer the capsules. Equipment that will perform this filling operation is available. Additionally, it might be possible to improve the flow of the chemical (drug) in these machines by making a simple formulation with a flow aid. Such simple formulations are within the time frame of the studies proposed above. The CICIB approach involves placing one or more CIC dosage forms in a solution, allowing the capsule shells to dissolve and administering the solution.

In recent years it has become clear that co-crystal formulations can achieve solubility and blood levels much higher than conventional formulations that involve the original, crystalline API. Thus, for insoluble compounds it might be judicious to explore the use of co-crystals to achieve higher exposure. A fast, abbreviated screen for co-crystals can be performed by adding approximately two weeks to the timeframe of the study. In this case, studies will need to be done to establish that the formulation is stable enough for clinical trials. An alternative strategy is to use the toxicological formulation or a soft-gel formulation.

Once the clinical trial material is manufactured, it is analyzed for purity, impurities, and solid form. The purity and impurity analyses are done using conventional high performance liquid chromatography (HPLC). The solid form can conveniently be determined using solid-state NMR or X-ray powder diffraction. Solid-state NMR is particularly attractive for dosage forms involving capsules since the entire capsule can be placed in the instrument.

6.6 Additional Development Factors for Co-crystals

6.6.1 Taste Masking

Taste masking often needs to be taken into consideration in the development of an oral dosage form to ensure patience compliance. Many active pharmaceutical ingredients and/or bulking agents possess undesirable bitter tastes, which need to be overcome during the formulation process. A variety of different formulation techniques are available to make orally administered, unpleasant tasting drugs palatable, for example, polymer coatings, microencapsulations, flavor enhancers, dispersions, and inclusion complexes. In this section we will briefly examine the use of co-crystals and inclusion complexes (which we consider to be a sub-group of co-crystalline materials) to mask undesirable taste efficiently.

Inclusion complexation occurs when one molecule (often referred to as the host) creates a cavity-type environment for a second molecule (the guest) to reside forming a stable complex. Complexing agents are capable of masking a bitter tasting drug by either decreasing its solubility upon ingestion or by decreasing the amount of drug–taste bud interactions, which reduces the perception of bitter taste.[39,40] β-cyclodextrin (a complexing agent) is a non-toxic, sweet tasting compound which is composed of seven sugar molecules bound in a ring. It is the most widely used inclusion compound and is commonly incorporated into drug formulations to mask bitter tasting drugs. For example, cetirizine (marketed as Zyrtec or Reactine) is an allergy relief drug with a very bitter taste. However, upon formulation with β-cyclodextrin, the masking of the bitter taste was dramatically improved and a palatable, chewable tablet was possible.[41] In another example, Ibuprofen (a very bitter tasting drug) was produced as a powder in sachet formulation due to elimination of its bitter taste through complexation with hydroxypropyl β-cyclodextrin.[42]

Co-crystallizing an API could also provide the means needed for taste altering especially if the dosage form is a fast disintegrating or chewable tablet. In fact, using co-crystals to mask taste was suggested almost 40 years ago by Higuchi and Pitman when they noticed the rates of dissolution in aqueous media of caffeine–gentisic acid co-crystals were dramatically decreased with respect to caffeine.[43] Thus, due to the slow release of caffeine, these materials offer a potentially viable path for formulating bitter tasting, caffeine in a chewable dosage form where interactions between API and taste buds are often of substantial duration. Another co-crystal example is between derivatized xanthines, propentofyllin and pentoxifyllin with acesulfame-H.[44] Acesulfame-H, is an artificial sweetener and successfully reduces the unpleasant taste of both caffeine-based molecules allowing for a chewable form. As a side, one might expect that a salt would form between the basic xanthine molecules and the acidic acesulfame, but through single-crystal X-ray diffraction studies, the molecules were characterized as neutral species.[45] From a formulation perspective, using a co-crystal or inclusion complex is a possible route to masking the taste of a bitter drug, but as stated earlier not the only one. In utilizing these solid forms, one must first determine if it is possible and/or profitable to co-crystallize the drug molecule with non-toxic, sweeteners or flavoring agents. Or, even if the formation of a co-crystal is possible, does the physicochemical properties of the API change, meaning the bitter taste might be masked, but the solubility/dissolution rate is now dramatically affected, leading to decreased bioavailability? To answer these questions and others, each system or API must be evaluated on a case by case basis, as co-crystal formation and physicochemical property modifications are not predictable and trends between molecules/systems are not easily drawn.

6.6.2 Scale-up

In order for any co-crystalline material to move from small scale, laboratory screening exercise to a possible option for development and ultimately a drug product, scalable (kilo to multi-kilo) processes offering high co-crystal purity and yields must be devised and established. Co-crystal quantities of milligrams to a few grams are typically necessary depending on the types of characterization and/or initial physicochemical studies (rate dissolution, solubility, bioavailability, *etc.*) which are to be performed. A recent survey of the open literature showed that slow evaporation and grinding (neat or liquid-assisted) are the two most common techniques for co-crystal generation; however, these approaches possess obvious limitations upon scale-up.[46] An extensive and thorough review of process development and scaling up co-crystals can be found in this book, although a few approaches will be briefly highlighted here.

From solution, large-scale production of crystalline material is often obtained through cooling crystallization with the option of anti-solvent addition potentially to increase the yield or seeding to promote crystallization of one form over another.[46,47] Recently it has also been established that co-crystals can be produced on large scale using spray drying techniques[19] and supercritical fluid technology.[21] One must consider the construction and use of ternary phase diagrams when designing a scale up process from solution because information about the relationship between equilibria of the solid phases and solvent choices can be obtained. It is critical to determine and map the solubilities of the individual components since the phase region where the most thermodynamically stable co-crystal is located will be altered based on whether or not they posses similar solubilities in a given solvent.[48,49]

In the end as co-crystalline materials move further into development and become viable options as marketed products, scalable co-crystal processes must be evaluated and optimized.

6.6.3 Polymorphism

Searching for polymorphs of a particular compound is commonplace in solid-state pharmaceutics. Since polymorphic compounds can potentially possess drastically different physical and chemical properties, considerable efforts should be taken to identify and characterize all forms during development. Polymorphism of single-component crystals is well documented and often occurs; however, cases of co-crystal polymorphism have been reported much less frequently. This is potentially due to the infancy of co-crystal screening and searching for polymorphs therein, not because it is less likely to occur. In fact, one could make the argument that it may even be more likely to occur, given the multifaceted nature of the typical co-formers used for screening purposes.

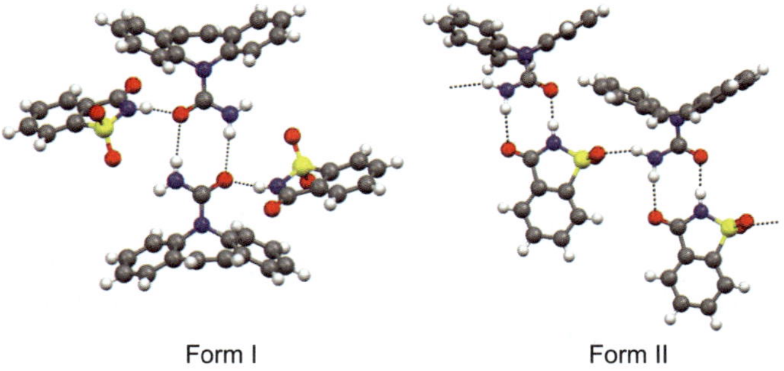

Figure 6.4 Form I and Form II of the 1:1 carbamazepine:saccharin co-crystal with the hydrogen bonds represented as black, dashed lines.

Furthermore, as for single-component systems, determining the thermodynamic relationship between polymorphs of co-crystals is very important, because, as it has been stated previously, changes in crystal form can lead to differences in physical and chemical properties. So conducting a thorough polymorph screen of a particular co-crystal of interest is highly valuable, just as it is for a single-component crystalline material.

In one example, an extensive polymorph screen on a 1:1 co-crystal of carbamazepine and saccharin (Form I) was carried out by Hickey *et al.*[36] Using high-throughput crystallization, 480 experiments including saccharin and carbamazepine in various solvents and solvent mixtures were performed, yielding 156 solid materials, which were characterized by XRPD and/or Raman spectroscopy. No polymorphic forms of the co-crystal were observed. Furthermore, manual liquid-assisted grinding experiments were tried using 24 different solvents and slurry conversion experiments utilizing seven different solvents; however, once again, only Form I resulted. Interestingly, a second polymorph of the 1:1 co-crystal of carbamazepine:saccharin (Form II) was found by Matzger and co-workers using a different set of crystallization techniques, polymer heteronuclei, which were not previously attempted when only Form I was continually observed.[50] The single crystal structures were determined for both co-crystal polymorphs and their differences can be viewed in Figure 6.4.

No doubt, polymorph screening of co-crystals will continue to attract more attention as these materials continue to gain momentum as another solid form choice in the development of new pharmaceuticals. As with salts, co-crystal polymorphs offer additional options to alter properties, increase patent protection, and improve marketed formulations.

6.7 Conclusions

This chapter has outlined the role of co-crystals within the pharmaceutical development continuum. It can clearly be seen that co-crystals have their rightful place as useful solid forms within the industry. Major solid-state properties to be considered for the development of a potential drug product such as stability, solubility/dissolution and bioavailability have all been adjusted and/or altered through forming co-crystals of APIs. Regardless of whether one feels a co-crystal is already available as a marketed pharmaceutical product or not, ongoing research in this field continues to excite and anticipation of an FDA approved co-crystal is surely just around the corner.

Acknowledgement

The authors would like to thank Dr. Ann Newman for her helpful contributions to this chapter.

References

1. M. Pulacki, J. D. Higgins, E. Kwong and A. C. Templeton, *J. Med. Chem.*, 2010, **53**, 5897–5905.
2. J.-O. Henck and S. R. Byrn, *Drug Discovery Today*, 2007, **12**, 189–199.
3. (a) K. R. Seddon and M. J. Zaworotko, *Crystal Engineering: The Design and Application of Functional Solids*, NATO-ASI Series, Kluwer, Dordrecht, 1999, vol. 539. (b) S. Datta and D. J. W. Grant, *Nat. Rev. Drug Discovery*, 2004, **3**, 42–57. (c) *Making Crystals by Design-from Molecules to Molecular Materials, Methods, Techniques, Applications*, ed. F. Grepioni and D. Braga, Wiley-VCH, Weinheim, 2007.
4. (a) *Handbook of Pharmaceutical Salts: Properties, Selection and Use*, ed. P. H. Stahl and C. G. Wermuth, Verlag Helvetica Chimica Acta, Zürich, 2002. (b) L. D. Bighley, S. M. Berge and D. C. Monkhouse, in *Encyclopedia of Pharmaceutical Technology,* ed. J. Swarbick and J. C. Boylan, Marcel Dekker, New York, 1996, vol. 13, pp. 453–499. (c) C. H. Gu and D. J. W. Grant, in *Handbook of Experimental Pharmacology: Stereochemical Aspects of Drug Action and Disposition*, ed. M. Eichelbaum, B. Testa and A. Somogyi, Springer, Berlin, 2003.
5. Molecules from the 'Generally Regarded As Safe' GRAS list http://www.cfsan.fda.gov, Code of Federal Regulations Title 21 - Food and Drugs, http://www.accessdata.fda.gov/scripts/cdrh/cfdocs/cfcfr/cfrsearch.cfm, R. C. Rave, P. J. Sheskey and M. E. Quinn, pharmaceutical excipients, in *Handbook of Pharmaceutical Excipients*, Pharmaceutical Press, New York, 2009.
6. (a) G. L. Amidon, H. Lennernäs, V. P. Shah and J. R. Crison, *Pharm. Res.*, 1995, **12**, 413–420. (b) *Drug Bioavailability: Estimation of Solubility, Permeability, Absorption and Bioavailability*, ed. H. Waterbeemd, H. Lennernäs and P. Artursson, Wiley-VCH, Weinheim, Germany, 2003.
7. R. Taylor and O. Kennard, *Acta Crystallogr., Sect. B*, 1983, **B39**, 133–138.
8. Newman, A. W., Childs, S. L. and Cowans, B. A. Salt co-crystal form selection, in *Preclinical Development Handbook*, John Wiley and Sons, Hoboken, 2008, pp. 455–481.

9. C. B. Aakeröy, D. J. Salmon, M. M. Smith and J. Desper, *Cryst. Growth Des.*, 2006, **6**, 1033–1042.
10. J. S. Stevens, S. J. Byard and S. L. M. Schroeder, *J. Pharm. Sci.*, 2010, **99**, 4453–4457.
11. (a) C. B. Aakeröy, A. Rajbanshi, Z. J. Li and J. Desper, *CrystEngComm*, 2010, **12**, 4231–4239. (b) C. A. Hunter, *Angew. Chem. Int. Ed.*, 2004, **43**, 5310.
12. Use of the Cambridge Structural Database (CSD) has become commonplace in examining the propensity of functional group complementarity and/or structural compatibility of molecules in the solid state. F. H. Allen, *Acta Crystallogr., Sect. B*, 2002, **58**, 380–388.
13. (a) GRAS Notice: http://www.cfsan.fda.gov/~rdb/opa-gras.html. (b) Food additive status list: http://www.cfsan.fda.gov/~dms/opa-appa.html.
14. One general example: D. R. Weyna, T. Shattock, P. Vishweshwar and M. J. Zaworotko, *Cryst. Growth Des.*, 2009, **9**, 1106–1123.
15. One general example: T. Friščić and W. Jones, *Cryst. Growth Des.*, 2009, **9**, 1621–1637.
16. One general example: T. Friščić, S. L. Childs, S. A. A. Rizvi and W. Jones, *CrystEngComm*, 2009, **11**, 418–426.
17. One general example: N. Blagden, D. J. Berry, A. Parkin, H. Javed, A. Ibrahim, P. T. Gavan, L. L. De Matos and C. C. Seaton, *New J. Chem.*, 2008, **32**, 1659–1672.
18. N. Rodríguez-Hornedo, S. J. Nehm, K. F. Seefeldt, Y. Pagán-Torres and C. J. Falkiewicz, *Mol. Pharmaceutics*, 2006, **3**, 362–367.
19. A. Alhalaweh and S. P. Velaga, *Cryst. Growth Des.*, 2010, **10**, 3302–3305.
20. C. Medina, D. Daurio, K. Nagapudi and F. Alvarez-Nunez, *J. Pharm. Sci.*, 2009, **99**, 1693–1695.
21. (a) L. Padrela, M. A. Rodriguez, S. P. Velaga, A. C. Fernandes, H. A. Matos and E. G. de Azevedo, *J. Supercrit. Fluids*, 2010, **53**, 156–164. (b) L. Padrela, M. A. Rodriguez, S. P. Velaga, H. A. Matos and E. G. de Azevedo, *Eur. J. Pharm. Sci.*, 2009, **38**, 9–17.
22. *Solid-State Chemistry of Drugs*, ed. S. R. Bryn, R. R. Pfeiffer and J. G. Stowell, SSCI, Inc., West Lafayette, Indiana, 2nd edn, 1999.
23. N. Schultheiss and A. Newman, *Cryst. Growth Des.*, 2009, **9**, 2950–2967.
24. FDA: Stability Testing of new Drug Substances and Products, International Conference on Harmonization (ICH), http://www.fda.gov/RegulatoryInformation/Guidances/UCM128179.htm.
25. S. M. Reutzel-Edens and A. W. Newman in *Polymorphism*, ed. R. Hilfiker, Wiley-VCH, Weinheim, 2006, pp. 235–258.
26. (a) A. V. Trask, W. D. S. Motherwell and W. Jones, *Cryst. Growth Des.*, 2005, **5**, 1013–1021. (b) A. V. Trask, W. D. S. Motherwell and W. Jones, *Int. J. Pharm.*, 2006, **320**, 114–123.
27. (a) A. M. Chen, M. E. Ellison, A. Peresypkin, R. M. Wenslow, N. Variankaval, C. G. Savarin, T. K. Natishan, D. J. Mathre, P. G. Dormer, D. H. Euler, R. G. Ball, A. Ye., Y. Wang and I. Santos, *Chem. Commun.*, 2007, 419–421. (b) N. Variankaval, R. Wenslow, J. Murry, R. Hartman,

R. Helmy, E. Kwong, S. Clas, C. Dalton and I. Santos, *Cryst. Growth Des.*, 2006, **6**, 690–700.
28. D. Good, C. Miranda and N. Rodriguez-Hornedo, *CrystEngComm*, 2011, **13**, 1181–1189.
29. J. F. Remenar, M. L. Peterson, P. W. Stephens, Z. Zhang, Y. Zimenkov and M. B. Hickey, *Mol. Pharmaceutics*, 2007, **4**, 386–400.
30. N. Huang and N. Rodríguez-Hornedo, *Cryst. Growth Des.*, 2010, **10**, 2050–2053.
31. E. K. Kalra, *AAPS PharmSci.*, 2003, **3**, e25.
32. I. Ivanisevic, M. C. Andres, K. Stephens, S. Bethune, M. Bower, S. Wolfe, J. Smit, A. Gilkinson and M. Roe, unpublished results.
33. (a) N. Schultheiss, S. Bethune and J.-O. Henck, *CrystEngComm,* 2010, **12**, 2436–2442. (b) S. Bethune, N. Schultheiss and J.-O. Henck, *Cryst. Growth Des.,* 2011, **11**, 2817–2823.
34. S. L. Childs, L. J. Chyall, J. T. Dunlap, V. N. Smolenskaya, B. C. Stahly and G. P. Stahly, *J. Am. Chem. Soc.*, 2004, **126**, 13335–13342.
35. D. McNamara, S. L. Childs, J. Giordano, A. Iarriccio, J. Cassidy, M. S. Shet, R. Mannion, E. O'Donnell and A. Park, *Pharm. Res.*, 2006, **23**, 1888–1897.
36. M. B. Hickey, M. L. Peterson, L. A. Scoppettuolo, S. L. Morisette, A. Vetter, H. Guzman, J. F. Remenar, Z. Zhang, M. D. Tawa, S. Haley, M. J. Zaworotko and O. Almarsson, *Eur. J. Pharm. Biopharm.*, 2008, **67**, 112–119.
37. A. Bak, A. Gore, E. Yanez, M. Stanton, S. Tufekcic, R. Sysed, A. Akrami, M. Rose, S. Surapaneni, T. Bostick, A. King, S. Neervannan, D. Ostovic and A. Koparkar, *J. Pharm. Sci.*, 2008, **97**, 3942–3956.
38. M. L. Cheney, N. Shan, E. R. Healey, M. Hanna, L. Wojtas, M. J. Zaworotko, V. Sava, S. Song and J. R. Sanchez-Ramos, *Cryst. Growth Des.*, 2010, **10**, 394–405.
39. H. Sohi, Y. Sultana and R. K. Khar, *Drug Dev. Ind. Pharm.*, 2004, **30**, 429–448.
40. J. Szejtli and L. Szente, *Eur. J. Pharm. Biopharm.*, 2005, **61**, 115–125.
41. K. Kasraian and T. Havlir, *EP* 20030745693, 2003.
42. S. Motola, G. R. Agisim and A. Mogavero, *US Patent* 5,024,997, 1991.
43. T. Higuchi and I. H. Pitman, *J. Pharm. Sci.*, 1973, **62**, 55–58.
44. Z. Ayenew, V. Puri, L. Kumar and A. K. Bansal, *Recent Pat. Drug Delivery Formulation*, 2009, **3**, 26–39.
45. A. Burgard, Patent application 101/65399, date issued 2/1/2005.
46. A. Y. Sheikh, S. Abd Rahim, R. B. Hammond and K. J. Roberts, *CrystEngComm,* 2009, **11**, 501–509.
47. J. W. Mullin, *Crystallization*, Butterworth-Heinemann, Oxford, 4th edn., 2001.
48. S. J. Nehm, B. Rodriguez-Spong and N. Rodriguez-Hornedo, *Cryst. Growth Des.*, 2006, **6**, 592–600.
49. R. A. Chiarella, R. J. Davey and M. L. Petterson, *Cryst. Growth Des.*, 2007, **7**, 1223–1226.
50. W. W. Porter III, S. C. Elie and A. J. Matzger, *Cryst. Growth Des.*, 2008, **8**, 14–16.

CHAPTER 7

Solid Forms and Pharmacokinetics

N. BISWAS

Piramal Life Sciences Limited, 1, Nirlon Complex, Goreagon (East), Mumbai, 400 063, India

7.1 Introduction

Pharmacokinetics, often abbreviated as PK, deals with the kinetics of a pharmaceutical dosage form and its metabolites after it has been administered into the body. It is the mathematical basis for assessing the time course of a drug *in vivo* and is essential for determining the drug regimen in patient treatment. Although in principle it encompasses all classes of xenobiotics, this chapter is limited only to drug molecules that are administered to prevent and/or for the treatment of different diseases.

In order to achieve therapeutic significance, it is critical that an "adequate" concentration of the drug reaches the specific targeted area in required time, that is, the concentration range of the drug in the target tissue should be above the minimum efficacy value but below the minimal toxicity levels (called "therapeutic window", Figure 7.1). This is ensured by biological processes that guide the drug from its site of administration to the target tissues.[1] These processes include:

- *Absorption* from the site of administration to be available for distribution by the circulatory system,

RSC Drug Discovery Series No. 16
Pharmaceutical Salts and Co-crystals
Edited by Johan Wouters and Luc Quéré

Published by the Royal Society of Chemistry, www.rsc.org

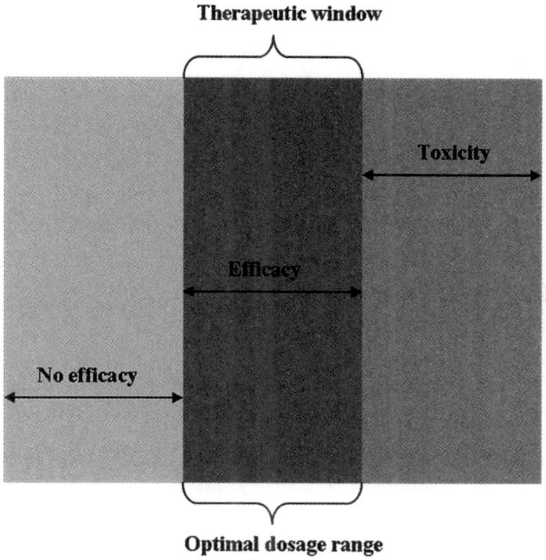

Figure 7.1 Relationship between therapeutic window, efficacy and toxicity.

- *Distribution* from the blood plasma into the interstitial and intracellular fluids,
- *Metabolism* in one or multiple organs (such as liver, small intestine, lungs, kidney *etc.*) to enable its enzyme-mediated biotransformation from lipophilic to more hydrophilic molecules to facilitate excretion,
- *Excretion* by urine, bile or feces to prevent any undesired accumulation that could lead to toxicity, and
- *Toxicity* caused by non-specific or non-selective interactions or undesired interference leading to cytotoxicity, mutagenicity and/or teratogenicity.

Together they are popularly known as the ADMET properties of a drug and determine the distribution of the drug both temporally and spatially in the body.

In the early days of drug discovery, the pharmaceutical industry was primarily focused on studying structure–activity relationships (SAR) aiming to improve the efficacy (and lowering the toxicity) of the potential drug candidate. However, with a significant number of drugs failing during clinical trials due to inadequate ADMET properties, the industry revolutionized the criterion for screening clinical candidates early in the discovery process. Now, proper understanding and achievement of "drug-like" properties is given as much importance in the early phases as the optimization of structure for activity. This has been attributed to the evolution of a new area of study called the "structure–property relationships (SPR)" which involves the optimization of the molecular structure in order to attain the desired "drug-like" properties with direct or indirect consequences for the ADMET properties (Figure 7.2).[1]

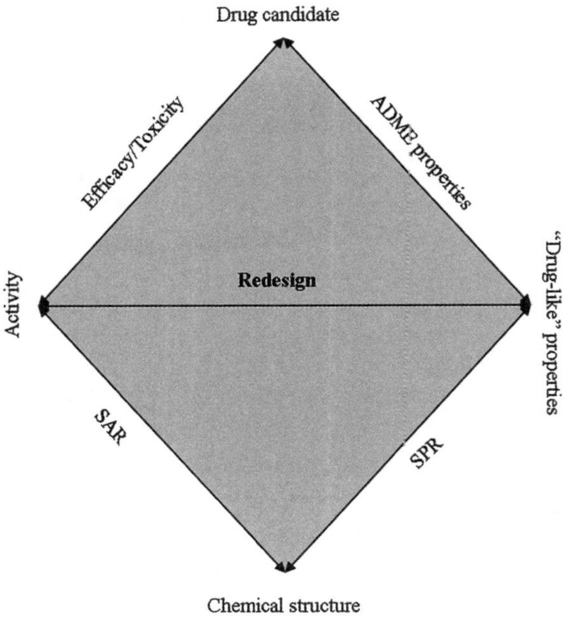

Figure 7.2 "Feedback loop" for optimizing the "drug-like" properties of a drug candidate.[1]

SPR study strategies involve the assessment of (a) the dissolution rate, solubility, permeability, lipophilicity, pK_a, stability metabolite screening, and protein and tissue binding using *in vitro* models and (b) the pharmacokinetic parameters indicating the exposure of the drug and bioavailability using *in vivo* models (please refer to Section 7.1.2 for details of the PK parameters).[1] Consequently, this field involves an interdisciplinary platform requiring close collaboration between multiple disciplines such as medicinal chemists, pharmacologists, analytical chemists, and formulation chemists, who work in harmony in the "feedback" loop (Figure 7.2).

Although the "drug like" properties of a developmental molecule can be classified into different categories, they are interlinked.[1] The property most fundamental to a molecule is its structural property. It arises from the scaffold used and the functional groups attached to it and directly influences its interaction with the environment (including solvent molecules, proteins and tissues). These interactions are clubbed together and classified as the physicochemical and biochemical properties of the drug depending on the nature of the interaction (physical or biological). They determine the kinetics of the drug *in vivo* and are responsible for its distribution in plasma and tissues. Different parameters such as the concentration of drug in plasma and its half-life provide information about its absorption and elimination profile and are therefore summed as its pharmacokinetics properties (Table 7.1). Together, all these properties dictate the behavioral pattern of the drug leading to its success or failure.

Table 7.1 Different classifications of properties of a drug.

Classification	*Properties*
Structural	Molecular weight, ionization, polarity, lipophilicity/hydrophilicity, shape, reactivity, polar surface area, H-bonding
Physicochemical	Solubility, dissolution rate, permeability, chemical stability
Biochemical	Protein/tissue/cell binding, metabolism, receptor mediated transport (influx and efflux)
Pharmacokinetic	Bioavailability, half-life, clearance, drug-drug interactions, toxicity, maximum concentrations in plasma

Often, in a pursuit of enhancing target binding, lipophilic groups are added which become detrimental for the development of the molecule due to adverse effects on its physicochemical characteristics. For example, improved lipophilic character may improve the permeability of the molecule through the lipid bilayer, but in the process its solubility could be compromised which can limit its absorption leading to lower bioavailability. Consequently, higher doses would be required to achieve the desired therapeutic effect, increasing the probability of toxic effects. Again, the addition of large hydrophobic groups could also reduce its permeability by passive diffusion (in addition to its reduced solubility) due to the increased bulk of the molecule. Therefore, an ideal candidate can only be obtained when these oppositely directed parameters are balanced (hydrophilic–lipophilic balance). It should also be noted at this point, that the pharmacokinetics of a drug is very complex and the above-mentioned properties are not the only factors that influence it.

7.1.1 Impact of Solid Form on the Pharmacokinetics

Every drug form, after its administration, has to travel from its site of administration specifically to the target site. In doing so, the drug encounters multiple barriers that limit its concentration at the target. For example, after oral dosing, the drug passes from the mouth to the gastrointestinal (GI) tract, where it is exposed to an aqueous environment with varying pH.[1] The pH of the GI tract varies not only along its length but is also dependent on the fasting condition. For example, the pH of the stomach under the fasted state is in the range of 1.4–2.1, while under fed conditions, it is less acidic with a pH ranging from 3 to 7. The pH in the different parts of the intestine (for example duodenum, jejunum and ileum) also varies from 4.4 to 8 in the fasted state to 5.2 to 8 in fed conditions. Details about the different barriers have been cited in most drug metabolism pharmacokinetic (DMPK) books and are only briefly discussed here.

The dissolution of a drug is a function of the pH of the environment it is exposed to and the pK_a of the molecule. It is facilitated by the pH gradient in the GI tract as predicted by the Henderson–Hasselbalch equation (equation (7.1)). For instance, basic drugs ionize in the acidic pH of the stomach and the upper parts of the intestine while acidic drugs remain neutral and have very limited solubility in these low pH regions. As the pH increases along the intestinal tract, they become ionized and dissolve in the basic gastric medium:

$$pH = pK_a + \log([\text{base}]/[\text{acid}]) \qquad (7.1)$$

Solubility of the lipophilic molecules, does not however, gain from the pH gradient of the GI tract. Their solubility is aided by the bile acids secreted from the gallbladder into the intestinal region.

Once the drug is solubilized, it is absorbed through the intestinal walls and enters the systemic circulation for distribution. Although there are different possibilities for absorption, permeability through passive diffusion is the most important route followed. Here lipophilicity and size of the molecule play a very important role. A smaller sized neutral molecule is more likely to pass through the hydrophobic phospholipid membrane of the intestinal wall than a large, charged molecule.

After absorption into the systemic circulation, the drug is circulated through the body. Here, it is exposed to the various plasma proteins such as albumin and erythrocytes which can bind to the drug and limit its passage through the tissue membrane. These affiliations are primarily influenced by the structure of the molecule and its lipophilicity.[1,2] They reduce the "free state" of the drug in plasma, thereby reducing its absorption into the target tissues.

Other major organs that act as barriers restricting the bioavailability of the drug are the liver and the kidney. The liver is the primary site of metabolism and biliary extraction, while the kidneys, along with the liver, play a major role in the elimination of the compound. The passage of the drug molecules through these barriers is also guided by their molecular structure and permeability by passive diffusion.

Most of the PK steps such as those dealing with metabolism, distribution and elimination, are governed by the molecular structure and the physicochemical properties of the active moiety and not the solid form of the drug. The impact of an altered solid form is restricted mostly to its dissolution properties and consequently its absorption.[3] This has a considerable impact on the bioavailability of the drug, which can expose the patient to varying drug concentrations leading to grave clinical consequences. The significance of such altered pharmacokinetics due to altered solid forms has led to the recall of multiple drugs, not only from clinical development, but also from marketing phases (please refer to Section 7.3 for case studies). The clinical significance and the enormous financial liabilities for the pharmaceutical companies have entailed detailed solid state screening at an early stage of drug development as part of routine analysis.

During development, it is common practice to perform salt screening in an attempt to increase the bioavailability of the drug candidate. Detailed solid state characterization of the salts along with the free acids or bases under *in vivo* and different manufacturing and processing conditions is also performed at the earliest time to ensure the reproducibility of a form and its clinical effect.

Another approach that is fast gaining popularity is the formation of co-crystals to alter the pharmacokinetics of a drug. The tendency of the system to salt or co-crystal formation is governed by the pK_a difference, steric and electrostatic properties of the molecules involved. The rule of thumb widely accepted is that there is salt formation if $\Delta pK_a > 2$, while $\Delta pK_a < 0$, results in co-crystal formation. For $0 < \Delta pK_a < 2$, the system could swing either way resulting in the formation of a salt or co-crystal or a complex with partial proton transfer.[4] Other approaches include formation of solvates, hydrates or variation in the crystal lattice through the formation of polymorphs.

In most cases, solid state screening is performed to overcome low solubility, difficulty of crystallization, low melting point, high hygroscopicity and chemical instability.[5] They can potentially alter not only the kinetics of the drug *in vivo* but are also of significance during manufacturing, processing and storage. Detailed knowledge of the forms and their kinetics can also be exploited to alter the dissolution/release profile and the solubility of the drug. They can be manipulated to cater for the specific clinical application for which the drug is intended, for example high release rate in quick-pain relief formulations *vs.* sustained release for basal diabetic therapy. A thorough understanding of the possible forms and their kinetics of interconversion is also important during formulation, manufacture and storage of the drug. It can help prevent any unexpected changes in the solid state profile that may have severe side effects in patients.

7.1.2 Important PK Parameters

Pharmacokinetic studies in drug discovery research deal with the measurement of the drug concentration in plasma with respect to time and the key parameters involved are volume of distribution (V_d), bioavailability (F), exposure (measured as the area under the curve or AUC), maximum plasma concentration attained (C_{max}), time of maximum drug concentration (T_{max}), clearance (Cl) and half-life ($t_{1/2}$). These parameters and their role in drug discovery have been discussed in detail in most text books related to drug metabolism and pharmacokinetics (DMPK), and have been discussed only briefly here.[1,6] Each of the above are defined as follows:

- Volume of distribution (V_d) is defined as the ratio of the total amount of drug in the body to that in plasma and is expressed as L kg^{-1}. In PK experiments it is calculated as shown in equation (7.2):

$$V_d = dose/C_0 \tag{7.2}$$

  where C_0 is the initial concentration of the drug in blood after an intravenous (IV) dose. V_d is an indicator for the distribution of the drug in the body, that is, preferential distribution in the bloodstream *vs.* tissues.[1,6]
- Bioavailability (BA or F) is the fraction of the administered dose of the drug that reaches the circulation unaltered and is expressed as %F. In PK

experiments, the bioavailability of the drug using a particular dosage route (e.g. oral) is calculated relative to the intravenous route as shown in equation (7.3):

$$\%F = (AUC_{\text{oral}}/AUC_{\text{IV}}) \times (dose_{\text{IV}}/dose_{\text{oral}}) \times 100\% \quad (7.3)$$

where AUC is the area under the curve in the plasma concentration *vs.* time curve (Figure 7.3).[1,6]

It is an indicator of the amount of drug in the bloodstream after passing through the different barriers in the body (*e.g.* enzymatic or pH-induced degradation of the molecule, first-pass metabolism *etc.*) and is directly related to its physicochemical properties.

- Exposure is the "area under the curve" (AUC) in the drug concentration in plasma *vs.* time plot (Figure 7.3).[1,6] For an adequate absorption and elimination profile, a higher AUC value usually indicates greater exposure of the drug. This parameter is widely used in the discovery program to measure the relative and absolute bioavailabilities of the drug and its clearance.
- Maximum concentration (C_{max}) is defined as the maximum concentration of the drug in systemic circulation after administration of the dose and expressed in concentration units (*e.g.* ng mL^{-1}). The highest concentration attained in the plasma concentration *vs.* time plot is assigned the value of C_{max} (Figure 7.3).[1,6]
- Time of maximum drug concentration (T_{max}) is defined as the time taken to achieve the maximum concentration of the drug in systemic circulation, that is, the time taken to achieve C_{max} and is expressed in hours (Figure 7.3).[1,6]

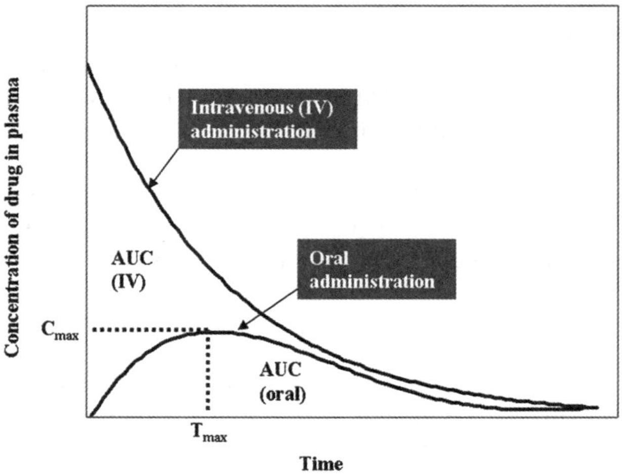

Figure 7.3 Determination of oral bioavailability of a drug.[1,6]

- Clearance (*Cl*) is defined as the efficiency of extraction of the drug in its unaltered form from the systemic circulation in irreversible elimination from the body. Since multiple organs facilitate this elimination, the total body clearance is the sum of all the different clearance processes (such as renal, hepatic) and is expressed as mL min^{-1} kg^{-1}. It is calculated as shown in equation (7.4):[1,6]

$$Cl = dose/AUC_{\mathrm{IV}} \tag{7.4}$$

- Half-life ($t_{1/2}$) is defined as the time required for the concentration of the drug in the systemic circulation to reduce to half of its value. Considering first-order kinetics, it is calculated as shown in equations (7.5) and (7.6):

$$t_{1/2} = 0.693/k \tag{7.5}$$

$$t_{1/2} = 0.693(V_{\mathrm{d}}/Cl) \tag{7.6}$$

where k is the slope of the drug concentration in the logarithmic scale *vs.* time curve (Figure 7.4).[1,6]
- Bioequivalence (BE) is defined as the similarity in the rate and extent of therapeutic and toxic effects observed by the administration of equivalent doses of the pharmaceutical formulations or pharmaceutical alternatives, under similar conditions. It is estimated by comparing the analytical and the above-mentioned pharmacokinetic parameters at different doses for the pharmaceutical preparation under study.

7.2 Dissolution and Solubility

Often during drug development, the terms "dissolution", "dissolution rate", "intrinsic dissolution rate" (IDR) and "solubility" are used interchangeably.

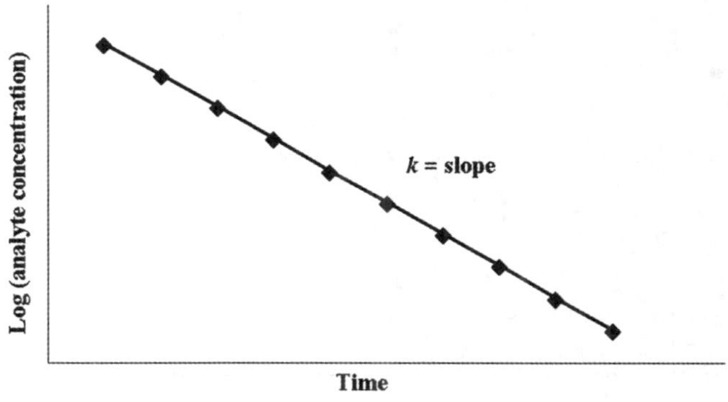

Figure 7.4 First-order rate constant (k) for elimination is obtained from the plot of log(analyte concentration) *vs.* time.[1,6]

This is incorrect as there are significant differences between the phenomenon they refer to. On many occasions however, their effects and/or alterations are related and this is responsible for the subtlety in their difference.

"Dissolution" of a drug refers to the overall process by which the solid form of the drug dissolves in a liquid (solvent) medium at a given temperature, while "dissolution rate" refers to the kinetics of dissolution, that is, the rate at which the drug dissolves. IDR is a special case referring to the dissolution rate of the pure compound under the condition of constant surface area.[7] The term "solubility" on the other hand, refers to the equilibrium state of the dissolution process, where the undissolved solid phase is in equilibrium with the drug in the solution phase in a saturated solution, at a given temperature. Manipulation of these physicochemical properties to achieve the desired pharmacokinetics is of paramount importance in the pharmaceutical industry.

7.2.1 Effect of dissolution on absorption

Of the many theories proposed for studying the dissolution of solids, the simple diffusion model mostly suffices for pharmaceutical applications.[3,8–10] The diffusion layer model assumes a thin stagnant diffusion layer (thickness $= h$) at the interface of the dissolving solid ($x = 0$) and the dissolution medium ($x = h$), referred to as bulk (Figure 7.5).[3]

Following Fick's first law, the rate of dissolution is expressed as shown in equation (7.7):

$$dW/dt = DA(C_S - C_B)/h \tag{7.7}$$

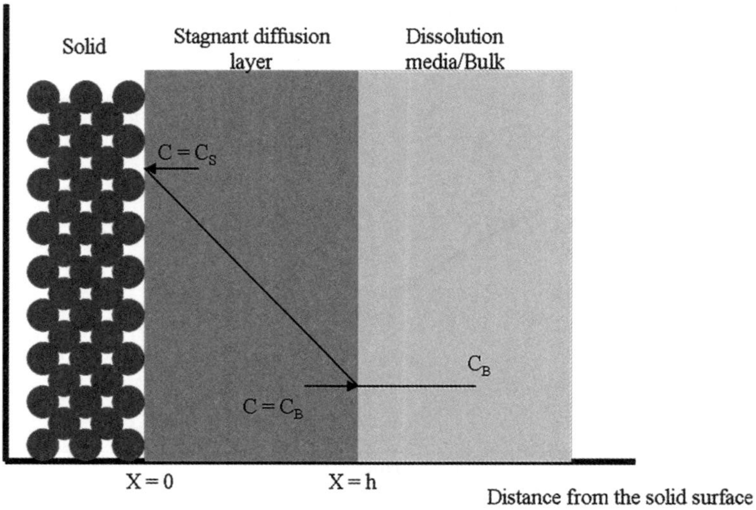

Figure 7.5 Drug dissolution by the diffusion layer model.[3]

where dW/dt is the flux or rate of mass transfer per unit time. It indicates the mass of solute (W) that has dissolved and has been transported across the diffusion layer in unit time (t) per unit surface area (A) of the dissolving solid, where D is the diffusion coefficient of the solute. C_B is the concentration of the drug in the bulk solution at time t, and C_S is the saturation solubility of the solute in the diffusion layer. The above equation is referred to as the Nernst–Brunner diffusion layer model of the Noyes–Whitney equation.[11]

As is evident from equation (7.7), dissolution is driven by ($C_S - C_B$), that is, the difference between the saturation solubility in the diffusion layer and the concentration of the drug in bulk solution.[12] This implies that if the rate of absorption of the drug is lower than its dissolution rate, the bulk concentration (C_B) increases and approaches C_S, thereby limiting further dissolution. The absorption of the drug is therefore said to be "absorption limited". On the other hand, if absorption rate is very high in comparison to its dissolution rate, then the accumulation of drug is prevented at the site of absorption (*e.g.* intestinal membrane) and ($C_s - C_B$) approaches C_s. This is usually the case for a very poorly soluble drug and the absorption is thus said to be "dissolution limited". Therefore, a change in the dissolution rate and/or solubility of the active pharmaceutical ingredient (API), triggers a cascading effect that has an impact on the absorption of the drug. It alters the concentration of the molecule at the site of absorption which is directly proportional to its cellular uptake, especially when the permeability of the compound is limited to passive diffusion. Therefore, the PK parameters most impacted by the alteration of the dissolution rate are the bioavailability and exposure represented by the AUC.

7.2.2 Impact of Dissolution Rate on Bioavailability

The intrinsic dissolution rate (IDR) and the solubility of a molecule are a function of its molecular structure and the physical condition (*i.e.* composition and pH of solvents) it is exposed to. The molecular structure dictates intra and/or intermolecular interactions governed by polar and non-polar interactions, pK_a, molecular weight and shape. As a result, structural modifications including (a) addition of ionizable groups, polar groups and H-bonding, (b) reduction in log P and molecular weight and (c) construction of prodrugs, are popular strategies employed to improve the solubility of the drug. However, medicinal chemists often fail to improve the solubility and IDR of the molecule while maintaining its efficacy. Many times the alteration of the structure leads to other complications such as its instability, manufacturing and processing difficulties and drug–drug interaction liabilities, which restrict its choice as a lead for drug development. In such cases, scientists venture into other alternatives that could potentially improve the pharmacokinetics of the molecule without altering its basic structure. These alternatives include complexation with cyclodextrin, use of carriers (*e.g.* nanoparticles, liposomes), co-solvency and emulsification.[1]

The above-mentioned approaches can enhance the amount of solid phase going into solution (*i.e.*, solubility), but in order to improve the kinetics of dissolution, other strategies are employed that reduce the time required for dissolution. These include (i) increasing the surface area by reducing the particle size, (ii) predissolving the molecule in gastric-simulated formulations, and (iii) increasing the wettability of the solid using surfactants during formulation. Another strategy that has gained considerable importance lately is the manipulation of the solid form, by salt, polymorph, hydrate, solvate or co-crystal formation. Some cases which highlight the relationship between solid forms, dissolution rate and bioavailability are discussed in Section 7.2.3.

7.2.3 Case Studies Demonstrating the Relationship Between Dissolution Rate and Bioavailability

7.2.3.1 Salt Forms of Piroxicam

There are abundant examples in the literature that demonstrate the direct relationship between dissolution rate and bioavailability. One such example is that of piroxicam (Figure 7.6), a potent non-steroidal anti-inflammatory drug that was launched in 1981 by Pfizer for multiple conditions such as arthritis (osteoarthritis and rheumatoid arthritis) and spondylitis.[13] It is a zwitterionic drug ($pK_{a1} = 1.86$ and $pK_{a2} = 5.46$) and is classified as a low solubility and high permeability drug (*i.e.*, class II) based on the Biopharmaceutics Classification System (BCS). In an attempt to improve its bioavailability after oral administration, the authors prepared three different ethanolamine salt forms (*i.e.*, mono-, di- and triethanolamine salts). These salt forms were selected based on prior literature precedence where they enhanced the percutaneous absorption of piroxicam.[14] PK studies of the ethanolamine salts revealed that both the exposure (AUC) and the C_{max} of piroxicam in plasma followed the same trend as the dissolution profile at pH 6.8, which varied in the order mono- > di- > tri-piroxicam.[13] The monoethanolamine salt showed the highest exposure with the relative bioavailability increasing almost 1.9-fold, while the di- and tri- salts were ~1.7 times better than piroxicam. The C_{max} values showed 2.14-, 1.6- and

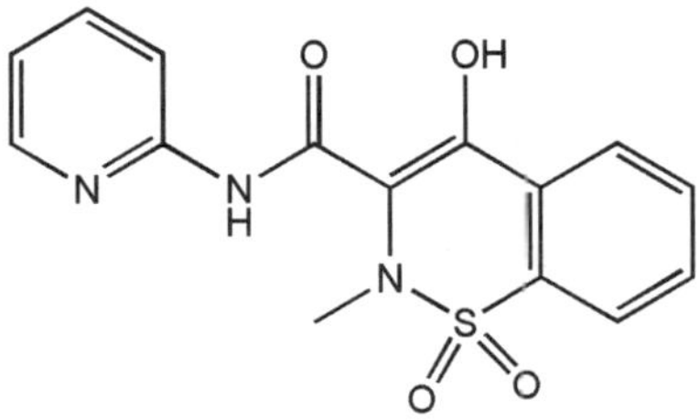

Figure 7.6 Piroxicam.

1.2-fold improvements in the mono-, di- and tri-salts, respectively, in comparison to piroxicam.

7.2.3.2 *Co-crystals of 2-[4-(4-chloro-2-fluorophenoxy)phenyl]pyrimidine-4-carboxamide with Glutaric Acid*

Sometimes, the traditional methods (such as salt formation) used for increasing the dissolution rate and thereby the bioavailability of the drug in the body may not be feasible. This was the case for 2-[4-(4-chloro-2-fluorophenoxy)-phenyl]pyrimidine-4-carboxamide, a sodium channel blocker.[15] It was a weak base with very low solubility of 0.1 μg mL^{-1} in water, simulated gastric fluid (SGF) and simulated intestinal fluid (SIF). The extremely low pK_a of -0.7, made it non-amenable to salt formation, even with strong acids such as hydrochloric acid, sulfuric acid and phosphoric acid. Also, its glass transition temperature (T_g) at 43 °C, was close to that at room temperature, which limited the formation of the amorphous phase reproducibly. It should also be noted that the amorphous phase, which might be an option in some cases, was not desirable either. This is because it could lead to the formation of a metastable state that could limit its shelf life. As a result, co-crystal formation was attempted in order to manipulate its physicochemical and pharmacokinetic properties.

Several commonly acceptable carboxylic acids were screened as co-crystal formers. Of these, the glutaric acid co-crystal was selected for development because of its lower melting point (64 °C lower than that of the free base) and almost 18-fold higher dissolution rate. Again, being a natural component of dietary foodstuffs and a metabolite product of fatty acids, tryptophan and lysine, glutaric acid was considered as a "safe" co-crystal former.

Glutaric acid co-crystals were formed by the H-bonded interaction of one carboxylic acid group of glutaric acid with the amide group of the API to form a self-complimentary 8-membered ring motif, while the second acid group interacted with the pyrimidine acceptor site (Figure 7.7). The result was monoclinic crystals belonging to the *P*2(1)/*n* space group, with a melting point of 142 °C. Oral PK studies in male beagle dogs with 5 and 50 mg kg^{-1} doses was in accordance with the improved intrinsic dissolution rate and showed a 2.5 to 3-fold increase in the plasma AUC for the co-crystals. Additionally, a 3-fold increase was also observed in the C_{max} values for the glutaric acid co-crystals in less than half the time required (T_{max}) for the free base.[15]

7.2.3.3 *Co-crystals of Aceclofenac with Chitosan*

In another example, Mutalik *et al.* demonstrated the use of natural polymers in preparing co-crystals that improve the pharmacokinetics and pharmacodynamics in animal models.[16] The biocompatibility and the biodegradability of these polymers provide an added advantage as co-crystal formers. In the

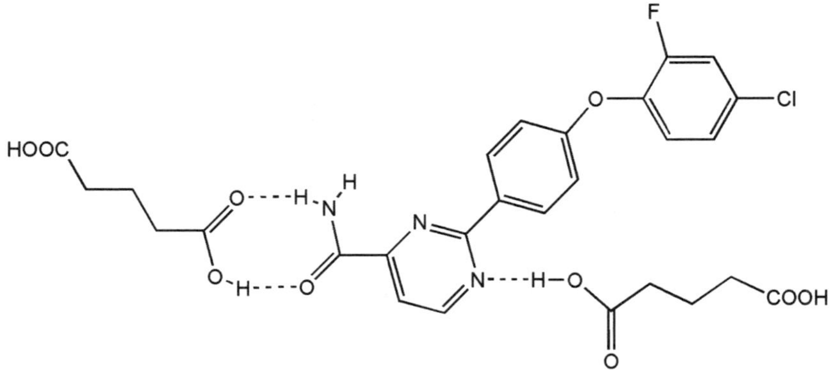

Figure 7.7 Co-crystal of 2-[4-(4-chloro-2-fluorophenoxy)phenyl]pyrimidine-4-carboxamide and glutaric acid.

present study, significant improvement was observed in the drug release rates and the bioavailability of the poorly water-soluble drug aceclofenac, when chitosan, a biopolymer was precipitated on the API crystals.

Aceclofenac (2-[2,6-dichlorophenyl)amine]phenylacetoxyacetic acid) is a non-steroidal anti-inflammatory drug (NSAID) of the phenyl acetic acid group with anti-inflammatory, analgesic and antipyretic properties. It has low oral bioavailability because of its low aqueous solubility, which has prompted many scientists to find alternatives. Mutalik *et al.* co-crystallized the API with a non-toxic natural polycationic polymer, called chitosan (β-(1-4)-2-amino-2-deoxy-D-glucose), a linear hydrophilic biopolymer of D-glucosamine. It is found abundantly in nature in the exoskeleton of crustaceans such as shrimps and is degraded by microflora in the colon. The presence of amino groups imparts a pH-dependent solubility profile, making it insoluble in neutral to basic pH and soluble in an acidic environment. It degrades to non-toxic amino sugars which are completely absorbed by the human body. This biocompatibility of chitosan makes it a potential co-crystal former in solid-state screening.

It has been hypothesized that interaction of chitosan with multivalent anions (such as sodium citrate which acts as anion cross-linker) leads to the formation of bridges between polymeric chains, resulting in cross-linking (due to electrostatic interaction) between the polymer molecules.[17] This facilitates efficient adsorption of the polymer on the drug molecule.

Different formulations were prepared using a varying aceclofenac:chitosan ratio by the solvent change method.[16] Of the series, the aceclofenac:chitosan crystals, with 40 mg of the API dispersed in 0.2% chitosan in 1% glacial acetic acid and crystallized from sodium citrate solution (referred to as C-3 crystals) were selected as the optimum ones based on their dissolution profile. These crystals demonstrated a release rate of about 99% in 1 hour in comparison to the 54% in 3 hours for the API alone. Their solubility in water and 0.1 N HCl was also 1.5- and 2-fold better than that of aceclofenac (Table 7.2).[16]

Table 7.2 Comparison of pure aceclofenac and C-3 crystals of aceclofenac with chitosan.

Parameter	*Pure Aceclofenac*	*C-3 crystals*
Particle size	$D_{10}=34.72$ μm, $D_{50}=62.62$ μm, $D_{90}=95.65$ μm, $D_{97+}=$ 96.25 μm	$D_{10}=0.11$ μm, $D_{50}=0.53$ μm, $D_{90}=0.95$ μm, $D_{97}=1.20$ μm
DSC thermogram		
(i) Melting point (°C)	(i) 154.49	(i)152.66
(ii) Enthalpy (J g^{-1})	(ii) –151.66	(ii) –97.07
Solubility (mg mL^{-1}) in		
(i) Water	(i) 0.053 ± 0.004	(i) 0.078 ± 0.009
(ii) 0.1 N HCl	(ii) 0.013 ± 0.001	(ii) 0.027 ± 0.002
Drug release in 0.1 N HCl with 2% Tween 80	54.13% (after 3 hr)	99.16% (after 1hr)

Improvements in the solubility and the dissolution profiles of the C-3 crystals were highlighted in the morphological differences with respect to the API. They were smaller and fluffy with a porous and rough surface. This was in contrast to the bigger and regular shaped crystals of aceclofenac which possessed a smooth surface.

The significant improvements in solubility and the dissolution rate for the C-3 crystals were demonstrated *in vivo* by their anti-inflammatory and analgesic properties. They achieved a higher percentage inhibition of writhe in the Writhing test (~82%) than aceclofenac (~65%) at the end of 30 min, proving their superiority in analgesic activity. They also induced faster anti-inflammatory effects in Wistar rats when tested for their ability to inhibit edema in the hind paw of these animals (~83% in 240 min compared with ~82% in 300 min for aceclofenac). Pharmacokinetic studies in Wistar rats also revealed an almost 1.7-fold increase in bioavailability of the C-3 crystals with respect to the pure drug.[16] They also achieved a higher concentration maxima in plasma of ~1.65 μg mL^{-1} (C_{max}) in an hour in comparison to 1.23 μg mL^{-1} in 1.08 hours for aceclofenac.

7.2.4 Adverse Effects

All of the examples discussed in Section 7.2.3, reveal the great potential of solid forms that can be exploited to achieve the desired pharmacokinetic properties. However, one has to also keep in mind the possibility that an increase in dissolution rate and solubility could sometimes even prove to be detrimental. It could lead to the accumulation of the drug, beyond the equilibrium solubility, creating a driving force for the nucleation and crystallization of some stable phases in the physiological medium.[18] To avoid any such accumulation, the drug should possess a sufficient absorption rate for its optimal distribution in the body.

7.3 Significance of Altered Dissolution Properties

With increased awareness of the effect of solid forms on clinical efficacy and toxicity, almost all drug discovery programs have integrated solid state screening and characterization in the early stages of development. This is aimed at reducing developmental costs and preventing any odd surprises during the clinical and marketing phases. This realization and its incorporation in the pharmaceutical industry have been triggered by a few high profile cases where inadequate knowledge of the marketed solid form and/or non-bioequivalence of generic products have led to the recall of specific lots or the product as a whole from the market. These undesired alterations in the drug effect, not only led to huge financial losses but also failed to improve patient life. As a result, regulatory bodies have become more vigilant and use more stringent guidelines for drug approval.[19]

7.3.1 Case Studies

7.3.1.1 Bioequivalence of the Generic Products of Carbamazepine

The variation in the PK parameters due to altered solid forms (such as salts/polymrophs/co-crystals), is especially highlighted for drugs whose oral absorption is limited by their dissolution properties. It is mostly observed for the BCS class II drugs, which have low aqueous solubility and high intestinal permeability. In these cases, if the low aqueous solubility and high permeability is coupled with a narrow therapeutic window of the drug, then the probability of altered BA/BE increases further with increased chances of toxicity. Carbamazepine, a sodium channel blocker used as an anticonvulsant antiepileptic drug, is a classic example that illustrates the problems of altered dissolution and bioavailability.

Carbamazepine was originally launched by Geigy Pharmaceuticals, after which many generic counterparts have been introduced in the market since 1986.[20] It has a narrow therapeutic range corresponding to 4–12 μg mL^{-1} of its concentration in plasma at steady state. Around the fall of 1988, one of the generic firms recalled 53 different lots of the 200 mg carbamazepine tablets amounting to approximately 70 million tablets. This was initiated based on several reports of clinical failures and changes in the dissolution characteristics of the marketed generic. Several patients who had earlier been stable with treatment using the innovator product experienced increased seizures with the generic product. Meyer and co-workers compared the bioavailability of three lots (Pharmaceutical Basics, Inc., Lot nos. K583-01, F844-07 and F915-03) of the recalled carbamazepine generic tablets to that of the innovator product (Geigy Pharmaceuticals, Inc., Lot no. 1T108912) using 24 healthy male subjects.[20] The relative bioavailability was observed to be 60% (in Lot. No. K583-01), 113% (in Lot. No. F844-07) and 78% (in Lot no. F915-03) of the original product from Geigy. The relative C_{max} values for these lots also followed a similar trend, being 61%, 142% and 74% of the innovator product for lot nos. K583-01, F844-07 and F915-03, respectively.

Statistical analysis of the data demonstrated that the three generic lots were not only non-bioequivalent to the innovator, they were also significantly different from each other ($P < 0.01$). They failed to meet the 80–120% requirement for the 90% confidence limits for the mean AUC and C_{max} with respect to the originally launched drug form.

Detailed analysis also revealed considerable differences in the dissolution of the four tested lots. Lot no. F844-07 dissolved more rapidly than its counterparts, which justified its higher C_{max} and $AUC_{0-\infty}$ values and thereby its clinical effect. Similar clinical effects have also been observed previously with rapidly absorbed carbamazepine tablets in Finland.[21]

Although the cause of these PK changes have not been determined with certainty, it has at least been partially attributed to the increase in the moisture content of the tablets during storage. The source of the raw materials used, along with their particle sizes, have also been thought to play a role in altering the clinical affect of these lots.

This example highlights the impact of altered dissolution properties, especially for drugs with a narrow therapeutic range (Figure 7.1). In these cases both a slight reduction and an enhancement in the drug concentration in plasma can easily cross the barriers of the therapeutic window and lead either to no efficacy or to toxicity. It is therefore extremely important to reproduce the physicochemical and pharmacokinetic properties of original drug product in the generic counterparts.

7.3.1.2 Polymorphism in Ritonavir

Another high profile case involving the recall of a drug due to polymorphic changes was that of ritonavir ((5*S*,8*S*,10*S*,11*S*)-10-hydroxy-2-methyl-5-(1-methylethyl)-1-[2-(1-methylethyl)-4-thiazoyl]-3,6-dioxo-8,11-bis(phenylmethyl)-2,4,7,12-tetraazatridecan-13-oic acid 5-thiazoylmethyl ester) (Figure 7.8). It was launched in 1996 by Abbott Laboratories as an HIV protease inhibitor for the treatment of AIDS. Ritonavir is a large lipophilic molecule with inadequate bioavailability in the solid form. As a result, it was marketed as Norvir oral liquid and Norvir semi-solid capsules, both formulations containing ethanol/water solutions.[22–24] At the time of development and launch, only one form, known as form I, which was a monoclinic crystal form, was known, with no stability issues. However, about two years after its introduction into the market, several lots of the capsules failed the dissolution specifications. Evaluation of the failed lots revealed the presence of a second, very stable polymorph of ritonavir, called form II, which appeared to have precipitated from the formulations during its storage.[22] Characterization of the new form revealed form II as a conformational polymorph of form I. It possessed a "*cis*" conformation around the carbamate linkages as opposed to the "*trans*" conformation in form I. The new form formed orthorhombic crystals with higher stability despite being slightly less dense than form I

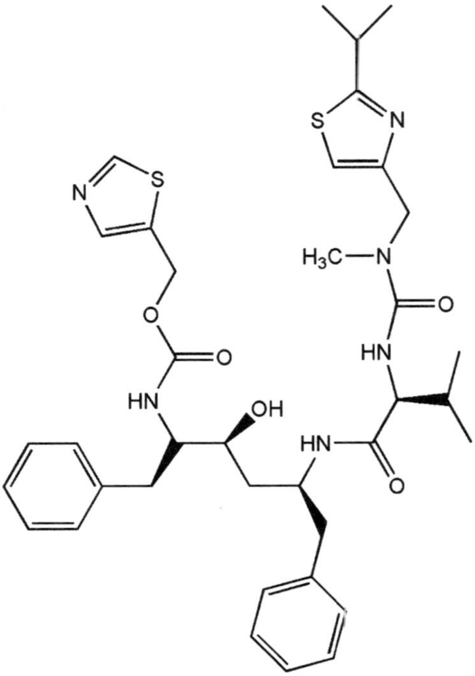

Figure 7.8 Ritonavir.

(density of form I = 1.28 g cm^{-3} and density of form II = 1.25 g cm^{-3}). This was primarily due to the synergistic effect caused by the hydrogen donor and acceptor properties of the alcohol group in form II, which resulted in very strong hydrogen bonds, providing more stability to the crystals. Its solubility was markedly lower (by about 50%) than that of form I, accounting for its failure during dissolution tests and inadequate bioavailability. Consequently, the drug was withdrawn from the market and huge losses were incurred.[23]

This unexpected change of form came as a surprise after its introduction to the market and rang a bell of caution. Its recall triggered major research in the area, resulting in the discovery of multiple polymorphs of ritonavir. Prior knowledge of these polymorphic systems would have been useful in averting the recall. This incident emphasizes the need for a thorough solid form screening and characterization with respect to the physicochemical and pharmacokinetic properties of the drug candidates at the early developmental stage, in order to prevent complications and life risks for patients and late stage developmental costs.

7.4 Selection of Solid Form Based on the Therapeutic Condition: Case Study of Indinavir

The effect of the solid form on the PK parameters is also a function of the disease state of the patient. This is well studied and demonstrated in the

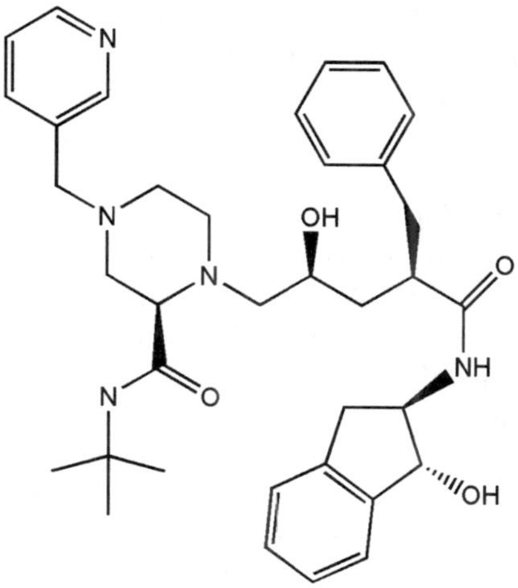

Figure 7.9 Indinavir.

development of indinavir by Merck Research Laboratories.[25,26] Indinavir (launched in 1996, brand name Crixivan®, see Figure 7.9) is a human immunodeficiency virus-1 (HIV-1) protease inhibitor that binds to the active site of the enzyme, rendering it inactive for proteolytic cleavage of viral polyprotein precursors. This inhibition prevents the formation of active proteins required for the viral infection, thereby interfering with the viral replication cycle.

The aqueous solubility of indinavir is pH dependent, with increased solubility at acidic pH.[27] During development, the compound was initially used in the free base monohydrate form and its absorption kinetics was tested in rats and dogs in two separate formulations of (i) 0.5% methocel suspension and (ii) 0.05 M citric acid solution, for an oral dose of 10 mg kg^{-1}.[25] In rats, indinavir attained a C_{max} value of 0.56 μM in 15 min in the methocel formulation, almost similar to that in citric acid (C_{max} 0.44 μM in 35 min). In dogs, there were more prominent differences in the two formulations. The C_{max} value in methocel suspension was 3.72 μM in 25 min as opposed to 11.4 μM in 30 min in the citric acid solution. It was also observed that the bioavailability of indinavir in both rats and dogs was quite similar (~16%) when administered as a 10 mg kg^{-1} dose in methocel suspension. However, in 0.05 M citric acid solution, the bioavailability of the drug in dogs was approximately 33% higher than that in rats.

It was also observed that the bioavailability of indinavir in rats for the two formulations was not significantly different. It varied from 15.8% in methocel to 23.9% in citric acid. However, in dogs, the bioavailability varied from 16% in methocel to 71.6% in citric acid.

The similarity in the PK parameters in rats, for the two formulations, has been attributed to first-pass metabolism. First-pass metabolism varies from species to species and is a significant contributor in rats. The blood clearance value (107 mL min^{-1} kg^{-1}) of indinavir in rats is higher than its hepatic blood flow rate (70–80 mL min^{-1} kg^{-1}), which results in its lower bioavailability for both the formulations. In dogs, however, the hepatic blood flow (30–35 mL min^{-1} kg^{-1}) is faster than that of its blood clearance value (11.4 mL min^{-1} kg^{-1}) and therefore the first-pass metabolism is not a significant contributing factor.[25]

The variation in oral bioavailability in dogs has been attributed to two factors, (i) the difference in gastric secretion which is poor in dogs but substantial in rats, and (ii) the pH-dependent solubility profile of the drug. When indinavir was administered in methocel in dogs, the lower bioavailability may have resulted due to insufficient solubility of the drug in the dog stomach. However, when administered as a citric acid solution, the formulation media provided the acidic pH (pH 2.5) necessary for dissolving the drug, leading to an increase in bioavailability. This rationale has been supported by the increase in C_{max} and AUC values in the fed state over the fasted state of the animal. Feeding the animal stimulated the secretion of gastric acid, which lowered the pH, aiding the solubility of the drug. Therefore, C_{max} increased 2.5-fold from 2.4 μM (fasted state) to 5.9 μM (fed state), while the exposure (AUC) increased by over 2-fold from 2.2 μM h in the fasted state to 5.1 μM h in the fed state.[25]

Humans, like rats, are good gastric acid secretors. Therefore, the oral bioavailability of indinavir in normal healthy human beings is expected to be less sensitive to pH, as in the rat model. However, for patients suffering from achlorhydria, the bioavailability of the drug would be significantly affected depending on the pH in the stomach, as in the dog model.[26] This is because achlorhydria refers to the condition where there is a lack of hydrochloric acid in the stomach and is unfortunately prevalent in patients suffering from HIV/AIDS, for whom the drug is intended. In such patients, the unpredictable variation in the concentration of the drug in plasma would not only be ineffective for patient treatment but could also lead to the development of new strains of the virus that would be resistant to the drug.[26]

In order to minimize these pharmacokinetic variations, indinavir was later developed as a sulfate salt, instead of the free base monohydrate. The sulfate salt had higher aqueous solubility (more than 500 mg mL^{-1}) and the pH of the resultant aqueous solution was also as low as 3, which facilitated its dissolution in the GI tract. As a result, the sulfate form of indinavir was favored more pharmacokinetically than the free base monohydrate form.

7.5 Manipulation of the Pharmacological Action of a Drug by Co-Crystallization: Case Study of Insulin

Tailoring the release and absorption profile of a drug to manipulate the site and the time of its release is of prime importance in the pharmaceutical industry. This is of special therapeutic significance in cases where slow and continuous release of the drug is required for prolonged treatments, such as for insulin treatment in diabetic patients.[28]

Insulin, a peptide hormone produced in the islets of Langerhans in the pancreas, is essential for glucose metabolism in the body and is the key for diabetes treatment. It is primarily administered as an injection, as it is inactivated by proteolytic enzymes by the oral route. Moreover, its high molecular weight and lower lipophilic character result in poor absorption of the drug *in vivo*, leading to lower oral bioavailability. This makes insulin unsuitable for oral administration.

There are many other strategies employed for the administration of insulin depending on the desired therapeutic effect. They include but are not limited to short-acting insulin, intermediate acting insulin and long-acting insulin. These formulations are categorized depending on their effective working hours, which vary from 30 min to 8 hours in short-acting insulin, from 1 to 14 hours for intermediate acting insulin and from 1 to 24 hours for long-acting insulin. In spite of their differences in the duration of action, it is essential in all insulin treatment to match the insulin time profile to that of the blood glucose time profile. Any mismatch in the two profiles can lead to hypoglycemic or hyperglycemic complications in the patients.

"Neutral protamine Hagedorn" (or NPH) marketed by Eli Lilly and Company as Humulin N is an intermediate-acting insulin and is commonly used for prolonged insulin action for diabetic patients. However, NPH fails to be in accordance with the physiological insulin profile leading to a mismatch in the blood glucose and insulin peaks. Brader *et al.* made an alternative formulation by co-crystallizing an insulin molecule with modified insulin to achieve the desired pharmacological activity.[28] They proposed prolonging the release and action of the drug by incorporating lipophilic characteristics that would decrease the absorption rate by slowing the dissolution process.

Presently, the existing strategy for controlled release of protein based drugs as injectables includes their encapsulation in exogenous matrices such as in liposomes, hydrogels and microspheres. However, stabilization and manufacture of the protein systems along with the prediction and modulation of the *in vivo* protein release is difficult, since these systems are physicochemically very complex. Brader *et al.*[28] attempted an alternative, by co-crystallizing human insulin (HI) with a lipophilically modified insulin derivative (octanoyl-N^{ε}-LysB29-human insulin (C8–HI) in the presence of Zn(II) and protamine. The steric requirement of the crystallizing unit of the insulin crystals was maintained in the co-crystals, allowing the insulin derivative to form the Zn-hexamer.[29] The co-crystals mimicked the pharmacological activity of the parent molecule and were analogous to the commonly used "neutral protamine Hagedorn" (or NPH) crystalline complex to the extent that it reproduced its favorable pharmacological action without the liability of the undesired pharmacodynamic profile.

The authors hypothesized that introduction of hydrophobicity into the crystal lattice in the form of fatty acid chain derivatives of human insulin (C_x–HI, where $x = 1,2,3....$) would reduce the aqueous solubility of the insulin molecules. This would facilitate slower release of the drug. Again, with increased lipophilicity due to the attachment of the fatty acid chain at the ε-amino group of LysB29, the affinity for albumin binding would also increase. This would help to extend the time-release profile of the drug.[2,28] In fact, the attachment of C14 and C16 long-acyl chains was able to reproduce the release profile of the NPH insulin. However, due to their increased size and lipophilic character, their biopotencies in humans were reduced. As a result, the shorter C8 chains were selected, which favored plasma protein binding optimally while retaining the biopotency of insulin.

To explore the utility of the lipophilically modified C8–HI–HI co-crystals, a series of co-crystals were prepared with varying proportions (0–100%) of C8–HI. The specific proportion of the insulin derivative in the co-crystals was easily controlled because of the quantitative nature of the co-crystallization process. As a result, the percentage of C8–HI incorporated into the co-crystal lattice could be directly related to its proportion in the initial co-crystallization solution. The lipophilic C8 modified C8–HI molecules appeared to be capable of replacing the unmodified HI molecules within the crystal lattice with relatively minor steric modification. The co-crystals formed were rod-like, bipyramid-tipped tetragonal crystals that resembled the NPH crystals.

In vitro dissolution studies demonstrated an inverse relationship between the C8–HI percentage in the co-crystals and the dissolution rate, confirming the authors' hypothesis that the hydrophobically modified insulin molecules reduce the solubility of the insulin co-crystal.[28]

The pharmacokinetics and the pharmacodynamics of these co-crystals were evaluated *in vivo* with 2 nmol kg^{-1} doses in somatostatin-treated beagle dogs, an animal model used for acute insulin deficiency. As a control, the time-action profile for the C8–HI molecule was compared to that of insulin. However, the former failed to demonstrate the protracted time action exhibited by the co-crystals, although their biopotencies in glucose reduction were similar. This underlines the necessity of the C8–HI, as part of the co-crystal lattice, for its controlled release action.

The glucodynamic profiles of the C8–HI–HI co-crystals were proportional to the concentration of C8–HI incorporated in the co-crystals. Plasma glucose levels were more effectively reduced over a period of 24 hours as the proportion of C8–HI increased from 58 to 92%. The co-crystals maintained more uniform plasma glucose levels than the NPH crystals which showed a steep rise between 15 and 24 hours.

The pharmacokinetic profiles for both the NPH crystals and the co-crystals also reflected a similar pattern. Co-crystals containing 75% C8–HI demonstrated a more uniform and favorable pharmacokinetic profile over a period of 24 hours than the NPH crystals. Studies measuring the insulin AUC indicated a much faster absorption rate for the NPH crystals (*i.e.* higher insulin AUC values) in the first 12 hours than the co-crystals, in accordance with the dissolution results. However, in the 12–24 hour period, there was a significant decrease in the AUC for the NPH insulin, while that for the co-crystals remained comparable to the 0–12 hour period. This uniformity and prolonged absorption rate made the co-crystals more suitable for basal therapy in glycemic control than the existing intermediate-acting NPH crystals.

7.6 Toxicity Caused by Co-crystal Formation

There is some evidence demonstrating the advantages of altering/controlling the solid form by co-crystal formation.[30–35] However, co-crystallization of otherwise non-toxic molecules can sometimes also prove to be fatal. One such case, involving the unexpected co-crystallization of melamine and cyanuric acid has acted as an eye-opener in this field. It further underlines the importance of detailed studies of the solid state characteristics of not only the individual components in a formulation but also their different combinations.

In 2004 and 2007, there were widespread reports of domestic dogs and cats dying due to nephrotoxic renal failure, in the geographically segregated continents of Asia and North America, respectively.[36–38] These outbreaks led to the recall of several pet food products, most of which were produced by the Canadian company Menu Foods.[39] Studies attributed these fatal incidents to the ingestion of pet food that was contaminated with melamine and cyanuric acid. These contaminants were thought to have been incorporated from the wheat gluten, rice protein and corn gluten used during their manufacture and marketing.[40] It was suspected that melamine was intentionally added to mislead the cost of the products. As, the protein concentration of the products were quantified based on their total nitrogen content, the addition of melamine would deceptively enhance the apparent protein content, paving the path for increased profits.[45] The source of cyanuric acid was however not known. It was speculated to be a by-product of melamine synthesis.[45]

Melamine (2,4,6-triamino-1,3,5-triazine) is a relatively non-toxic compound with acute toxicity observed in rodents, with oral lethal doses (LD_{50}) of 3100 mg kg^{-1} and 3900 mg kg^{-1} in male rats and male mice, respectively.[41] In another study, a 125 mg kg^{-1} oral dose of melamine showed only diuretic effects in dogs.[42] Toxicity studies of cyanuric acid derivatives (*e.g.* sodium cyanurate) at 700 mg kg^{-1} and 2200 mg kg^{-1} doses in rats and mice, respectively, resulted in bladder calculi and bladder epithelial hyperplasia, with no other adverse effects.[43] In other studies, it was observed that dietary monosodium cyanurate administered in dogs for 6 months at 0.8% concentration produced no adverse effects, although at 8% for a longer period of 16 to 24 months it produced kidney fibrosis and fical dilation.[44] This was attributed to the low solubility of sodium cyanurate, due to which it precipitated at higher concentrations in urine, leading to renal dysfunction.[45]

The melamine and cyanuric acid levels in the recalled products, as analyzed by the California Animal Health and Food Safety (CAHFS) toxicology laboratory, showed 10 mg kg^{-1} (0.001%) to 3200 mg kg^{-1} (0.32%) of melamine and more than 10 mg kg^{-1} of cyanuric acid.[45] Studies in cats by the authors revealed acute renal failure at 32 mg kg^{-1} doses of melamine and cyanuric acid. At higher doses of 2000 mg kg^{-1}, which is the minimum dietary concentration, clinical signs of vomiting, weakness and anorexia were observed within 24 hours of a single dose. The effects were however not dose dependent. In contrast to the above clinical and clinicopathologic findings, the extent of renal

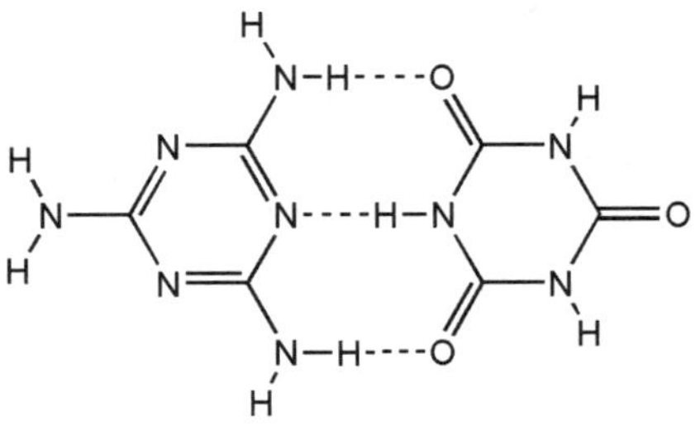

Figure 7.10 Co-crystal of melamine and cyanurate.

lesions observed in these cases, including tubular damage, interstitial edema and visibility of crystals, were dose dependent.

The above study proved that although the two components were non-toxic in animals when administered separately, they could prove to be lethal in combination. The renal failures were associated with the formation of highly insoluble birefringent melamine–cyanuric acid co-crystals, catalyzed by the acidic nature of the urine especially in dogs and cats.[45] The morphology and location of these co-crystals were similar to those observed in animals with renal failure during the 2004/2007 outbreaks. The conclusions of the study are supported by results from other work that demonstrate the formation of co-crystals between melamine and cyanuric acid (Figure 7.10).[46] The two components form an extensive two-dimensional network where they are superimposed to form a hexameric rosette structure. They are stabilized by a three-point molecular recognition pattern by forming H-bonds between N–H....O and N–H....N. These co-crystals possess very low aqueous solubility which leads to their precipitation in the renal tubules, causing failure of the organ.

7.7 Conclusion

Optimal pharmacokinetic features are one of the key parameters responsible for the success of a drug candidate in clinic. These parameters are greatly influenced by the solid form of the drug, which can potentially alter its dissolution and absorption properties, thereby affecting the bioavailability of the compound. Thorough investigation of the possible solid forms and their kinetics of interconversion can prevent unexpected and undesired alterations in the pharmacokinetic properties, which have a significant impact on its therapeutic effect. Solid state properties can also be manipulated to achieve suitable physicochemical properties depending on the application of the drug. For example, an altered solid form can potentially alter the release profile of the drug (*e.g.* fast release or slow release formulations) for enhanced therapeutic significance. This realization of the great impact of solid forms on the physicochemical and pharmacokinetic properties of a drug has led to the incorporation of detailed solid state screening at an early development phase by most companies, in an attempt to prevent unwarranted delays and cost liabilities and ensure quality treatment for patients.

References

1. E. H. Kerns and L. Di, *Drug-Like Properties: Concepts, Structure Design and Methods from ADME to Toxicity Optimization*, Academic Press, Burlington, MA, 2008.
2. J. F. Douglas, W. H. Bradshaw, B. J. Ludwig and D. Powers, *Biochem. Pharmacol.*, 1964, **13**, 537.
3. P.H. Stahl and C.G. Wermuth, *Handbook of Pharmaceutical Salts Properties, Selection and Use*, ed. P.H. Stahl and C.G. Wermuth, Wiley-VCH, Zürich and Weinheim, 2008.
4. N. Variankaval, R. Wenslow, J. Murry, R. Hartman, R. Helmy, E. Kwong, S-D. Clas, C. Dalton and I. Santos, *Cryst. Growth Des.*, 2006, **6**, 690.
5. R. Hilfiker, *Polymorphism in the Pharmaceutical Industry*, ed. R. Hilfiker, Wiley-VCH, Weinheim, Germany, 2006.
6. D. J. Birkett, *Pharmacokinetics Made Easy*, McGraw Hill Australia Pty Ltd., NSW, Australia, 2002.
7. http://www.pharmacopeia.cn/v29240/usp29nf24s0_c1087.html (last accessed Nov 2010).
8. D. J. W. Grant and T. Higuchi, *Solubility Behavior of Organic Compounds*, John Wiley & Sons, New York, 1990.
9. L. J. Lesson and J.T. Carstensen, *Dissolution Technology*, American Pharmaceutical Association, Washington DC, 1974.
10. J. T. Carstensen, *Solid Pharmaceutics: Mechanical Properties and Rate Phenomena*, Academic Press, New York, NY, 1980.
11. A. A. Noyes and W.R. Whitney, *J. Am. Chem. Soc.*, 1897, **19**, 930.
12. S. M. Berge, L. D. Bighley and D. C. Monkhouse, *J. Pharm. Sci.*, 1977, **66**, 1.
13. H-S Gwak, J-S Choi and H-K Choi, *Int. J. Pharm.*, 2005, **297**, 156.
14. H-A. Cheong and H-K. Choi, *Pharm. Res.*, 2002, **19**, 1375.
15. D. P. McNamara, S. L. Childs, J. Giordano, A. Iarriccia, J. Cassidy, M.S. Shet, R. Mannion, E. O'Donnell and A. Park, *Pharm. Res.*, 2006, **23**, 18888.
16. S. Mutalik, P. Anju, K. Manoj and A. N. Usha, *Int. J. Pharm.*, 2008, **350**, 279.
17. Z. Lu, J. H. Steenekamp and J. H. Hamman, *Drug Dev. Ind. Pharm.*, 2005, **31**, 311.
18. Y. Qiu, Y. Chen and G. G. Z. Zhang, *Developing Solid Oral Dosage Forms Pharmaceutical Theory and Practice*, ed. Y. Qiu, Y. Chen and G. G. Z. Zhang, Academic Press, Burlington, MA, 2009.
19. D. L. Azarnoff and D. H. Huffman, *Annu. Rev. Pharmacol. Toxicol.*, 1976, **16**, 53.

20. M. C. Meyer, A. B. Straughn, E. J. Jarvi, G. C. Wood, F.R. Pelsor and V. P. Shah, *Pharm. Res.,* 1992, **9**, 1612.
21. P. J. Neuvonen, *Int. J. Clin. Pharmacol. Therapy Toxicol.*, 1985, **23**, 226.
22. J. Bauer, S. Spanton, R. Henry, J. Quick, W. Dziki, W. Porter and J. Morris, *Pharm. Res.*, 2001, **18**, 859.
23. S. L. Morissette, S. Soukasene, D. Levinson, M. J. Cima and Ö. Almarsson, *Proc. Natl. Acad. Sci. U.S.A.*, 2003, **100**, 2180.
24. G. Zhang, D. Law, E. A. Schmitt and Y. Qiu, *Adv. Drug Deliv. Rev.*, 2004, **56**, 371.
25. J. H. Lin, *Adv. Drug Deliv. Rev.*, 1999, **39**, 33.
26. C. R. Gardner, C. T. Walsh and Ö. Almarsson, *Nat. Rev. Drug Discovery*, 2004, **3**, 926.
27. J. H. Lin, I-W. Chen, K. J. Vastag and D. Ostovic, *Drug Metab. Dispos.*, 1995, **23**, 730.
28. M. L. Brader, M. Sukumar, A. H. Pekar, D. S. McClellan, R. E. Chance, D. B. Flora, A. L. Cox, L. Irwin and S.R. Myers, *Nat. Biotechnol.*, 2002, **20**, 800.
29. E. N. Baker, T. L. Blundell, J. F. Cutfield, S. M. Cutfield, E. J. Dodson, G. G. Dodson, D. M. Hodgkin, R. E. Hubbard, N. W. Isaacs, C. D. Reynolds, K. Sakabe, N. Sakabe and N. M. Vijayan, *Philos. Trans. R. Soc. Lond. B. Biol. Sci.*, 1988, **319**, 369.
30. M. B. Hickey, M. L. Peterson, L. A. Scoppettuolo, S. L. Morisette, A. Vetter, H. Guzmán, J.F. Remenar, Z. Zhang, M. D. Tawa, S. Haley, M. J. Zaworotko and Ö. Almarsson, *Eur. J. Pharm. Biopharm.*, 2007, **67**, 112.
31. S. G. Fleischman, S. S. Kuduva, J. A. McMahon, B. Moulton, R. D. Bailey Walsh, N. Rodríguez-Hornedo and M. J. Zaworotko, *Cryst. Growth Des.*, 2003, **3**, 909.
32. S. L. Childs, L. J. Chyall, J. T. Dunlap, V. N. Smolenskaya, B. C. Stahly and G. P. Stahly, *J. Am. Chem. Soc.*, 2004, **126**, 13335.
33. J. F. Remenar, S. L. Morissette, M. L. Peterson, B. Moulton, J. M. MacPhee, H. R. Guzmán and Ö. Almarsson, *J. Am. Chem. Soc.*, 2003, **125**, 8456.
34. J. F. Remenar, M. MacPhee, M. Peterson, S. Morissette and Ö. Almarsson, *USP* 2005/0070551A1, 2004.
35. M. Zegarac, E. Mestrovic, A. Dumbovic, M. Devcic and P. Tudja, *WO* 2007/080362 A1, 2007.
36. Anonymous, *New York Times*, March 28, 2007, p. 13.
37. W.-I.I. Jeong, S. H. Do, D.-H. Jeong, J.-Y. Chung, H.-J. Yang, D.-W. Yuan, I.-H. Hong, J.-K. Park, M.-J. Goo and K.-S. Jeong, *J. Vet. Sci.*, 2006, **7**, 299.
38. C. A. Brown, K.-S. Jeong, R. H. Poppenga, B. Puschner, D. M. Miller, A. E. Ellis, K.-I. Kang, S. Sum, A. M. Cistola and S. A. Brown, *J. Vet. Diagn. Invest.*, 2007, **19**, 525.
39. http://www.usatoday.com/money/industries/2007-04-30-chinese-imports-usat_N.htm (last accessed Nov 2010).

40. T. McPheron, *Melamine and cyanuric acid interaction may play part in illness from recalled pet food* (press release), American Veterinary Medical Association, Schaumburg, IL, May 1, 2007 (Retrieved May 4, 2007).
41. R. L. Melnick, G. A. Boorman, J. K. Haseman, R. J. Montali and J. Huff, *Toxicol. Appl. Pharmacol.*, 1984, **72**, 292.
42. W. L. Lipschitz and E. Stokey, *J. Pharmacol. Exp. Ther.*, 1945, **83**, 235.
43. B. G. Hammond, S. J. Barbee, T. Inoue, N. Ishida, G. J. Levinskas, M. W. Stevens, A. G. Wheeler and T. Cascieri, *Environ. Health Perspect.*, 1986, **69**, 287.
44. H. C. Hodge, B. J. Panner, W. L. Downs and E. A. Maynard, *Toxicol. Appl. Pharmacol.*, 1965, **7**, 639.
45. B. Puschner, R. H. Poppenga, L. J. Lowenstine, M. S. Filigenzi and P. A. Pesavento, *J. Vet. Diagn. Invest.*, 2007, **19**, 616.
46. A. Ranganathan, V. R. Pedireddi and C. N. R. Rao, *J. Am. Chem. Soc.*, 1999, **121**, 1752.

CHAPTER 8

Application of Mechanochemistry in the Synthesis and Discovery of New Pharmaceutical Forms: Co-crystals, Salts and Coordination Compounds

TOMISLAV FRIŠČIĆ AND WILLIAM JONES

Department of Chemistry, University of Cambridge, Lensfield Road, Cambridge CB2 1EW, UK

In the past decade mechanochemical[1] methodologies, such as neat grinding and liquid-assisted grinding,[2] have been demonstrated to be a highly efficient means of both synthesising and screening for pharmaceutical co-crystals. This chapter provides an overview of these mechanochemical methodologies, their applications to date, an overview of mechanistic details and a brief comparison with more conventional co-crystallisation methods based on solution crystallisation and slurrying. In addition to the use of mechanosynthesis in the context of pharmaceutical co-crystals,[3] its application for pharmaceutical salt synthesis[4] and screening is also described, as well as recent applications in the synthesis of metal–organic derivatives.[5]

RSC Drug Discovery Series No. 16
Pharmaceutical Salts and Co-crystals
Edited by Johan Wouters and Luc Quéré

Published by the Royal Society of Chemistry, www.rsc.org

8.1 Introduction

A particularly important aspect of pharmaceutical materials science is the discovery and development of new solid forms of active pharmaceutical ingredients (APIs).[6] Over the past decade mechanochemical methods, such as grinding,[7] have demonstrated outstanding efficiency in this context, especially for the synthesis of single-[8] and multi-component crystalline solids,[9] but also for the synthesis of salts[4] as well as amorphous materials.[10] The success of mechanochemistry in the synthesis and discovery of pharmaceutical co-crystals can be ascribed to two principal reasons. The first is the ease with which supramolecular interactions responsible for co-crystal formation, that is, hydrogen bonds,[11] halogen bonds,[12] $\pi \cdots \pi$ stacking interactions[13] and other weaker types of intermolecular interactions,[14] can be broken and reformed under mild mechanical agitation.[2] The second reason is the recent introduction of a modified mechanochemical methodology of liquid-assisted grinding[15] (LAG, also known as kneading or solvent-drop grinding, SDG[16,17]). The LAG methodology provides considerable advantages over the simpler conventional approaches based on neat grinding (*i.e.* dry grinding) of solid components. Specifically, LAG utilises small amounts of a liquid phase to combine the synthetic advantages inherent in mechanochemistry, such as simplicity and lack of bulk solvent, with high yields and crystallinity of product, fast reaction rates and the ability in some cases to direct the polymorphic composition of the reaction product. This set of advantageous properties has made LAG a co-crystal screening method in several aspects superior to both simple neat grinding as well as to other methodologies of co-crystallisation, such as solution crystallisation, co-crystallisation from the melt and co-crystallisation using sonication methods.[18–20] This contribution provides a brief overview of the use of mechanochemical methodologies for the formation and discovery of pharmaceutical co-crystals. While this application of mechanochemistry is now well-established and LAG is acknowledged as a highly efficient screening methodology,[18–21] we have every reason to believe that the importance of mechanosynthesis in pharmaceutical materials science will continue to grow. This will result from the future development of the still emerging areas of mechanochemical synthesis of pharmaceutical salts[4,22] and of metal–organic pharmaceutical derivatives,[23–26] which are described at the end of this contribution. The development of techniques for solid-state characterisation, such as structure determination from X-ray powder diffraction data,[27,28] solid-state NMR spectroscopy,[29,30] terahertz time-domain spectroscopy[31,32] and even solid-state NMR crystallography[33] will be key in such development and the broader use of mechanochemistry in pharmaceutical research.

8.2 Historical Overview

The simplest mechanochemical method for pharmaceutical co-crystallisation is by neat grinding of two or more co-crystal components.[1] The first examples of neat grinding co-crystallisation of pharmaceutically relevant molecules were reported independently in 1995 by Caira *et al.*[34] and by Etter *et al.*[11] Caira *et al.* explored the co-crystallisation of the drug sulfadimidine by grinding with a variety of carboxylic acid co-crystal formers, including benzoic, anthranilic, salicylic and acetylsalicylic acids (aspirin) (Figure 8.1(a)).[34] Co-crystals were obtained in all cases and were identical to those that have been previously obtained by solution methods. First order kinetic behaviour was observed for the co-crystallisation of sulfadimidine with anthranilic and salicylic acids. This was interpreted as an indication of a random nucleation reaction mechanism. The exceptional stability of the sulfadimidine–anthranilic acid co-crystal (Figure 8.1(b)) was observed through two types of mechanochemical competition experiments. First, mechanochemical co-crystallisation of sulfadimidine was conducted with equimolar mixtures of two different acids. In all cases where one of the acids was anthranilic, only the formation of the sulfadimidine–anthranilic acid co-crystal was observed. Powder X-ray diffraction revealed that the other acid remained separate in the reaction mixture, either in a crystalline or in an amorphous form (Figure 8.1(c)). The second type of competition experiment involved neat grinding of the previously prepared sulfadimidine co-crystals with a different potential carboxylic acid co-crystal former. The addition of anthranilic acid repeatedly resulted in a solid-state substitution to form the sulfadimidine–anthranilic acid co-crystal. Using the sulfadimidine–salicylic acid co-crystal as the starting material, the formation of a salicylic acid co-crystal with anthranilic acid was also observed, along with the co-crystal of sulfadimidine with anthranilic acid.[34]

Etter *et al.* reported the solid-state co-crystallisation of 9-methyladenine and 1-methylthymine,[11] driven by the formation of hydrogen-bonded Hoogsteen

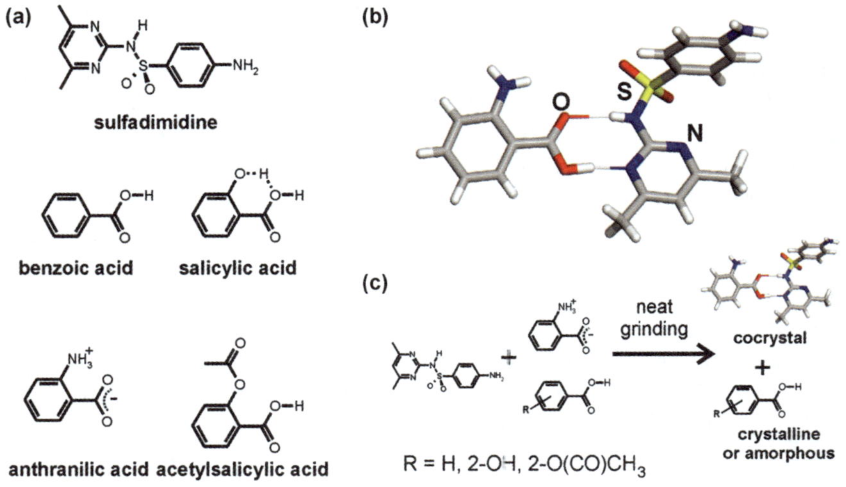

Figure 8.1 (a) Molecules used in the mechanochemical synthesis of pharmaceutical co-crystals by Caira *et al.*, (b) fragment of the crystal structure of the co-crystal of sulfadimidine with anthranilic acid and (c) competition between carboxylic acids as co-formers for sulfadimidine typically resulted in the formation of the co-crystal with anthranilic acid.[34]

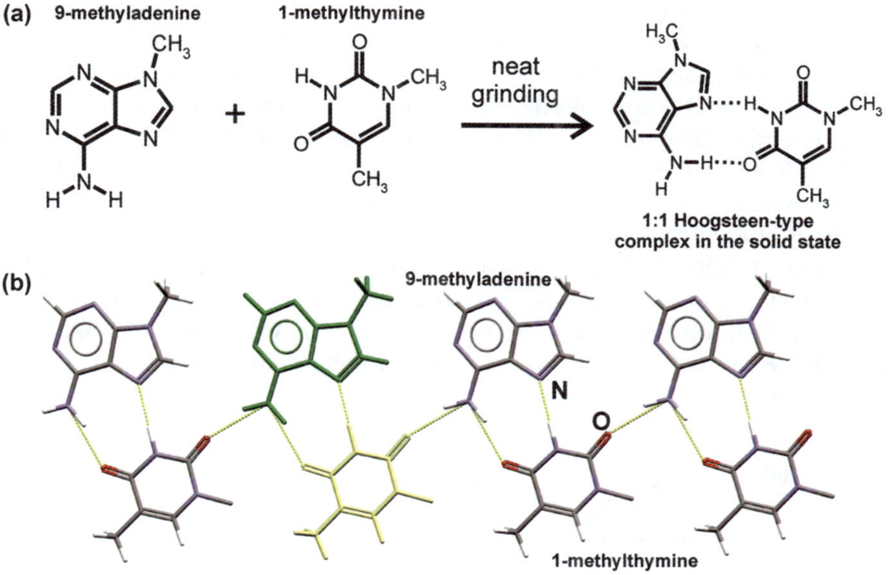

Figure 8.2 (a) Mechanochemical reaction of 9-methyladenine and 1-methylthymine to form a co-crystal[11] and (b) fragment of the (9-methyladenine)–(1-methylthymine) co-crystal, displaying Hoogsteen-type dimers (a single dimer is highlighted in green and yellow) assembled into hydrogen-bonded tapes.[35]

complexes between the base pairs (Figure 8.2).[35] Although 9-methyladenine and 1-methylthymine are not themselves pharmaceutical ingredients, they are derivatives of biologically active molecules and, consequently, are of importance for the field of pharmaceutical co-crystal science.

8.3 Overview of Mechanochemical Methodologies

8.3.1 Comparison with Solvent-based Co-crystallisation Techniques

While certain co-crystals are readily obtained by simple neat grinding of the mixture of co-crystal components,[9,36] the addition of small amounts of a liquid phase has been demonstrated to bring about numerous advantages for mechanochemical reactivity. First, liquid-assisted grinding (LAG)[2,15,16] enables a much broader range of reactions than neat grinding. Moreover, LAG usually provides hydrogen-, as well as halogen-bonded co-crystals rapidly and in quantitative yield. Finally, the products of LAG are highly crystalline even after extended grinding periods, in contrast to neat grinding which inevitably leads to partial amorphisation of the product. The latter difference between neat grinding and LAG was clearly demonstrated through the combined use of powder X-ray diffraction (PXRD) and terahertz time-domain spectroscopy (THz-TDS).[31]

The efficiency of LAG was established in a comparative study involving neat grinding, LAG, slurry sonication and crystallisation from solution.[37] Model co-crystals in this study involved theophylline and caffeine as the model APIs and either the L- or DL-form of tartaric acid as the co-former: (theophylline)$_2$ · (L-tartaric acid), (theophylline)$_2$ · (D-tartaric acid) and (caffeine) · (L-tartaric acid) (Figure 8.3(a)). The liquid phases for LAG, slurry sonication and solution growth were selected on the basis of the solubility of the co-crystal components – synthesis of each co-crystal was achieved using a solvent that was either a poor or a good solvent for both co-crystal components, or was a good solvent for only one of the co-crystal components. In all cases, quantitative formation of the co-crystal was observed only in the case of LAG, while the other methods provided either no reaction or gave a mixture of the co-crystal and individual components.

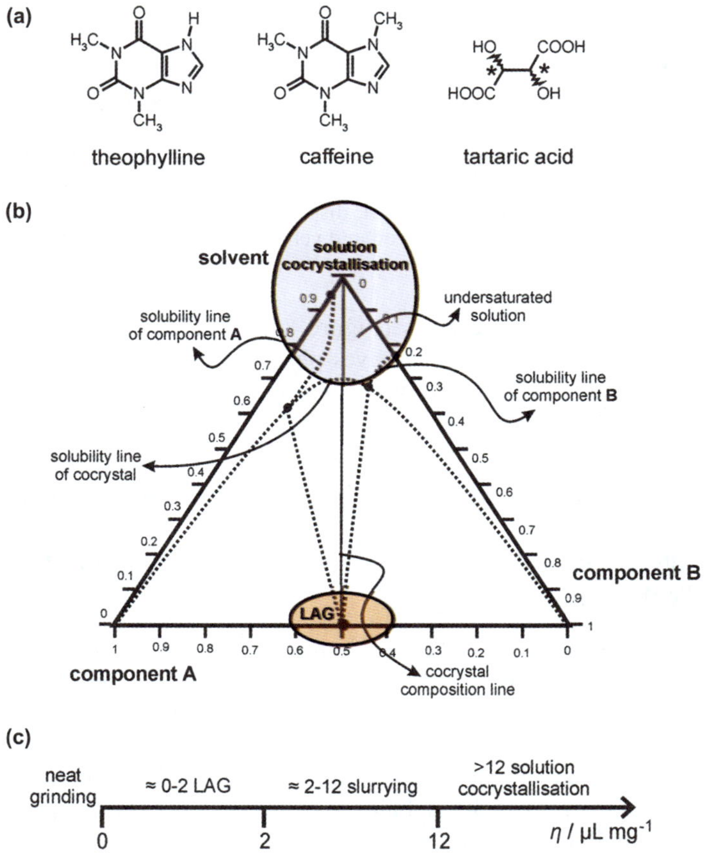

Figure 8.3 (a) Molecular diagrams of model API molecules theophylline, caffeine and tartaric acid, (b) a typical three-component phase diagram relevant for co-crystal formation in the presence of a liquid phase, with indicated compositions of solution growth and liquid-assisted grinding, (c) the application of the η-parameter to define co-crystallisation methodologies.[37]

The superior efficiency of LAG over neat grinding was explained by kinetic effects, specifically by an enhanced rate of reaction in the presence of the liquid. The distinction between LAG and either slurry sonication or solution crystallisation was explained through thermodynamic factors, specifically through a difference in solubilities of the co-crystal and its individual components. Consequently, LAG was identified as a methodology in which the liquid phase is used to enhance the rate of mechanochemical co-crystal formation, but in a sufficiently small amount to avoid the effects of relative solubilities of co-crystal components. This conclusion is fully consistent with the model of liquid-assisted grinding proposed by Davey and co-workers.[38,39] who described LAG reaction conditions in terms of a three-component phase diagram involving the two co-crystal components and the solvent (Figure 8.3(b)). In particular, the characteristic reaction conditions place LAG in the part of the phase diagram where the relative solubilities of co-crystal components are no longer relevant: the amount of the liquid phase in the system is sufficiently small for it to become continuously saturated with respect to all co-crystal components. In this way co-crystallisation conditions resembling those in solution-based reaction co-crystallisation are obtained.[40] The comparative study[37] also provided an empirical parameter, η, which may be used to distinguish between different mechanochemical, slurry- and solution-based methodologies (Figure 8.3(c)). In addition, η, which corresponds to the ratio of the liquid phase and the mass of the reactants in the co-crystallisation experiment, can be used to optimise the design of solvent-mediated co-crystal screening experiments.

8.3.2 Screening for Co-crystals Using Neat and Liquid-assisted Grinding

Since the structure of a pharmaceutical co-crystal is based on hydrogen bonds or weaker interactions which are likely to be disrupted by interactions with solvent molecules that may be present, screening for pharmaceutical co-crystals using conventional solution-based methods is of limited applicability. For example, solution methods are likely to fail in screening for co-crystals of low-solubility APIs which typically separate as a solid while the more soluble co-former is retained in solution.[41,42] In such cases mechanochemical methods represent obvious alternatives for co-crystal screening. In particular, LAG is preferred to neat grinding as it is more general and provides improved kinetics and more crystalline products. As an example, while caffeine and citric acid do not form a co-crystal upon neat grinding, LAG with either water or organic solvents results in the formation of the pharmaceutical solid (caffeine)–(citric acid) (Figure 8.4).[43] The advantage of LAG over neat grinding was also observed in screening for co-crystals of piroxicam.[44] The study by Childs *et al.* established that LAG was of comparable efficiency to solution-based and thermal methods for co-crystallisation of carbamazepine.[19] The study by Karki *et al.* demonstrated that mechanochemistry was more effective than solution- and melt-based methods in screening for co-crystals of nicotinamide.[20]

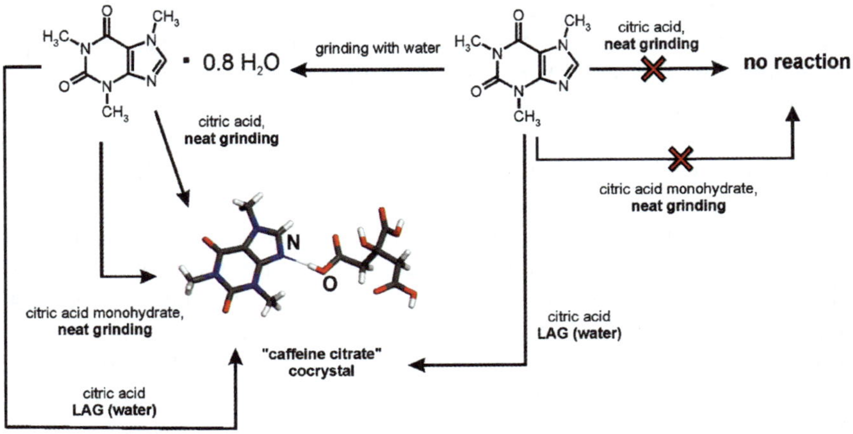

Figure 8.4 Schematic representation of the mechanochemical reactivity of caffeine, citric acid and their respective hydrates under neat and liquid-assisted grinding conditions.[43]

The application of LAG to form co-crystals of low-solubility APIs was further demonstrated using the model compound theobromine (Figure 8.5(a)). Grinding theobromine with trifluoroacetic and malonic acid resulted in co-crystals, while none were obtained by crystallisation from solution.[27] Additionally, because of the high melting point of theobromine (>400 °C) the two co-crystals could not be obtained from the melt. The inability to obtain the two co-crystals from solution also prevented their structural characterisation through single crystal X-ray diffraction.

Consequently, characterisation was achieved using powder diffraction data (Figure 8.5(a)).[27,28] The combined approach of solid-state synthesis and powder structure analysis was also applied to the study of co-crystals of theobromine with acetic acid,[45] and of theophylline with chiral and racemic malic acids (Figure 8.5(b)).[46] Further pharmaceutically interesting molecules that were recently explored through LAG co-crystal screening include the cases of dihydrocarbamazepine,[47] indomethacin[48] and the drug candidate AMG-517 (Figure 8.5(c)).[49,50]

Since co-crystallisation by LAG is typically achieved within 20 minutes of grinding, mechanochemical screening for pharmaceutical co-crystals is also faster than conventional slow evaporation and requires substantially less solvent. The latter advantage was demonstrated by Zaworotko and co-workers who explored the construction of 25 model co-crystals previously obtained from solution.[18] In each case, the co-crystal was successfully constructed mechanochemically, using only 4-20 μL of the liquid per 100 mg of the solid. Furthermore, LAG was also shown to be advantageous to co-crystal screening from the melt, as it avoids exposing thermally-sensitive APIs or co-formers to high temperatures. This was demonstrated in screening for co-crystals of the model API nicotinamide with dicarboxylic acids: screening from the melt was not possible with thermally sensitive oxalic acid,[20] or with the high melting fumaric acid.[51]

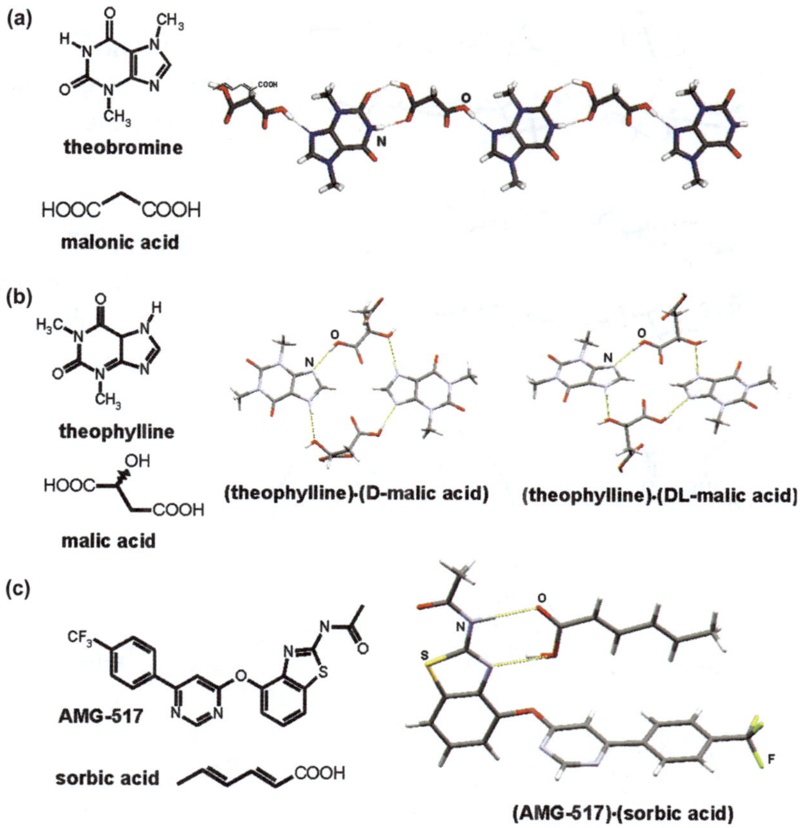

Figure 8.5 Pharmaceutical co-crystal components and a fragment of the solid-state co-crystal structure for: (a) theobromine and malonic acid,[27] (b) theophylline and D- and DL-malic acids and[46] (c) AMG-517 and sorbic acid.[49,50]

8.3.3 Control of Polymorphism Using Mechanochemistry

The ability to vary the nature of the liquid phase in LAG allows control over the polymorphic behaviour of mechanochemically obtained co-crystals. This was demonstrated by Trask *et al.* in mechanochemical co-crystallisation of caffeine with glutaric acid.[52] Co-crystallisation of the two components in chloroform solution provided the co-crystal (caffeine)–(glutaric acid) as two concomitant polymorphs. However, LAG with either chloroform or cyclohexane resulted in the selective formation of each form separately (Figure 8.6(a)).[52] This control of polymorphic behaviour was also observed in LAG screening for co-crystals of piroxicam.[44] The application of LAG to the control of polymorphism is not limited to co-crystals, however and has also been shown to be applicable to single-components systems, for example anthranilic acid (Figure 8.6(b)).[8]

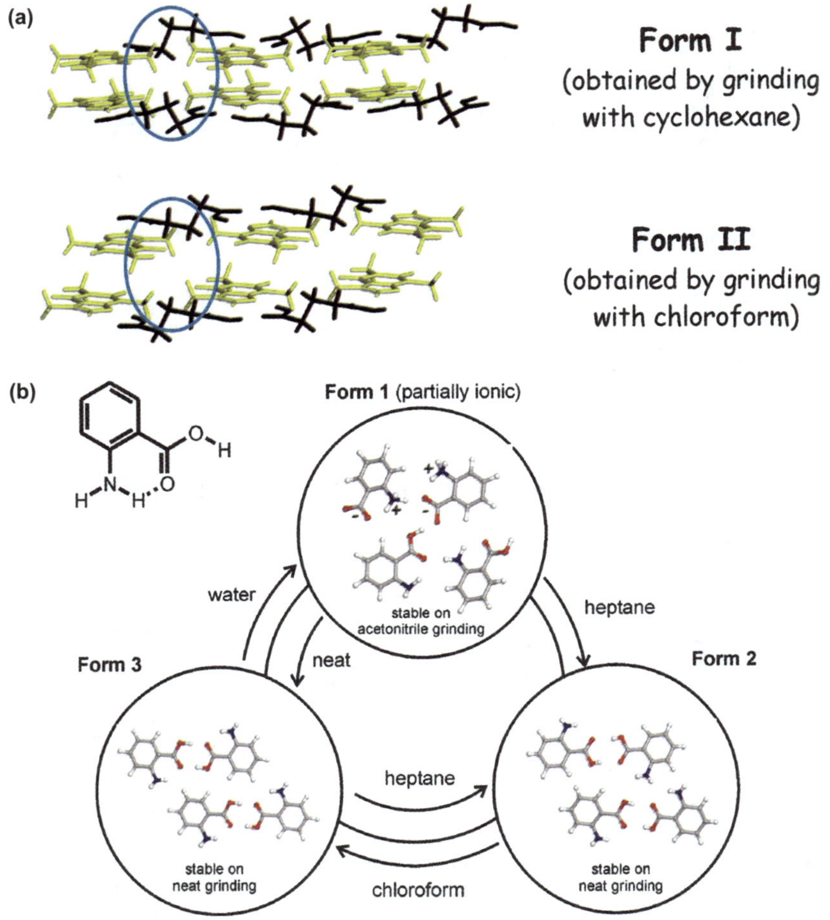

Figure 8.6 (a) Comparison of crystal structures of two polymorphs of the co-crystal of caffeine (yellow) with glutaric acid (black) obtained by LAG with cyclohexane and chloroform, respectively.[52] The most important structural difference between the two polymorphs is in the glutaric acid conformation (highlighted in blue). (b) Polymorphic transformations of anthranilic acid achieved through neat and liquid-assisted grinding.[8]

8.3.4 Control of Co-crystal Stoichiometry Using Mechanochemistry

An attractive aspect of co-crystallisation in the synthesis of new API forms is the possibility of forming co-crystals containing identical constituents in different stoichiometric ratios.[53] Mechanochemical methods provide a particularly simple procedure for constructing such stoichiometric variations by simply grinding different amounts of starting materials. This was first demonstrated for the model API caffeine upon co-crystallisation with acetic

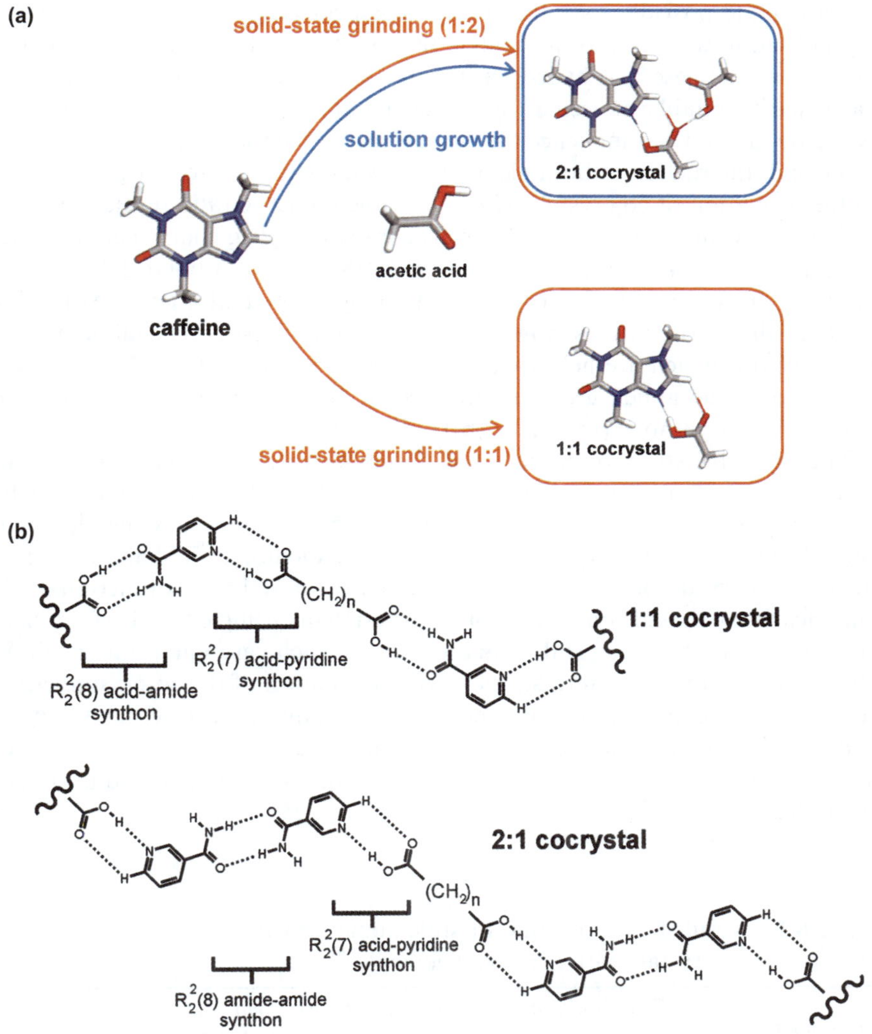

Figure 8.7 (a) Mechanochemical synthesis[53] of the co-crystals (caffeine)–(acetic acid) and (caffeine)–(acetic acid)$_2$ and (b) schematic structures of hydrogen-bonded chains in the 1:1 and 2:1 co-crystals of terminal dicarboxylic acids with the model API nicotinamide.[20]

acid (Figure 8.7(a)).[53] Crystallisation of caffeine from liquid acetic acid results in the formation of co-crystals with the composition (caffeine)–(acetic acid)$_2$. The same product is observed by grinding together the two components in the appropriate stoichiometric ratio. Grinding of equimolar amounts of caffeine and acetic acid, however, resulted in the formation of a different co-crystal, with the composition (caffeine)–(acetic acid).[53]

The synthesis of stoichiometric variations was systematically investigated for co-crystals of the model API nicotinamide with various dicarboxylic acids.[20]

Nicotinamide provides two distinct hydrogen bonding sites available for co-crystallisation with carboxylic acids – the pyridine and the amide group. The availability of these two hydrogen-bonding sites allows nicotinamide to be a particularly suitable molecule for the deliberate study of how the choice of synthetic method can influence the stoichiometric composition of the resulting co-crystal. Interaction of both nicotinamide sites with a dicarboxylic acid leads to the formation of co-crystals with a 1:1 stoichiometric ratio of nicotinamide and the co-former. These 1:1 co-crystals consist of zigzag chains held together by pyridine-acid $R_2^2(7)$ and amide–acid $R_2^2(8)$ hydrogen-bonded heterosynthons.[54] The formation of co-crystals containing nicotinamide and the diacid in a 2:1 stoichiometric ratio is, however, also possible. These co-crystals consist of zigzag chains which are held together by the pyridine–acid $R_2^2(7)$ heterosynthons, while the amide group is involved in amide–amide $R_2^2(8)$ hydrogen-bonded homosynthons (Figure 8.7(b)).[20,55]

The extensive study by Karki *et al.*[20] demonstrated that the formation of either 1:1 or 2:1 nicotinamide–diacid co-crystals is efficiently controlled by the composition of the reaction mixture during mechanochemical co-crystallisation (Table 8.1). In particular, the stoichiometric composition of the final co-crystal product closely resembled the initial ratio of nicotinamide and dicarboxylic acid solids in 16 of 20 experiments using either LAG or neat grinding. Notably, co-crystal formation was in each case quantitative within 20 min of grinding. In contrast, the partial formation of the co-crystal with a stoichiometric composition corresponding to the initial ratio of nicotinamide and dicarboxylic acid was observed in 12 of 20 cases when using co-crystallisation from the melt and in 10 of 20 experiments when using solution co-crystallisation.[20,51,56]

Table 8.1 Results[a] of screening for stoichiometric variations of nicotinamide co-crystals with dicarboxylic acids.[20]

Stoichiometric ratio nicotinamide: acid	*Co-crystallisation from solution*		*Co-crystallisation from the melt*		*Neat grinding*		*LAG*	
	1:1	*2:1*	*1:1*	*2:1*	*1:1*	*2:1*	*1:1*	*2:1*
Oxalic acid	+	−	−	−	+	+	+	+
Malonic acid	−	+	−	+	−	+	−	+
Succinic acid	−	+	−	+	−	+	−	+
Glutaric acid	+	−	+	−	+	−	+	−
Adipic acid	−	+	−	+	+	+	+	+
Pimelic acid	+	−	+	−	+	+	+	+
Suberic acid	+	−	+	−	+	+	+	+
Azelaic acid	+	−	+	−	+	−	+	−
Sebacic acid	−	+	−	+	+	+	+	+
Fumaric acid	+	−	−	−	+	−	+	−

[a]The symbols ' + ' and ' − ' indicate whether the co-crystal with the composition corresponding to the initial stoichiometric ratio of nicotinamide and the carboxylic acid was or was not obtained, respectively.

8.4 Examples of Pharmaceutical Co-crystal Synthesis and Screening Using Mechanochemistry

In principle, the formation of pharmaceutical co-crystals can be guided utilising the concept of supramolecular synthons. However, the physicochemical properties of the resulting materials cannot be readily predicted. Furthermore, since the synthon-based approach considers only specific recognition between selected functional groups, its usefulness in controlling the three-dimensional architecture of the co-crystal is limited. Additionally, the synthon approach itself is recognised as being sensitive to competition between different functional groups and the overall lattice energy.[57–61] Consequently, pharmaceutical co-crystals with improved properties must be discovered in a trial-and-error process which is strongly assisted by efficient screening methods such as LAG. We now consider examples of pharmaceutical co-crystals with improved properties that were first obtained by mechanochemical screening.

The use of co-crystals to enhance the hydration stability of a solid API was first demonstrated for model APIs caffeine and theophylline by forming co-crystals with dicarboxylic acids.[62,63] Co-crystallisation with oxalic, malonic, maleic and glutaric acids provided co-crystals based on the expected $R_2^2(7)$ carboxylic acid–imidazole[54] heterosynthon. In both cases, the co-crystal with oxalic acid demonstrated enhanced hydration stability compared to the solid APIs and did not transform into a hydrate even after seven weeks exposure to 98% relative humidity. Similar observations were also made in the case of analogous co-crystals of theophylline.[62,63]

Liquid-assisted grinding was used to construct co-crystals of nicotinamide with the low melting API molecules S-ibuprofen and RS-ibuprofen (Figure 8.8(a)).[64] It was anticipated that co-crystallisation would lead to solid forms with higher melting points, as a result of extended hydrogen bonding involving the amide–amide $R_2^2(8)$ homosynthon. Indeed, LAG screening provided co-crystals with significantly higher melting points compared to the parent APIs. Single crystals of these co-crystals were subsequently grown from solution and structurally characterised using synchrotron radiation, which confirmed the formation of the expected hydrogen-bonded networks (Figure 8.8(b,c)).[65]

Mechanochemical co-crystallisation has also been exploited in the synthesis of readily compressible and thermodynamically stable forms of the API paracetamol (Figure 8.8(b)).[66] While tablet formation using the thermodynamically stable polymorph of paracetamol is difficult, the metastable orthorhombic polymorph yields tablets much more readily due to its layered crystal structure. Consequently, it was expected that co-crystals of paracetamol with a similar layered structure would also be readily compressible. Screening by LAG revealed four co-crystals of paracetamol with improved ability to compress into tablets. Structural characterisation and Density Functional Theory (DFT) calculations revealed that enhanced compressibility was related to the formation of a sheet-like structure reminiscent of the second polymorph of paracetamol (Figure 8.8(c)–(e)).[66]

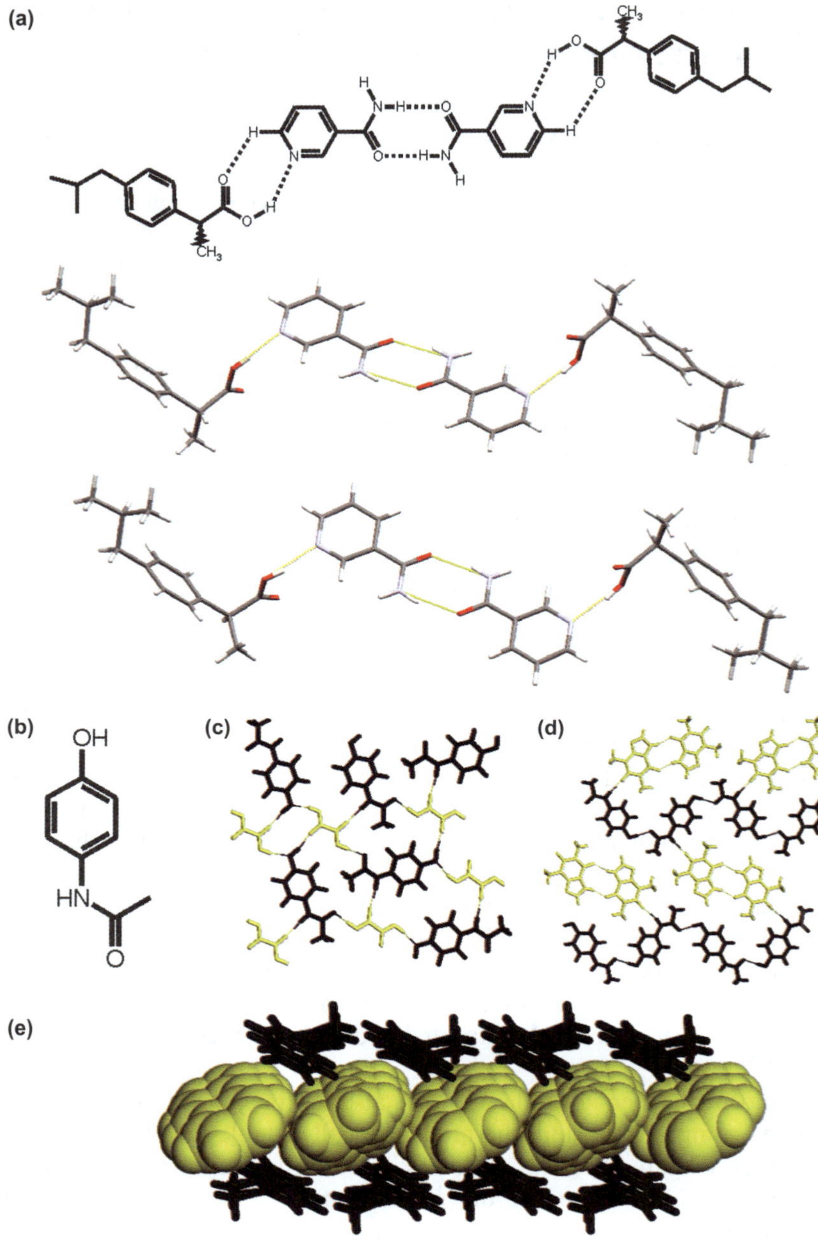

Figure 8.8 (a) Expected hydrogen-bonded assembly[64] (top) and corresponding fragments in the observed[65] crystal structures of nicotinamide co-crystals with *RS*- (middle) and *S*-ibuprofen (bottom), (b) molecular diagram of paracetamol, (c) single layer in the crystal structure of the paracetamol co-crystal with oxalic acid, (d) single layer in the crystal structure of the paracetamol co-crystal with theophylline and (e) stacked layers of paracetamol and naphthalene in the co-crystal. Molecules of paracetamol are shown in black and the molecules of the co-former in yellow.[66]

One of the advantages of co-crystal formation over salt formation is the ability to generate new solid forms of APIs regardless of the presence of ionisable groups. In particular, the design of co-crystals has largely focused on the use of functional groups capable of participating in hydrogen or halogen bonding. However, co-crystallisation is also possible in the absence of complementary hydrogen- or halogen-bonding groups. In such cases, however, there are no obvious guidelines for the selection of a suitable co-crystal former and the availability of a rapid methodology for co-crystal screening is a necessity.

The use of mechanochemical liquid-assisted grinding in this context was demonstrated using the antimalarial drug artemisinin as a model API (Figure 8.9(a)).[67] Artemisinin was selected as a particularly challenging co-crystallisation target, as its molecular structure does not offer any obvious sites for the formation of strong hydrogen-bonded synthons. The only potential hydrogen-

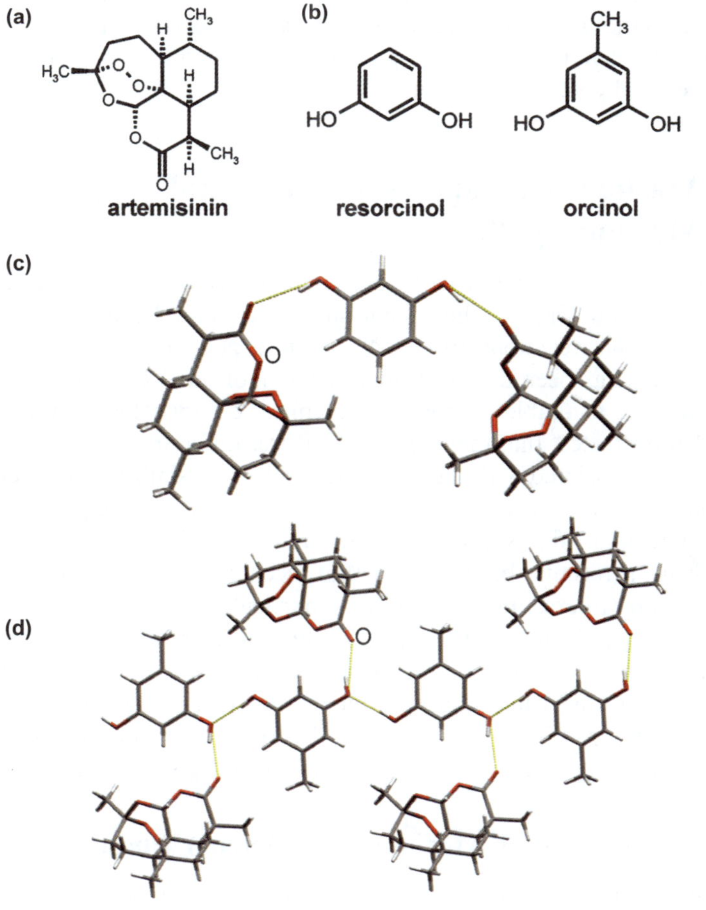

Figure 8.9 Molecular diagrams of (a) artemisinin and (b) resorcinol and orcinol. Fragments of the crystal structures of (c) (artemisinin)$_2$–(resorcinol) and (d) (artemisinin)–(orcinol) co-crystals.[67]

bonding functionalities on the artemisinin molecule are an ester, an ether and a peroxide group. There are no reliable supramolecular synthons involving any of these functionalities. Consequently, LAG was employed to screen over 75 different co-former molecules. This extensive screen revealed only two molecules which formed a co-crystal with artemisinin – resorcinol and orcinol (Figure 8.9(b)).[67] In both cases, co-crystal formation was based on the formation of O–H···O hydrogen bonds between the ester group of artemisinin and the phenolic group of the co-former.

However, aside from the formation of O–H···O hydrogen bonds, the structures of these two co-crystals exhibit very little in common. The hydrogen-bonding patterns are different, leading to different stoichiometric ratios of artemisinin to co-former: $(artemisinin)_2$–(resorcinol) and (artemisinin)–(orcinol) (Figure 8.9(c) and (d)). Whereas certain guidelines for the selection of co-formers for artemisinin could be subsequently derived using the molecular descriptor approach,[68,69] the driving force behind co-crystal formation involving orcinol and resorcinol remains largely unknown, thus demonstrating the value of mechanosynthesis as an efficient co-crystal screening approach in pharmaceutical co-crystallisation.

8.5 Mechanistic Aspects of Co-crystal Mechanosynthesis

The mechanisms associated with mechanochemical reactions are, to a large extent, still unknown.[2,70] The exceptions are a limited number of mechanochemical organic reactions for which a liquid generated by an intermediate eutectic phase has been demonstrated to be involved.[71] However, more extensive studies are available for mechanochemical co-crystallisation and a recent review demonstrates[2] that there are at least three different types of mass transfer mechanisms associated with mechanochemical co-crystallisation. These include molecular diffusion through the vapour phase, liquid eutectic phase and via an intermediate amorphous phase. In addition to differences in mass transfer mechanisms, it recently became clear that mechanochemical co-crystallisations can also exhibit interesting kinetic effects, particularly the stepwise formation of different co-crystal forms during the grinding process. It should be noted that each mechanochemical reaction is most likely to be a combination of all three types of mass transfer mechanisms which constitute, along with kinetic effects, the complicated process which has been described by Kaupp in the context of phase rebuilding, phase transformation and crystal disintegration steps.[72–74]

8.5.1 Mass Transfer Mediated by a Vapour Phase

This is probably the most studied mechanism of mechanochemical co-crystal formation. Co-crystallisation mediated by molecular diffusion through the vapour is likely to occur in cases where one or both reactants exhibit significant vapour pressures. Product formation is expected then to occur readily upon

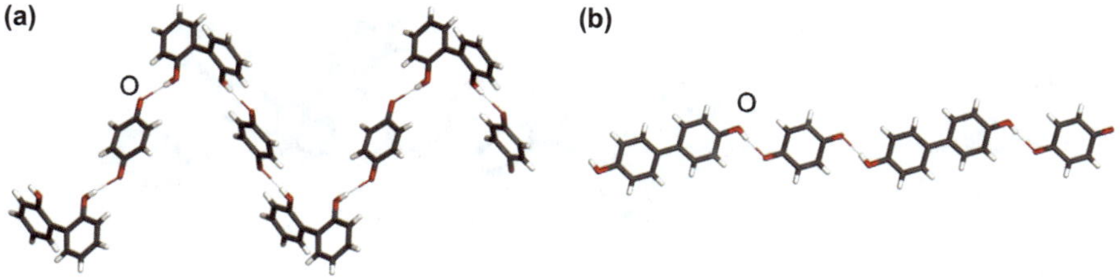

Figure 8.10 Hydrogen-bonded chains in the co-crystal of *p*-benzoquinone with (a) 2,2'-biphenol and (b) 4,4'-biphenol.[81]

contact between reactant solids, even in the absence of mechanochemical agitation. Rastogi *et al.* established that surface migration and diffusion through the vapour phase are likely to be the principal mechanisms for the formation of picric acid co-crystals with aromatic hydrocarbons.[75] These are also the likely mechanisms for the formation of two- and three-component co-crystals by grinding the volatile solid *p*-benzoquinone with various phenol derivatives.[76–80]

Kuroda *et al.* have demonstrated that co-crystal formation with *p*-benzoquinone can be mediated by molecular diffusion.[81]Mixing, in the solid state, crystals of *p*-benzoquinone with either 2,2'-biphenol or 4,4'-biphenol crystals rapidly results in the formation of highly coloured co-crystals, in which the components are held together through charge-transfer interactions and hydrogen bonds (Figure 8.10).

Co-crystal formation is also observed upon exposing solid 2,2'- or 4,4'-biphenol to vapours of *p*-benzoquinone, again demonstrating that co-crystallisation is aided by gas phase diffusion.[81] The rate of co-crystal formation by mixing in the solid state was reported to be much slower in the case of 4,4'-biphenol, than for the 2,2'-analogue. This difference in reactivity between the two phenols was explained by the extensive network of hydrogen bonds in solid 4,4'-biphenol which hinders surface diffusion of molecules. Co-crystallisation of solid powdered *p*-benzoquinone and 2,2'-biphenol was found to proceed for approximately 60 hours in a non-agitated mixture of solid reactants. After this time, the reaction could be initiated again and brought to completion by brief milling, suggesting that the co-crystallisation is limited by the availability of fresh surface for molecular diffusion of solid reactants. This interpretation is supported by vapour diffusion experiments, where co-crystal formation was observed only on the surface of solid biphenol reactant exposed to *p*-benzoquinone vapours. The role of grinding in mechanochemical co-crystallisation of *p*-benzoquinone with 2,2'- or 4,4'-biphenol appears to be the enhancement of the reaction rate by exposing fresh reactant surfaces and mixing the solid reaction mixture.[81]

Exposure of crystals of racemic bis-β-naphthol (Figure 8.11(a)) to *p*-benzoquinone vapour or solid-state mixing with *p*-benzoquinone does not provide a co-crystal.[81] In contrast, mechanochemical grinding provides a co-crystal composed of finite hydrogen-bonded molecular assemblies (Figure 8.11(b)).

A possible explanation is that mechanochemical grinding can play a more important role in co-crystallisation than simply mixing volatile reactants.

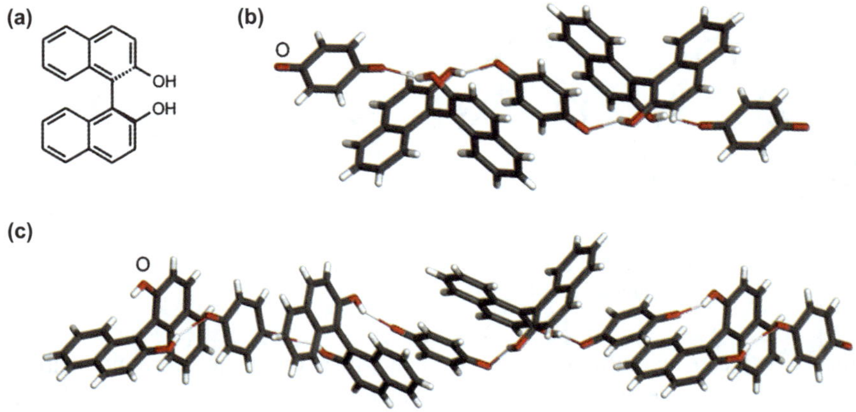

Figure 8.11 (a) Molecular diagram of bis-β-naphthol, (b) and (c) finite and infinite assemblies of *p*-benzoquinone with racemic bis-β-naphthol obtained by grinding and from solution, respectively.[81]

Grinding was suggested to enable co-crystallisation by overcoming the strong intermolecular forces that hold together the molecules in solid bis-β-naphthol.[81] This is believed to involve forming disordered molecular layers on the surface of reactant crystals through shearing. The change from a vapour diffusion-aided mechanism of co-crystal formation in case of 2,2'-biphenol and 4,4'-biphenol to a mechanism that requires mechanical activation of the reactant surface has been explained by strong intermolecular forces in solid β-naphthol. In contrast to grinding, co-crystallisation of the racemic bis-β-naphthol and *p*-benzoquinone from solution yields a different co-crystal of 1:1 composition, consisting of hydrogen-bonded tapes (Figure 8.11(c)).

The relevance of surface diffusion in co-crystallisation by grinding was first recognized by Rastogi and Singh who monitored the reaction rates of picric acid with different aromatic co-formers (Figure 8.12).[82,83]

These results suggest a model in which both vapour and surface diffusion aid solid-state co-crystallisation. Significant participation of vapour diffusion in the co-crystal forming mechanism was suggested in the case of the naphthalene–picric acid co-crystal, while surface diffusion was deemed more relevant in the co-crystallisation with heavier arenes. Experiments suggest a mechanism of co-crystallisation predominantly based on surface diffusion, assisted to varying degrees by gas phase diffusion. The role of grinding in such systems is to enhance surface diffusion by mixing the reactants, provide fresh reactive surfaces by removing the product, and activate the reactant surface through introducing strain and defects.

8.5.2 Mechanochemical Co-crystallisation Mediated by a Eutectic

It is very likely that understanding of the role of eutectic formation in co-crystal synthesis will become increasingly important in the context of pharmaceutical

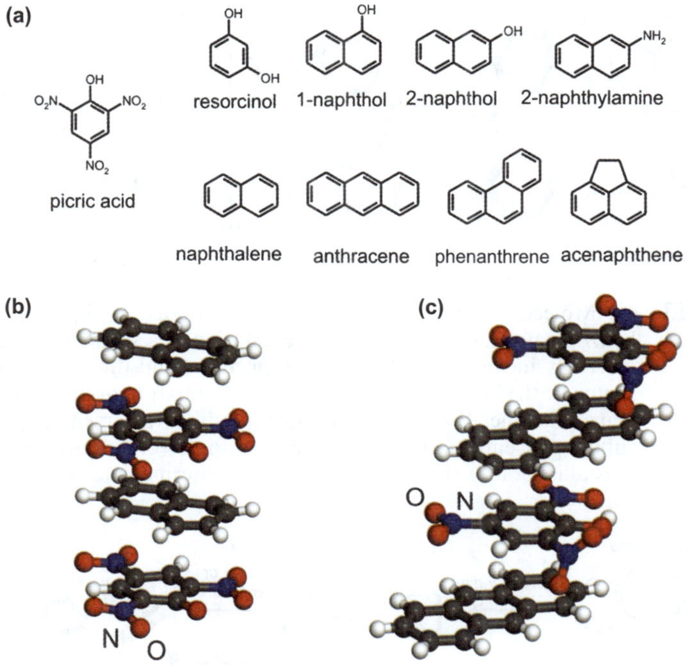

Figure 8.12 (a) Molecular diagrams of co-crystal components investigated by Rastogi and Singh.[82,83] Stacking of molecules in the co-crystals of picric acid with (b) naphthalene and (c) anthracene.

materials, as demonstrated by the recent development of a eutectic-based thermal methodology for co-crystal synthesis by Rodríguez-Hornedo and coworkers.[84] Co-crystallisation preceded by the formation of a metastable eutectic phase was investigated by Davey and co-workers in the case of the co-crystal of diphenylamine and benzophenone (Figure 8.13(a)).[85] Co-crystallisation from solution yields a co-crystal consisting of discrete assemblies held by an N–H···O hydrogen bond (Figure 8.13(b)).[86,87]

Upon mixing of the two colourless co-crystal constituents in the solid state, the formation of a yellow co-crystal phase is rapidly (within 1 min) observed at the interface of the two solids.[85] Microscopic observation of the interface between two macroscopic single crystals of diphenylamine and benzophenone revealed melting of the surface which proceeds until most of the material is converted to a liquid at room temperature. Subsequent nucleation of the co-crystal phase results in the solidification of the entire melt to form the solid co-crystal.[85] Consequently, the role of grinding in a eutectic-mediated co-crystallization is most likely two-fold: first, it provides agitation to expose fresh reactant surfaces for eutectic formation and second, it enhances co-crystal formation by inducing co-crystal nucleation in the eutectic phase (Figure 8.13(c)).

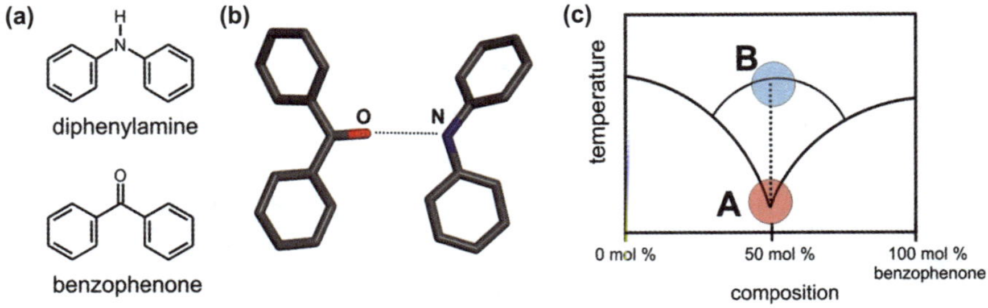

Figure 8.13 (a) Molecular diagrams of co-crystal components diphenylamine and benzophenone, (b) wireframe representation of a single hydrogen-bonded assembly in the co-crystal of benzophenone and diphenylamine and (c) sketch of the binary phase diagram for the mixture of diphenylamine and benzophenone. Point A (red) corresponds to the melting point of the submerged eutectic and path from A to B (blue) corresponds to the crystallization of the co-crystal from the intermediate eutectic phase.[85]

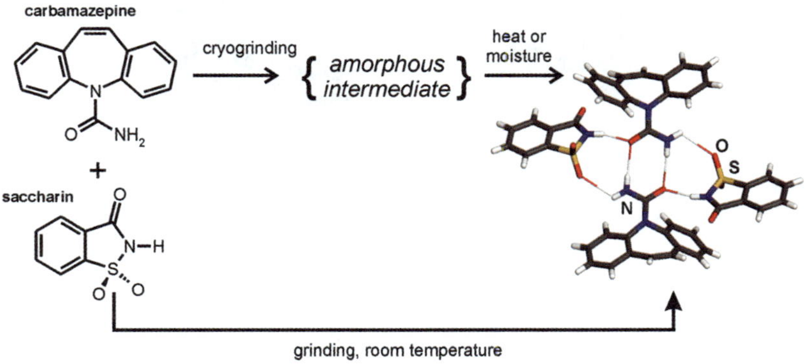

Figure 8.14 Mechanochemical co-crystallisation by neat grinding of carbamazepine and saccharin[88–90] involving an amorphous intermediate.

8.5.3 Mechanochemical Co-crystallisation Mediated by an Amorphous Phase

The co-crystal of the API carbamazepine with saccharin was used by the group of Rodríguez-Hornedo as a model to study the mechanism of neat grinding co-crystallisation (Figure 8.14).[88–90] Co-crystallisation of saccharin and carbamazepine by neat grinding proceeds through an intermediate amorphous phase. The mechanochemical reaction was accelerated at temperatures close to the glass transition temperature (T_g) of the ground mixture. Cryogenic grinding, that is, grinding at low temperatures, allowed the observation of the intermediate amorphous phase which recrystallised on warming. This behaviour is consistent with the work of Descamps and co-workers and the mechanochemical behaviour of single-component solids, which become amorphous upon grinding at low temperatures or undergo polymorphic transformations if ground above their T_g.[91–94] Mechanochemical co-crystallisation was also accelerated by exposing the intermediate amorphous phase to water, which effectively lowers the T_g of the reaction mixture. Exposure to water was achieved either by storing the ground reaction mixture under conditions of high relative humidity or by using carbamazepine dihydrate as the reactant.[88–90] Higher reactivity of hydrated carbamazepine compared to the anhydrous form was also observed by Rades and co-workers in the mechanosynthesis of the (carbamazepine)–(nicotinamide) co-crystal.[95] Similarly, neat grinding of either citric acid or its monohydrate with caffeine did not provide the pharmaceutical co-crystal (caffeine)–(citric acid), while using caffeine hydrate as the reactant provided the co-crystal of both citric acid and its hydrate (Figure 8.4).[43]

The ability to construct pharmaceutical co-crystals on a large scale is important for the industrial application of mechanosynthesis. This aspect of mechanochemical co-crystallisation was recently addressed by Medina *et al.* who described a scalable solvent-free process for pharmaceutical co-crystallisation by using twin screw extrusion.[96]

8.5.4 Kinetic Effects in the Mechanosynthesis of Co-crystals

The mechanism of mechanochemical co-crystal formation has been considered in the macroscopic sense of transformations between bulk reactant, intermediate and product phases. Cinčić *et al.*, however, have recently addressed the mechanism of co-crystallization by neat grinding at the microscopic level, that is, from the standpoint of forming and cleaving individual supramolecular bonds.[97] In particular, the formation of halogen-bonded co-crystals of the halogen bond acceptor 1,4-thiomorpholine with linear or bent halogen bond donors was observed to proceed via a stepwise mechanism. In case of the linear 1,4-diiodotetrafluorobenzene donor, the first step in the grinding reaction is the formation of three-component assemblies held together by N···I halogen bonds.[97] The assemblies further react with additional halogen bond donor, if available, to yield co-crystals composed of infinite halogen-bonded chains that are held together via N···I and S···I interactions (Figure 8.15(a)).[98]

Analogous behaviour was observed in the non-linear donor 1,2-diiodotetrafluorobenzene, wherein the early step in grinding co-crystallisation was the formation of five-membered assemblies held by three N···I and one S···I bonds. The assemblies again crosslink with additional halogen bond donor to form infinite chains held by alternating N···I and S···I halogen bonds (Figure 8.15(b)).[97]

The stepwise mechanism of co-crystal formation has been interpreted in terms of kinetic and thermodynamic co-crystallisation products, wherein the finite assemblies are the kinetic product whose formation is driven by the formation of stronger N···I halogen bonds.[97] The formation of the thermodynamic product is subsequently followed by the formation of enthalpically less favourable S···I halogen bonds.

This stepwise mechanism is most likely the consequence of the hierarchy[99–101] of strong and weak halogen bonds in the reaction mixture. The stepwise mechanism is likely to be a general characteristic of co-crystal formation by grinding reactant molecules that have significantly different halogen- or

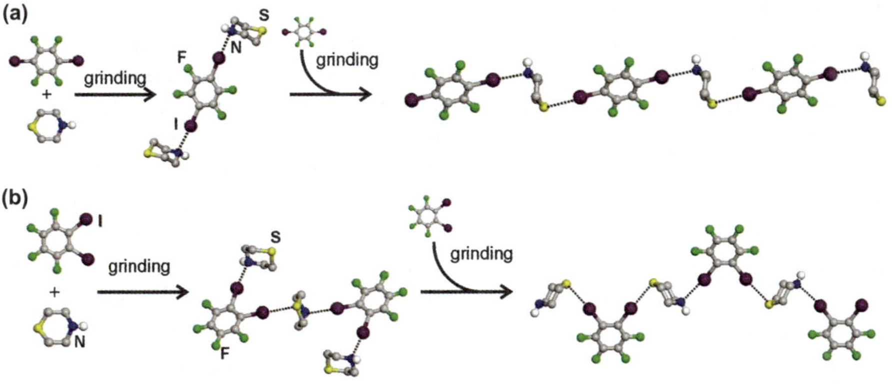

Figure 8.15 (a) The stepwise formation of linear infinite halogen-bonded chains of the co-crystal of 1,4-thiomorpholine and tetrafluoro-1,4-diiodobenzene, via a finite assembly intermediate and (b) stepwise formation of infinite zig zag halogen-bonded chains of the co-crystal of 1,4-thiomorpholine and tetrafluoro-1,2-diiodobenzene.[97]

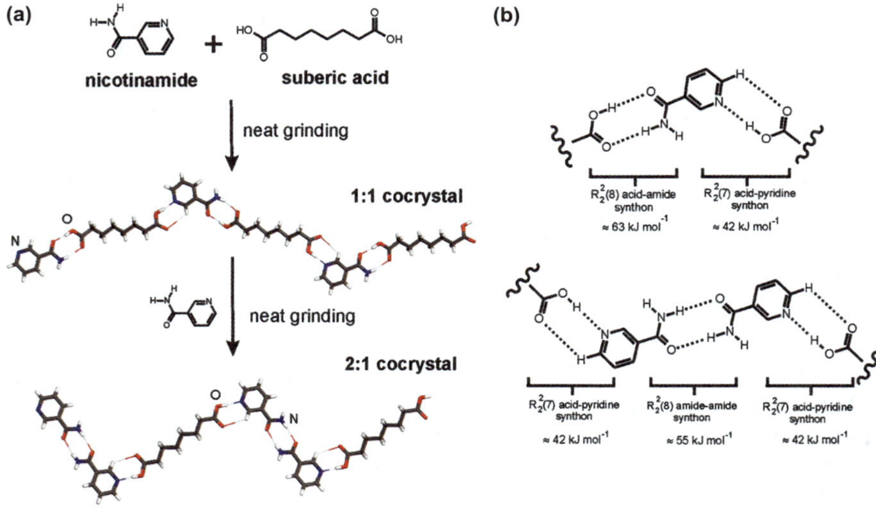

Figure 8.16 (a) Stepwise synthesis of the 2:1 co-crystal of nicotinamide and suberic acid, via the intermediate 1:1 co-crystal and (b) relevant hydrogen-bonded synthons for each co-crystal, along with approximate enthalpies of formation.[20]

hydrogen-bonding sites. Indeed, Karki *et al.* have reported that the neat grinding formation of a 2:1 co-crystal of nicotinamide and suberic acid proceeds through an intermediate co-crystal that has a 1:1 stoichiometry (Figure 8.16).[20] Considering the approximate enthalpies of hydrogen-bonded supramolecular synthons that could form by combining amide, pyridine and carboxylic acid moieties of nicotinamide and suberic acid provided a tentative explanation of this stepwise process. The initial formation of the 1:1 co-crystal was explained by the kinetically controlled formation of the strongest possible supramolecular synthons: the pyridine–carboxylic acid $R_2^2(7)$ ($\approx$42 kJ mol^{-1}) and amide–carboxylic acid $R_2^2(8)$ ($\approx$63 kJ mol^{-1}) heterosynthons. Subsequent formation of the 1:2 co-crystal by grinding with excess nicotinamide results in the dismantling of the strongest amide–carboxylic acid $R_2^2(8)$ heterosynthon, compensated by the formation of a pair of weaker pyridine–carboxylic acid $R_2^2(7)$ heterosynthons and an amide–amide ($\approx$55 kJ mol^{-1}) $R_2^2(8)$ homosynthon.

8.6 Three-component Pharmaceutical Solids

8.6.1 Chiral and Racemic Co-crystals and Co-crystal-Co-crystal Reactions

Chirality is relevant in the formation of pharmaceutical co-crystals for two reasons. The first is that a large number of pharmaceutically active components are chiral and such co-crystallisation is illustrated by the co-crystal formation involving ibuprofen,[64] artemisinin[67] or quinidine.[102] However, the formation of pharmaceutical co-crystals involving chiral co-crystal formers is also of considerable interest in pharmaceutical materials science. On one hand, the use of chiral co-formers may allow the resolution of racemic forms of an API.[103] On the other hand, the stability of a pharmaceutical co-crystal can be significantly modified by selecting either the pure enantiomeric form or the racemic form of the co-crystal former.[46] This choice also brings about an interesting distinction in the structure of the resulting co-crystal; an enantiomerically pure co-former leads to a two-component chiral co-crystal, whereas a racemic co-former is expected to provide a three-component centrosymmetric solid. Consequently, the use of racemic co-formers provides a simple entry into the synthesis of three-component solids.

The absence of bulk solvent in mechanochemical synthesis allows qualitative assessment of the relative stabilities of chiral and racemic co-crystals. In one example, co-crystallisation of the non-chiral model API caffeine with enantiomerically pure L- or D-tartaric acid was found to provide the chiral co-crystal (caffeine)–(L-tartaric acid) or (caffeine)–(D-tartaric acid), respectively.[104] However, grinding with racemic DL-tartaric acid led to no reaction, suggesting that DL-tartaric acid was more stable towards caffeine compared to L- or D-tartaric acid. This stability was additionally demonstrated by conducting a mechanochemical reaction between the two enantiomeric co-crystals. This co-crystal–co-crystal reactivity is unique to mechanosynthesis and cannot be achieved in solution due to the obvious reason of co-crystal dissolution.[104] Grinding together equal amounts of (caffeine)–(D-tartaric acid) and (caffeine)–(L-tartaric acid) resulted in the dismantling of both co-crystals to yield the solid API along with the racemic form of tartaric acid (Figure 8.17(a)). The same study[104] established different behaviour if theophylline was used as the model API. In this case both L-, D- as well as DL-tartaric acids readily provided a co-crystal with theophylline upon grinding. In contrast to the experiments involving caffeine, grinding together of the L- and D-co-crystals of theophylline with

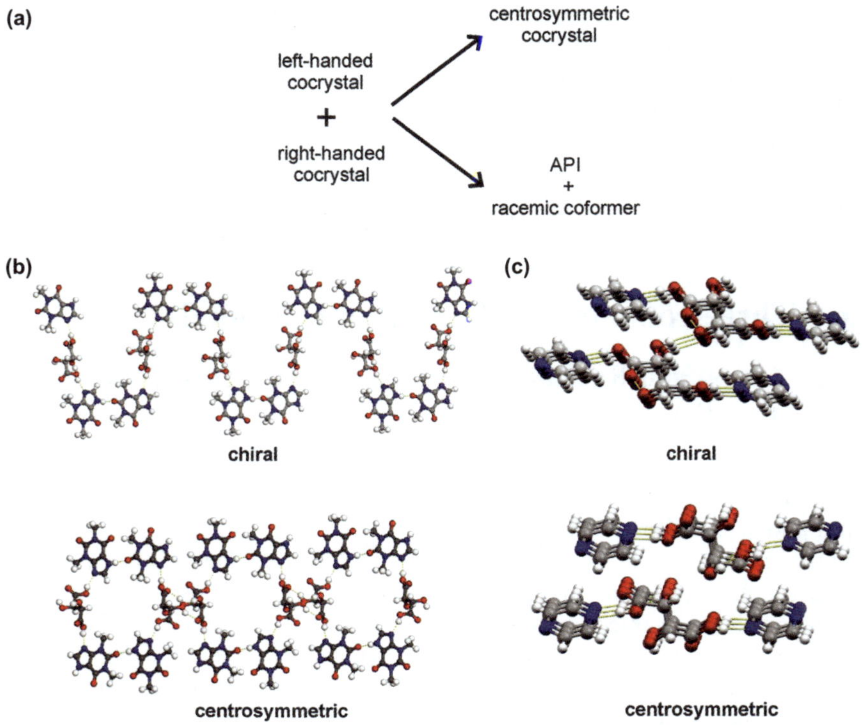

Figure 8.17 (a) Two different modes of co-crystal-co-crystal reactivity involving non-centrosymmetric co-crystals.[104] Fragments of crystal structures of chiral and centrosymmetric co-crystals of (b) theophylline with tartaric acid[104] and (c) pyrazine with tartaric acid.[105]

tartaric acid resulted in the formation of a centrosymmetric three-component co-crystal containing molecules of theophylline, L- and D-tartaric acids (Figure 8.17(b)). A similar observation was made by Braga *et al.* who conducted mechanochemical grinding experiments involving co-crystals of pyrazine with tartaric acid.[105] In particular, grinding the two enantiomeric co-crystals resulted in the formation of a centrosymmetric solid (Figure 8.17(c)). The described experiments have established two pathways of co-crystal–co-crystal reactivity involving chiral co-crystal formers: (1) the formation of a centrosymmetric co-crystal or (2) decomposition of the co-crystals with the separation of the API and the racemic co-former (Figure 8.17(a)).

8.6.2 Co-crystal Hydrates

The addition of a liquid to a grinding mixture of two co-crystal components can sometimes enable co-crystallisation through the formation of a product that incorporates the molecules of the liquid as a constituent. For this reason, grinding in the presence of liquids is a versatile technique for searching for three-component co-crystals (*e.g.* co-crystal hydrates,[43,106] inclusion compounds[15,107,108]). In some cases, the formation of a three-component co-crystal hydrate can be achieved either by using liquid water as the reactant or by using reactants in the form of crystalline hydrates. For example, neat grinding of anhydrous theophylline and anhydrous citric acid (Figure 8.18(a)) with a small amount of water provides a co-crystal hydrate that consists of theophylline, citric acid and water in equal stoichiometric proportions (Figure 8.18(b)).[43]

The same co-crystal hydrate is also obtained by the neat grinding of theophylline hydrate with anhydrous citric acid, the neat grinding of anhydrous theophylline with citric acid monohydrate, as well as the neat grinding of theophylline monohydrate with citric acid monohydrate. Neat grinding of anhydrous theophylline and anhydrous citric acid, by contrast, leads to the formation of an anhydrous co-crystal of 1:1 stoichiometry (Figure 8.18(c)). The similarity of the supramolecular motifs within the co-crystal hydrate to those observed in theophylline monohydrate and citric acid monohydrate crystals (Figure 8.19) suggests that, in the presence of water, the co-crystal hydrate could

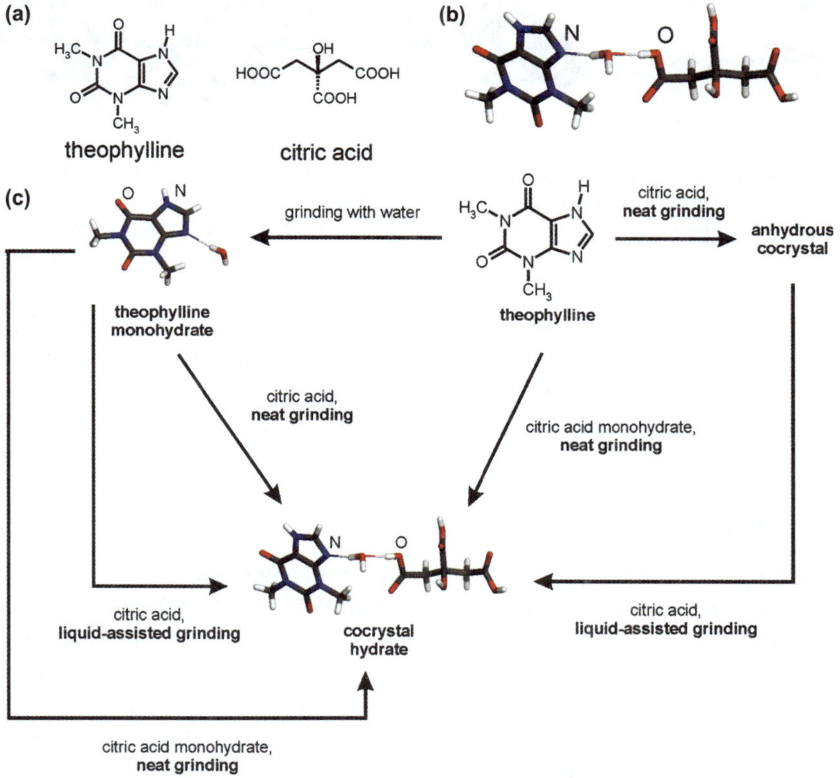

Figure 8.18 (a) Molecular diagrams of theophylline and citric acid, (b) wireframe representation of the asymmetric unit of the co-crystal hydrate involving theophylline and citric acid[43] and (c) an overview of solid-state reactions of hydrated and non-hydrated forms of theophylline and citric acid.

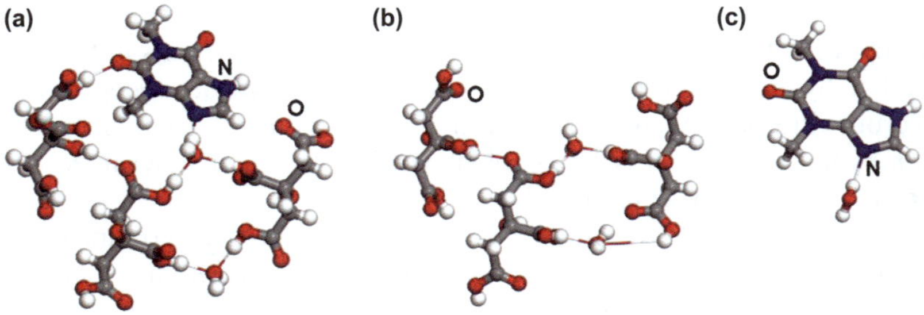

Figure 8.19 Similar structural motifs between (a) a co-crystal hydrate of theophylline and citric acid, (b) citric acid monohydrate and (c) theophylline monohydrate.[43]

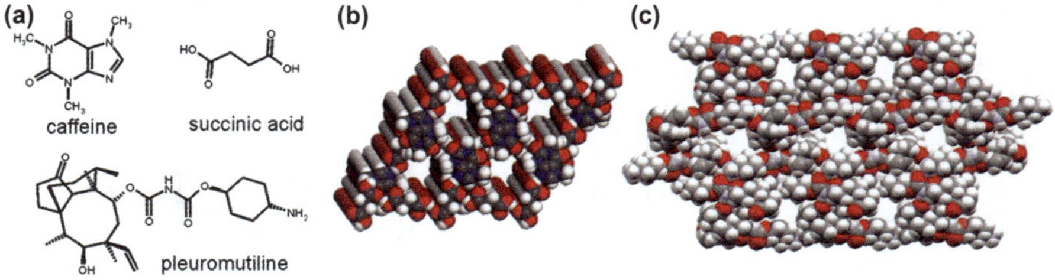

Figure 8.20 (a) Molecular diagrams of caffeine, succinic acid and pleuromutilin, (b) two-component host of caffeine and succinic acid[15] and (c) host structure of protonated pleuromutilin.[108–110] For clarity, included guests are not shown.

be preferred to the formation of an anhydrous co-crystal.[43] Interestingly, a set of similar experiments involving anhydrous β-caffeine and caffeine hydrate, instead of theophylline and theophylline monohydrate, led to an anhydrous co-crystal, perhaps through the formation of the intermediate caffeine hydrate which is more reactive towards citric acid than anhydrous β-caffeine.[43]

8.6.3 Host–guest systems

The ability to circumvent solubility limitations using neat grinding or LAG enables the synthesis of co-crystals of pharmaceutically relevant molecules which contain more than two components. This was first demonstrated in the systematic study of three-component (ternary) inclusion compounds involving a guest molecule within an open hydrogen-bonded host consisting of caffeine and succinic acid (Figure 8.20(a)).[15] The formation of ternary solids was attempted with 25 potential guest molecules using solution crystallisation, neat grinding and LAG. Solution crystallisation provided ternary inclusion compounds in four cases, neat grinding in 15 and LAG in 18 cases (Figure 8.20(b)).[15] The formation of ternary phases of the antibiotic pleuromutilin with succinic acid and methanol or water was studied by Clawson and co-workers.[109,110] Crystallisation from solution or a slurry produced a range of ternary solids containing succinic acid and pleuromutilin in ratios between 1:2 and 1.4:1, while LAG led to the synthesis of materials containing up to two succinic acid molecules per molecule of pleuromutilin. Combined studies using solid-state NMR and X-ray crystallography revealed that the material exhibiting the 1:2 ratio of succinic acid to pleuromutilin consisted of a host lattice of protonated pleuromutilin with included succinate anions and solvent (Figure 8.20(c)). Increasing the relative amount of succinic acid resulted in the progressive replacement of solvent guests with neutral succinic acid, accompanied by a change in crystallographic space group.[111]

8.7 Mechanochemical Synthesis of Pharmaceutical Salts

Besides synthesis of pharmaceutical co-crystals, mechanosynthesis has been successfully applied to the synthesis of pharmaceutical salts. Trask *et al.* have demonstrated mechanochemical screening for pharmaceutical salts by grinding the model API molecules trimethoprim and pyrimethamine with seven pharmaceutically acceptable carboxylic acids: formic, acetic, maleic, fumaric, succinic, glutaric and salicylic.[4] The results of screening experiments (Table 8.2) demonstrate that LAG is more versatile than neat grinding in screening for pharmaceutical salts. More importantly, the application of mechanochemical techniques not only led to the formation of already reported salt forms but also enabled the discovery of 14 new salt forms.

André and co-workers[22] have utilised kneading and neat grinding in the investigation of solvate and salt formation involving the small heterocycles dioxane, morpholine or piperazine and the API 4-aminosalicylic acid. In particular, the use of mechanochemical approaches has led to the simple formation of solvate and salt forms that were otherwise obtained using more complex solution-based routes.

Table 8.2 Results of mechanochemical salt screening for pyrimethamine.[a]

CCDC[b] code of the known crystalline salt	*Neat Grinding*	*LAG*
UHAYIL	√	√
CIVDIU	+	++
ULAXOU	x	√
ULAXIO	x	+
ULAYAH	x	√
UHAYEH	+	+
CIVDOA	x	+

[a]The symbols indicate the following observations based on powder X-ray diffraction: '√' indicates the formation of the already known form of the API salt; '+' the formation of a new form of the API salt, different from the already known one; '++' indicates a new form of the API salt, different from the already known one as well as from the one obtained by neat grinding; 'x' indicates no reaction.
[b]Cambridge Crystallographic Data Centre.

8.8 Mechanochemical Synthesis of Metal–Organic Pharmaceutical Derivatives

Compared to pharmaceutical co-crystals, salts and solvates, coordination complexes of pharmaceutical molecules are a much less developed family of pharmaceutical forms. These complexes can be roughly divided into two groups: (1) metallodrugs in which the metal ion is also the biologically active component and (2) metallopharmaceuticals in which the metal ion plays the role of a carrier of the API molecule, similar to the counterion in a pharmaceutical salt or the co-former in a pharmaceutical co-crystal. Well-known examples of metallodrugs involve the bismuth subsalicylate complex, marketed as Pepto-Bismol[112] and the platinum complexes used in cancer treatment, such as cisplatin, carboplatin or oxaliplatin.[113] A potential design for metallopharmaceuticals has been proposed by Ma and Moulton who utilised the copper(II) carboxylate paddlewheel motif to enhance the lypophyllicity of carboxylate APIs.[114,115]

Whereas the application of mechanochemistry for the synthesis of metal complexes and open metal–organic frameworks is well known, there are very few examples involving the mechanosynthesis of pharmaceutical derivatives. Notably, Braga and co-workers have used mechanical grinding to construct two derivatives of the neuroleptic drug gabapentin with zinc and copper(II) chlorides.[23] In an extension of this work, silver nitrate and nickel chloride were used to generate metal–organic derivatives with APIs 4-aminosalicylic acid and piracetam.[24] The synthesis involving silver nitrate and 4-aminosalicylic acid is particularly interesting both because it combines a known API with a known antimicrobial agent, and because it illustrates the potential of coordination chemistry in generating new solid forms of APIs. Complexation of $AgNO_3$ and 4-aminosalicylic acid in solution provides a hydrated product as a one-dimensional polymer based on O–H···O hydrogen bonds and Ag···O coordination bonds (Figure 8.21(a)). In contrast, mechanochemical synthesis provides an anhydrous product consisting of two-dimensional sheets held via Ag···O and Ag···N coordination bonds as well as O–H···O hydrogen bonds (Figure 8.21(b)). The anhydrous material can be recrystallised from acetone without converting to the one-dimensional hydrate polymer, most likely and importantly due to surface templating caused by sub-micrometre seeds that are present after dissolution of the bulk material.

The application of transition metal ions to the synthesis of pharmaceutical derivatives is hindered by the toxicity associated with many transition metal ions. As a potential alternative, Chow *et al.* have proposed the synthesis of API coordination complexes involving the use of the biologically benevolent magnesium ions (Figure 8.22(a)).[25] Spontaneous formation of metal–organic complexes upon ageing of mixtures involving the excipient magnesium oxide and carboxylic acid drugs was originally investigated by Byrn *et al.*[116] The same type of reactivity was subsequently applied in a more rapid mechanochemical process to synthesise several derivatives of the non-steroidal anti-inflammatory drugs (NSAIDs) *S*- and *RS*-ibuprofen, salicylic acid and naproxen (Figure 8.22(b)–(f)).[25,26] In particular, the reactions involving either the chiral or the

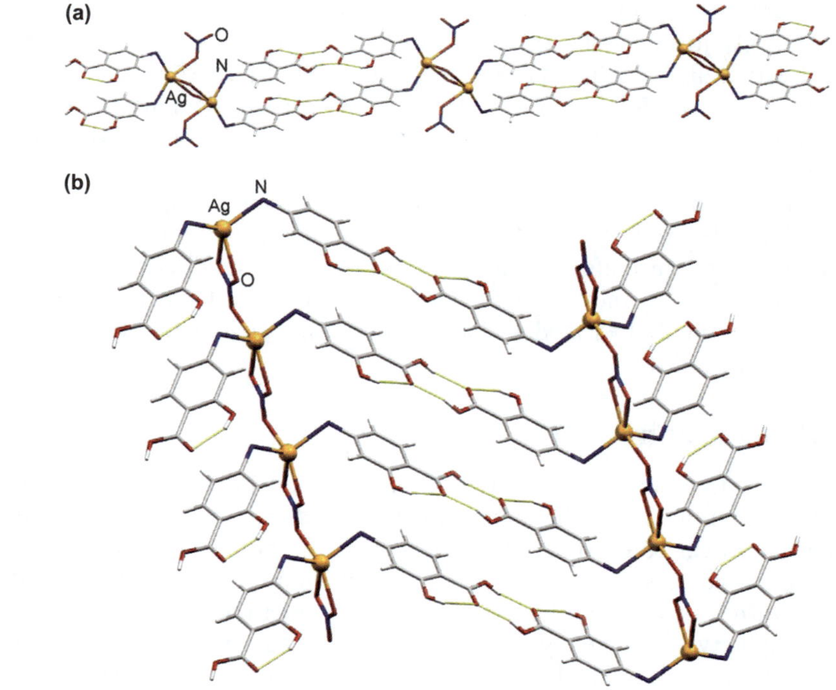

Figure 8.21 Fragments of crystal structures of complexes involving silver nitrate and 4-aminosalicylic acid: (a) the monohydrate 1-D coordination polymer and (b) a 2-D sheet of the anhydrous compound.[24]

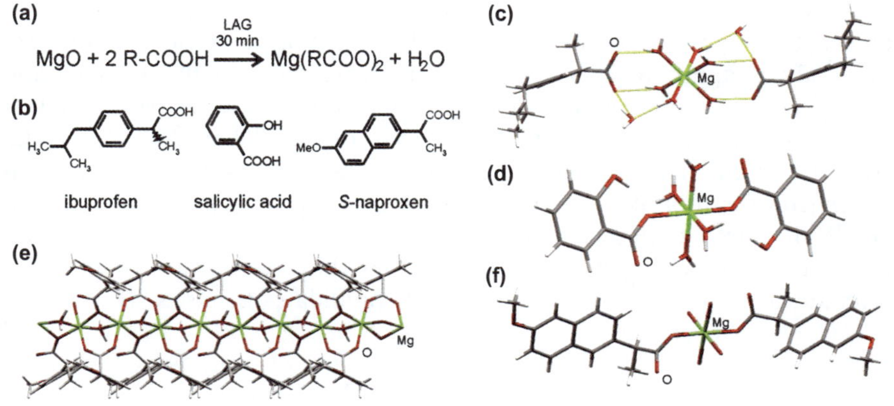

Figure 8.22 (a) Mechanochemical reactivity between the excipient MgO and carboxylic acid NSAID molecules,[25,26] (b) NSAID molecules explored in mechanochemical reactions with MgO, (c) fragment of the crystal structure of mechanochemically obtained magnesium ibuprofen, (d) fragment of the crystal structure of mechanochemically obtained magnesium salicylate, (e) fragment of the crystal structure of mechanochemically obtained magnesium naproxen monohydrate and (f) fragment of the crystal structure of mechanochemically obtained magnesium naproxen tetrahydrate.

racemic form of ibuprofen provided an explanation for the greater activity of ibuprofen when formulated with MgO, due to the formation of a metal–organic material with a higher solubility than the neutral NSAID.

A subsequent study[26] by Strobridge and co-workers demonstrated how variation of water content in mechanochemical reactions involving MgO and naproxen can be exploited to investigate the phase space of solid forms of magnesium naproxen systematically. In particular, low amounts of water in the grinding liquid led to the formation of a one-dimensional coordination polymer monohydrate, intermediate amounts of water provided the discrete units of a tetrahydrate complex and, finally, grinding with pure water resulted in the formation of a highly hydrated salt which was formulated as the octahydrate (Figure 8.22(e) and (f)).

8.9 Conclusions

We have attempted to provide a brief but nevertheless broad overview of the diverse applications of mechanochemical synthesis in research areas related to pharmaceutical co-crystals. We believe that the best way to summarise such a contribution is by stressing the advantages of mechanochemical liquid-assisted synthesis over the other co-crystallisation methodologies in terms of speed, simplicity, product crystallinity, control of co-crystal stoichiometric composition, as well as in the ability to generate co-crystal materials independent of the relative solubilities of the co-crystal components. Beyond these synthetic and screening advantages, however, mechanochemistry also brings considerable improvements over solution- or thermally-based co-crystallisation approaches in the context of environmentally friendly methodologies. Recent work in mechanosynthesis of small organic molecules[117] and metal–organic materials[5] suggests that mechanochemical methods can bring about approximately 10000-fold improvements in the solvent- and energy-usage. We believe that this advantage, along with the growing availability of characterisation methods based on solid-state NMR, crystal structure prediction or powder X-ray diffraction, will make mechanochemistry the tool of choice in the development of pharmaceutical materials. While such development is already taking place in the context of new pharmaceutical forms, we look forward to the broader use of covalent mechanochemistry[118,119] to synthesise API candidates themselves.

Acknowledgements

We are extremely grateful to the Pfizer Institute for Pharmaceutical Materials Science for support. Tomislav Friščić thanks the Herchel Smith Fund for the award of a Fellowship. Numerous present and former members of our group have contributed to various aspects of the work described here. They and colleagues at Pfizer Laboratories in Sandwich (UK) and Groton (USA) are also thanked for their substantial contributions.

References

1. A. V. Trask and W. Jones, *Topics Curr. Chem.*, 2005, **254**, 41–70.
2. T. Friščić and W. Jones, *Cryst. Growth Des.*, 2009, **9**, 1621–1637.
3. N. Schultheiss and A. Newman, *Cryst. Growth Des.*, 2009, **9**, 2950–2967.
4. A. V. Trask, D. A. Haynes, W. D. S. Motherwell and W. Jones, *Chem. Commun.*, 2006, 51–53.
5. T. Friščić, *J. Mater. Chem.*, 2010, **20**, 7599–7605.
6. T. Friščić and W. Jones, *J. Pharm. Pharmacol.*, 2010, **62**, 1547–1559.
7. Sonication is also considered to be a mechanochemical methodology, see J. M. J. Paulusse and R. P. Sijbesma, *Angew. Chem. Int. Ed.*, 2004, **116**, 4560–4562.
8. A. V. Trask, N. Shan, W. D. S. Motherwell, W. Jones, S. Feng, R. B. H. Tan and K. J. Carpenter, *Chem. Commun.*, 2005, 880–882.
9. V. R. Pedireddi, W. Jones, A. P. Chorlton and R. Docherty, *Chem. Commun.*, 1996, 986–987.
10. J. F. Willart and M. Descamps, *Mol. Pharmaceutics*, 2008, **5**, 905–920.
11. M. C. Etter, S. M. Reutzel and C. G. Choo, *J. Am. Chem. Soc.*, 1993, **115**, 4411–4412.
12. D. Cinčić, T. Friščić and W. Jones, *CrystEngComm.*, 2011, **13**, 3224–3231.
13. P. J. Nichols, C. L. Raston and J. W. Steed, *Chem. Commun.*, 2001, 1062–1063.
14. T. Friščić, R. W. Lancaster, L. Fábián and P. G. Karamertzanis, *Proc. Natl. Acad. Sci.*, 2010, **107**, 13216–13221.
15. T. Friščić, A. V. Trask, W. Jones and W. D. S. Motherwell, *Angew. Chem. Int. Ed.*, 2006, **45**, 7546–7550.
16. N. Shan, F. Toda and W. Jones, *Chem. Commun.*, 2002, 2372–2373.
17. S. A. Myz, T. P. Shakhtschneider, K. Fucke, A. P. Fedotov, E. V. Boldyreva, V. V. Boldyrev and N. I. Kuleshova, *Mendeleev Commun.*, 2009, **19**, 272–274.
18. D. R. Weyna, T. Shattock, P. Vishweshwar and M. J. Zaworotko, *Cryst. Growth Des.*, 2009, **9**, 1106–1123.
19. S. L. Childs, N. Rodríguez-Hornedo, L. S. Reddy, A. Jayasankar, C. Maheshwari, L. McCausland, R. Shipplett and B. C. Stahly, *CrystEngComm.*, 2008, **10**, 856–864.
20. S. Karki, T. Friščić and W. Jones, *CrystEngComm.*, 2009, **11**, 470–481.
21. S. Karki, T. Friščić, L. Fábián and W. Jones, *CrystEngComm.*, 2010, **12**, 4038–4041.
22. V. André, D. Braga, F. Grepioni and M. T. Duarte, *Cryst. Growth Des.*, 2009, **9**, 5108–5116.
23. D. Braga, F. Grepioni, L. Maini, R. Brescello and L. Cotarca, *CrystEngComm.*, 2008, **10**, 469–471.
24. D. Braga, F. Grepioni, V. André and M. T. Duarte, *CrystEngComm.*, 2009, **11**, 2618–2621.

25. E. H. H. Chow, F. C. Strobridge and T. Friščić, *Chem. Commun.*, 2010, **46**, 6368–6370.
26. T. Friščić, I. Halasz, F. C. Strobridge, R. E. Dinnebier, R. S. Stein, L. Fábián and C. Curfs, *CrystEngComm.*, 2011, **13**, 3125–3129.
27. S. Karki, L. Fábián, T. Friščić and W. Jones, *Org. Lett.*, 2007, **9**, 3133–3136.
28. S. H. Lapidus, P. W. Stephens, K. K. Arora, T. R. Shattock and M. J. Zaworotko, *Cryst. Growth Des.*, 2010, **10**, 4630–4637.
29. Z. J. Li, Y. Abramov, J. Bordner, J. Leonard, A. Medek and A. V. Trask, *J. Am. Chem. Soc.*, 2006, **128**, 8199–8210.
30. E. Salager, G. M. Day, R. S. Stein, C. J. Pickard, B. Elena and L. Emsley, *J. Am. Chem. Soc.*, 2010, **132**, 2564–2566.
31. K. L. Nguyen, T. Friščić, G. M. Day, L. F. Gladden and W. Jones, *Nature Mater.*, 2007, **6**, 206–209.
32. J. A. Zeitler, K. Kogermann, J. Rantanen, T. Rades, P. F. Taday, M. Pepper, J. Aaltonen and C. J. Strachan, *Int. J. Pharm.*, 2007, **334**, 78–84.
33. R. K. Harris, S. Cadars, L. Emsley, J. R. Yates, C. J. Pickard, R. K. R. Jetti and U. J. Griesser, *Phys. Chem. Chem. Phys.*, 2007, **9**, 360–368.
34. M. R. Caira, L. R. Nassimbeni and A. F. Wilderwanck, *J. Chem. Soc. Perkin Trans.*, 1995, **2**, 2213–2216.
35. K. Hoogsteen, *Acta Cryst.*, 1963, **16**, 907–916.
36. D. Braga, S. L. Giaffreda, M. Curzi, L. Maini, M. Polito and F. Grepioni, *J. Therm. Anal. Cal.*, 2007, **90**, 115–123.
37. T. Friščić, S. L. Childs, S. A. A. Rizvi and W. Jones, *CrystEngComm.*, 2009, **11**, 418–426.
38. R. A. Chiarella, R. J. Davey and M. L. Peterson, *Cryst. Growth Des.*, 2007, **7**, 1223–1226.
39. K. Chadwick, R. Davey, G. Sadiq, W. Cross and R. Pritchard, *CrystEng-Comm.*, 2009, **11**, 412–414.
40. N. Rodríguez-Hornedo, S. J. Nehm, K. F. Seefeldt, Y. Pagán-Torres and C. J. Falkiewicz, *Mol. Pharmaceutics*, 2006, **3**, 362–367.
41. D. J. Good and N. Rodríguez-Hornedo, *Cryst. Growth Des.*, 2009, **9**, 2252–2264.
42. D. J. Good and N. Rodríguez-Hornedo, *Cryst. Growth Des.*, 2010, **10**, 1028–1032.
43. S. Karki, T. Friščić, W. Jones and W. D. S. Motherwell, *Mol. Pharmaceutics*, 2007, **4**, 347–354.
44. S. L. Childs and K. I. Hardcastle, *Cryst. Growth Des.*, 2007, **7**, 1291–1304.
45. A. J. Cruz-Cabeza, S. Karki, T. Friščić, G. M. Day and W. Jones, *Chem. Commun.*, 2010, **46**, 2224–2226.
46. T. Friščić, L. Fábián, J. C. Burley, D. G. Reid, M. J. Duer and W. Jones, *Chem. Commun.*, 2008, 1644–1646.
47. A. J. Cruz-Cabeza, G. M. Day, W. D. S. Motherwell and W. Jones, *J. Am. Chem. Soc.*, 2006, **128**, 14466–14467.
48. S. Basavoju, D. Boström and S. P. Velaga, *Pharm. Res.*, 2008, **25**, 530–541.

49. M. K. Stanton, S. Tufekcic, C. Morgan and A. Bak, *Cryst. Growth Des.*, 2009, **9**, 1344–1352.
50. A. Bak, A. Gore, E. Yanez, M. Stanton, S. Tufekcic, R. Syed, A. Akrami, M. Rose, S. Surapaneni, T. Bostick, A. King, S. Neervannan, D. Ostovic and A. Koparkar, *J. Pharm. Sci.*, 2008, **97**, 3942–3956.
51. L. Orola and M. V. Veidis, *CrystEngComm.*, 2009, **11**, 415–417.
52. A. V. Trask, W. D. S. Motherwell and W. Jones, *Chem. Commun.*, 2004, 890–891.
53. A. V. Trask, J. van de Streek, W. D. S. Motherwell and W. Jones, *Cryst. Growth Des.*, 2005, **5**, 2233–2241.
54. M. C. Etter and J. C. MacDonald, *Acta Crystallogr. B*, 1990, **46**, 256–262.
55. P. Vishweshwar, A. Nangia and V. M. Lynch, *Cryst. Growth Des.*, 2003, **3**, 783–790.
56. Different co-crystal stoichiometries can be predicted, but controlling their formation in solution requires the knowledge of the phase diagram of the system: (a) A. J. Cruz-Cabeza, G. M. Day and W. Jones, *Chem. Eur. J.*, 2008, **14**, 8830–8836; (b) C. C. Seaton and A. Parkin, C. C. Wilson and N. Blagden, *Cryst. Growth Des.*, 2009, **9**, 47–56.
57. C. B. Aakeröy, M. Fasulo, N. Schultheiss, J. Desper and C. Moore, *J. Am. Chem. Soc.*, 2007, **129**, 13772–13773.
58. C. B. Aakeröy, N. Schultheiss, J. Desper and C. Moore, *Cryst. Growth Des.*, 2007, **7**, 2324–2331.
59. C. B. Aakeröy, A. M. Beatty and B. A. Helfrich, *J. Am. Chem. Soc.*, 2002, **124**, 14425–14432.
60. L. S. Reddy, S. K. Chandran, S. George, N. J. Babu and A. Nangia, *Cryst. Growth Des.*, 2007, **7**, 2675–2690.
61. N. J. Babu, L. S. Reddy and A. Nangia, *Mol. Pharmaceutics*, 2007, **4**, 417–434.
62. A. V. Trask, W. D. S. Motherwell and W. Jones, *Int. J. Pharm.*, 2006, **320**, 114–123.
63. A. V. Trask, W. D. S. Motherwell and W. Jones, *Cryst. Growth Des.*, 2005, **5**, 1013–1021.
64. T. Friščić and W. Jones, *Faraday Discuss.*, 2007, **136**, 167–178.
65. D. J. Berry, C. C. Seaton, W. Clegg, R. W. Harrington, S. J. Coles, P. N. Horton, M. B. Hursthouse, R. Storey, W. Jones, T. Friščić and N. Blagden, *Cryst. Growth Des.*, 2008, **8**, 1697–1717.
66. S. Karki, T. Friščić, L. Fábián, P. R. Laity, G. M. Day and W. Jones, *Adv. Mater.*, 2009, **21**, 3905–3909.
67. S. Karki, T. Friščić, L. Fábián and W. Jones, *CrystEngComm.*, 2010, **12**, 4038–4041.
68. L. Fábián, *Cryst. Growth Des.*, 2009, **9**, 1436–1443.
69. J. Fayos, *Cryst. Growth Des.*, 2009, **9**, 3142–3153.
70. I. A. Tumanov, A. F. Achkasov, E. V. Boldyreva and V. V. Boldyrev, *CrystEngComm.*, 2011, **13**, 2213–2216.
71. G. Rothenberg, A. P. Downie, C. L. Raston and J. L. Scott, *J. Am. Chem. Soc.*, 2001, **123**, 8701–8708.

72. G. Kaupp, *CrystEngComm.*, 2003, **5**, 117–133.
73. G. Kaupp, J. Schmeyers and U. D. Hangen, *J. Phys. Org. Chem.*, 2002, **15**, 307–313.
74. G. Kaupp, *CrystEngComm.*, 2006, **8**, 794–804.
75. R. P. Rastogi, P. S. Bassi and L. S. Chadha, *J. Phys. Chem.*, 1963, **67**, 2569–2573.
76. Y. Imai, N. Tajima, T. Sato and R. Kuroda, *Org. Lett.*, 2006, **8**, 2941–2944.
77. R. Kuroda, Y. Imaiv and N. Tajima, *Chem. Commun.*, 2002, 2848–2849.
78. F. Toda, M. Senzakiv and R. Kuroda, *Chem. Commun.*, 2002, 1788–1789.
79. Y. Imai, N. Tajima, T. Satov and R. Kuroda, *Chirality*, 2002, **14**, 604–609.
80. E. Y. Cheung, S. J. Kitchin, K. D. M. Harris, Y. Imai, N. Tajima and R. Kuroda, *J. Am. Chem. Soc.*, 2003, **125**, 14658–14659.
81. R. Kuroda, K. Higashiguchi, S. Hasebe and Y. Imai, *CrystEngComm.*, 2004, **6**, 463–468.
82. R. P. Rastogi and N. B. Singh, *J. Phys. Chem.*, 1968, **72**, 4446–4449.
83. R. P. Rastogi and N. B. Singh, *J. Phys. Chem.*, 1966, **70**, 3315–3324.
84. E. Lu, N. Rodríguez-Hornedo and R. Suryanarayanan, *CrystEngComm.*, 2008, **10**, 665–668.
85. K. Chadwick, R. J. Davey and W. Cross, *CrystEngComm.*, 2007, **9**, 732–734.
86. H. H. Lee and J. C. Warner, *J. Am. Chem. Soc.*, 1933, **55**, 209–214.
87. C. Brassy and J. Mornon, *C. R. Seances Acad. Sci. Ser. C*, 1972, **274**, 1728–1730.
88. A. Jayasankar, A. Somwangthanaroj, Z. J. Shao and N. Rodríguez-Hornedo, *Pharm. Res.*, 2006, **23**, 2381–2392.
89. A. Jayasankar, D. J. Good and N. Rodríguez-Hornedo, *Mol. Pharmaceutics*, 2007, **4**, 360–372.
90. C. Maheshwari, A. Jayasankar, N. A. Khan, G. E. Amidon and N. Rodríguez-Hornedo, *CrystEngComm.*, 2009, **11**, 493–500.
91. A. De Gusseme, C. Neves, J. F. Willart, A. Rameau and M. Descamps, *J. Pharm. Sci.*, 2008, **97**, 5000–5012.
92. N. Dujardin, J. F. Willart, E. Dudognon, A. Hédoux, Y. Guinet, L. Paccou, B. Chazallon and M. Descamps, *Solid State Comm.*, 2008, **148**, 78–82.
93. K. J. Crowley and G. Zografi, *J. Pharm. Sci.*, 2002, **91**, 492–507.
94. M. Otsuka, T. Matsumoto and N. Kaneniwa, *Chem. Pharm. Bull.*, 1986, **34**, 1784–1793.
95. N. Chieng, M. Hubert, D. Savillev T. Rades and J. Aaltonen, *Cryst. Growth Des.*, 2009, **9**, 2377–2386.
96. C. Medina, D. Daurio, K. Nagapudiv and F. Alvarez-Nunez, *J. Pharm. Sci.*, 2010, **99**, 1693–1696.
97. D. Cinčić, T. Friščić and W. Jones, *J. Am. Chem. Soc.*, 2008, **130**, 7524–7525.

98. D. Cinčić, T. Friščić and W. Jones, *Chem. Eur. J.*, 2008, **14**, 747–753.
99. C. B. Aakeröy, M. Fasulo, N. Schultheiss, J. Desper and C. Moore, *J. Am. Chem. Soc.*, 2007, **129**, 13772–13773.
100. C. B. Aakeröy, J. Desper, B. Helfrich, P. Metrangolo, T. Pilati, G. Resnati and A. Stevenazzi, *Chem. Commun.*, 2007, 4236–4238.
101. C. B. Aakeröy, N. Schultheiss, J. Desper and C. Moore, *CrystEngComm.*, 2007, **9**, 421–426.
102. M. Khan, V. Enkelmann and G. Brunklaus, *J. Am. Chem. Soc.*, 2010, **132**, 5254–5263.
103. E. Székely, G. Bánsághi, P. Thorey, P. Molnár, J. Madarász, L. Vida and B. Simándi, *Ind. Eng. Chem. Res.*, 2010, **49**, 9349–9354.
104. T. Friščić, L. Fábián, J. C. Burley, W. Jones and W. D. S. Motherwell, *Chem. Commun.*, 2006, 5009–5011.
105. D. Braga, F. Grepioni and G. I. Lampronti, *CrystEngComm.*, 2011, **13**, 3122–3124.
106. H. D. Clarke, K. K. Arora, H. Bass, P. Kavuru, T. T. Ong, T. Pujari, L. Wojtas and M. J. Zaworotko, *Cryst. Growth Des.*, 2010, **10**, 2152–2167.
107. T. Friščić, A. V. Trask, W. D. S. Motherwell and W. Jones, *Cryst Growth Des.*, 2008, **8**, 1605–1609.
108. N. Shan and W. Jones, *Green Chem.*, 2003, **5**, 728–730.
109. J. S. Clawson, F. G. Vogt, J. Brum, J. Sisko, D. B. Patience, W. Dai, S. Sharpe, A. D. Jones, T. N. Pham, M. N. Johnson and R. C. P. Copley, *Cryst. Growth Des.*, 2008, **8**, 4120–4131.
110. Y. Andemichael, J. Chen, J. S. Clawson, W. Dai, A. Diederich, S. V. Downing, A. J. Freyer, P. Liu, L. M. Oh, D. B. Patience, S. Sharpe, J. Sisko, J. Tsui, F. G. Vogt, J. Wang, L. Wernersbach, E. C. Webb, J. Wertman and L. Zhou, *Org. Process Res. Dev.*, 2009, **13**, 729–738.
111. J. S. Stevens, S. J. Byard and S. L. M. Schroeder, *Cryst. Growth Des.*, 2010, **10**, 1435–1442.
112. R. Ge and H. Sun, *Acc. Chem. Res.*, 2007, **40**, 267–274.
113. Z. Guo and P. J. Sadler, *Angew. Chem. Int. Ed.*, 1999, **38**, 1512–1531.
114. Z. Ma and B. Moulton, *Cryst. Growth Des.*, 2007, **7**, 196–198.
115. Z. Ma and B. Moulton, *Mol. Pharmaceutics*, 2007, **4**, 373–385.
116. S. R. Byrn, W. Xu and A. W. Newman, *Adv. Drug Deliv. Rev.*, 2001, **48**, 115–136.
117. T. Szuppa, A. Stolle, B. Ondruschka and W. Hopfe, *Green Chem.*, 2010, **12**, 1288–1294.
118. A. M. Belenguer, T. Friščić, G. M. Day and J. K. M. Sanders, *Chem. Sci.*, 2011, **2**, 696–700.
119. A. Bruckmann, A. Krebs and C. Bolm, *Green Chem.*, 2008, **10**, 1131–1141.

CHAPTER 9

Co-crystallization in Solution and Scale-up Issues

E. GAGNIÈRE,[a] D. MANGIN,[a] S. VEESLER[b] AND F. PUEL*[a]

[a] Université de Lyon F-69622, Lyon, France; Université Lyon 1; Villeurbanne; CNRS, UMR5007, Laboratoire d'Automatique et de Génie des Procédés (LAGEP), CPE-Lyon, 43 bd du 11 Novembre 1918, 69100 Villeurbanne, France; [b] CNRS, Aix-Marseille University, CINaM (Centre Interdisciplinaire de Nanoscience de Marseille), Campus de Luminy, Case 913, 13288 Marseille Cedex 09, France

9.1 Introduction

Co-crystallization is an increasingly popular research topic in the pharmaceutical industry, since it offers the opportunity to improve and tailor the physico-chemical properties of pharmaceuticals.[1] Moreover the contribution of a generic pharmaceutical company to the discovery of other solid forms of active pharmaceutical ingredients (APIs) such as salts and co-crystals is becoming more habitual when the basic molecule patent has expired. Pharmaceutical companies would like to protect their new branded pharmaceutical product when it reaches the market and to acquire extra years for the product beyond the expiry of the basic molecule patent by patenting new solid forms.[2] This is why the screening of all possible solid phases has now become compulsory in the earlier stages of the pharmaceutical development process. When a co-crystal form is chosen from among the different solid forms identified, the development of the co-crystallization process is begun on the lab scale in a thermostatted

RSC Drug Discovery Series No. 16
Pharmaceutical Salts and Co-crystals
Edited by Johan Wouters and Luc Quéré

Published by the Royal Society of Chemistry, www.rsc.org

mechanically stirred reactor, since it is representative of the most popular apparatus used in the chemical and pharmaceutical industry. The co-crystallization process is carried out in batch mode in order to meet the requirements stipulated by the regulatory authorities in the manufacture of API solid forms.

In this chapter we focus on the co-crystallization process in organic compounds. Even though it is possible to obtain co-crystals using mechanochemical methods, melt or ultrasonic techniques, the industrial manufacture of organic co-crystals is carried out in solution for economical reasons. The main mechanisms involved in crystallization of a mono-component crystal also act in the co-crystallization of a multi-component crystal. This is why we first recall these concepts and phenomena in solution. Mediated solid phase transition is also introduced since it may be involved. Preferential formation of a co-crystal may depend on the different solid phases which might appear and compete against each others during the experiments. This is why thermodynamic (phase diagram) and kinetic (phase transition) aspects of co-crystal formation are essential. These two aspects are discussed and illustrated with case studies. The final section deals with the development of a co-crystallization process to obtain a given co-crystal. Prerequisites are recalled and experimental techniques used for the screening of co-crystals are presented. This section ends by describing how to define the operating conditions in batch mode.

9.2 Concepts and Phenomenon Involved in a Co-crystallization Process in Solution

9.2.1 Supersaturation

Supersaturation is the driving force in nucleation and growth. After dissolving the chemical species in a solvent, whether or not of a predetermined nature, the solution must be supersaturated in order to observe nucleation or growth. Supersaturation is the difference between the chemical potential of the solute molecules in the supersaturated (μ) and saturated (μ_s) states, respectively. For one molecule the expression of this difference is:

$$\Delta\mu = \mu - \mu_s = kT \ln \beta \tag{1}$$

where k is the Boltzmann constant and T is the temperature. For simplification and in order to introduce the activities, we write here without specifying units:

$$\beta = C_i / C_s \tag{2}$$

where β is the supersaturation ratio, C_i is the concentration of the solute in solution and C_s its saturated concentration. Obviously, this ratio is dimensionless. Moreover, if $\beta > 1$, the crystals grow, if $\beta < 1$ the crystals dissolve, and if $\beta = 1$ crystals and solution are at the equilibrium.

9.2.2 Nucleation

When a solution is supersaturated, the solid phase forms more or less rapidly depending on the growth conditions: temperature, supersaturation, medium (chemical conditions) and hydrodynamics. Primary nucleation occurs in a solution that is clear, without crystals of its own type. It is called homogeneous nucleation if the nuclei form in the bulk of the solution. It is called heterogeneous if the nuclei preferentially form on substrates such as the wall of the crystallizer, the stirrer, or solid particles (such as dust particles or other solid phases). Conversely, secondary nucleation is induced by the presence of existing crystals of the same phase.

The nucleation rate or nucleation frequency, J, is the number of crystals that form in a supersaturated solution per unit of time and unit of volume.[3–6] Here, we only need to recall that:

$$J = nN_0\nu \exp\left(-\frac{f\Omega^2\gamma^3}{(kT)^3\ln^2\beta}\right) \qquad (3)$$

where f is the nuclei form factor and γ is the crystal-solution interfacial free energy (J m^{-2}).

J is proportional to n times the solubility expressed in number of molecules per unit of volume, N_0. ν is the frequency with which nuclei of critical size r^* become supercritical by addition of a molecule and develop into crystals. The term $nN_0\nu$ can be simply described as a pre-exponential factor K_o.

Equation (9.3) shows that the frequency of nucleation depends not only on the supersaturation β but also on the concentration of molecules nN_0. All things being equal, supersaturation included, the higher the probability of intermolecular contact, the easier is nucleation. Systems with high solubility, and thus low γ, meet this condition. For systems with low solubility, the solute molecules are separated by larger distances and by a greater number of solvent molecules. The probability that the molecules will come into contact and form a nucleus is thus lower.

9.2.3 Crystal Growth

Once the nuclei are formed and exceed the critical size, they become crystals; hereafter, we recall the basic principles of crystal growth.

9.2.3.1 Growth Form

A crystal is limited by its faces. The set of equivalent faces resulting from the crystal symmetry is called a crystalline form in crystallography. All the forms present in a crystal represent the morphology of the crystal. However, the concept of morphology alone does not fully cover the external form of the crystal, which is contained in the notion of crystal habit. The concept of habit includes the notion of face extension. However, it is important to emphasize that the growth form of

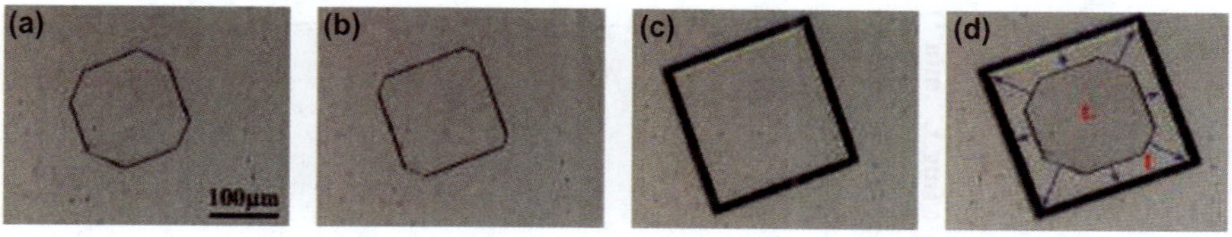

Figure 9.1 Growth of BPTI (bovine pancreatic trypsin inhibitor) protein crystal in 350 mM KSCN at pH = 4.9. (a)–(c) Frames of a time sequence obtained at different temperatures showing the evolution of the growth form as illustrated in (d), in which arrows indicate the face displacement with time (reproduced with permission from Astier and Veesler, ref. 7).

the crystal only includes the faces with the slowest growth rates. This is shown by Figure 9.1 which represents the growth of a protein crystal; experimental details are given by Astier and Veesler.[7] From time t_0 to time t, all the faces will have migrated parallel to themselves and crossed distances proportional to their growth rates (vectors in Figure 9.1(d)). Obviously, the growth form is different at time t and time t_0. The slowest faces develop at the expense of the fastest faces, which entirely disappear. Moreover, some very slow faces appear because their growth rates are slower than the others. The growth form thus depends on kinetic factors, that is to say, crystallization conditions.

9.2.3.2 Growth Medium and Kinetics

Growth kinetics and mechanisms depend on external factors (medium, temperature, supersaturation, and hydrodynamics) and on internal factors (structure, bonds and defects). The growth medium influences the growth kinetics of the faces in different ways. First of all, the solvent is more or less adsorbed by the faces and selectively slows down their growth rates. Solubility also plays a role: the higher the solubility, the higher the growth rate. The growth medium also influences solvation, desolvation, and complex formation.[8] If not predetermined by the process, variations in temperature also produce extremely different growth rates. Lastly, hydrodynamics, or more precisely the relative velocity of the solution compared to the crystal, is an important parameter.[9] When the solution is quiescent, the face grows slowly at a rate determined by the molecular diffusion of the solute towards the crystal. The growth rate of the face increases with the flow velocity of solution to the crystal. However, there is still a diffusion limitation; this growth rate tends very quickly towards a plateau and thus reaches an upper limit determined by the phenomenon of solute integration in the crystal lattice at the crystal surface.

9.2.4 Dissolution

Dissolution is a phenomenon that occurs when the crystal is located in an undersaturated solution. There is a loss of species (molecules, ions or atoms

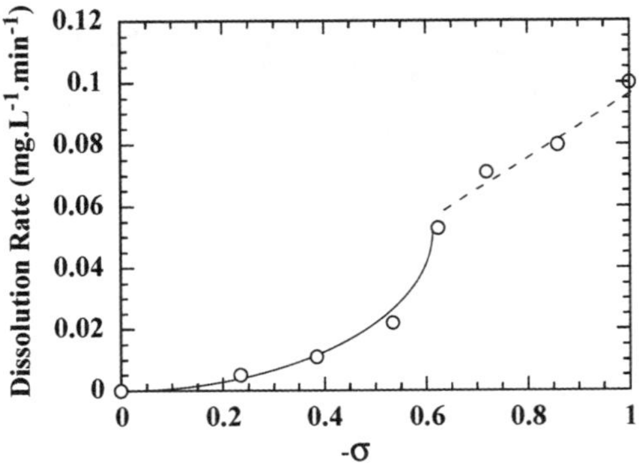

Figure 9.2 Dissolution rate of 5 mg L^{-1} of irbesartan form A as a function of undersaturation (denoted $-\sigma$) at 40 °C and at $N = 300$ rpm (reproduced with permission from Garcia *et al.*, ref. 13).

depending on the nature of the crystal) from the crystal lattice, leading to a decrease in the particle size and to an increase in the solute concentration. As in the case of crystal growth, dissolution globally involves two main steps: (i) surface reaction and detachment of the surface species followed by (ii) mass transfer of this species towards the bulk solution across the diffusion layer which surrounds the crystals.[10,11] Like crystal growth rate, the dissolution rate is controlled by the slowest step. The dissolution kinetics of a readily soluble compound is limited by mass transfer (step (ii)), while the dissolution of a sparingly soluble compound is found to be controlled by the kinetics of the events occurring at the crystal surface (step (i)). Moreover, for a mineral and an API, several authors have evidenced different behaviours depending on undersaturation.[12,13] The initial dissolution rate was found to vary linearly with the initial undersaturation at high undersaturation, and quadratically at lower undersaturation (Figure 9.2). The authors concluded that the mechanism changed from a diffusion-controlled process at high undersaturation to a process controlled by a surface mechanism at lower undersaturation. Moreover, disaggregation can occur during the initial stage of dissolution leading to an acceleration of the overall dissolution kinetics at the beginning of the dissolution process.[14]

9.2.5 Metastable Phases and their Transition: Solution-mediated Phase Transition (SMPT)

Unstable phases can nucleate for kinetic reasons. These unstable phases may stay in a metastable state for a few seconds or several centuries. The transformation of a metastable phase into a stable phase, corresponding to the minimal free enthalpy of the system, is called phase transition.

During solid processing like crystallization in solution, the presence of a liquid phase surrounding the crystals often promotes phase transition phenomena. This modification in the presence of a solvent is called solution-mediated phase transition (SMPT). In the case of a system presenting a co-crystal phase, the co-crystal phase can be either a metastable form or a stable form, depending on the domain in the phase diagram in which the co-crystallization process is carried out (see Section 9.3.1).

The basic phenomena involved in SMPT have already been described.[15] For a system consisting of a metastable phase and a stable phase, this transformation requires at least three mechanisms:

- primary nucleation, often heterogeneous, of the more stable solid phase (this step can be replaced by seeding the stable solid phase) and growth of both phases until solubility of the metastable phase is reached,
- dissolution of the metastable solid phase,
- and growth of the more stable solid by mass transfer of solute in the solution until solubility of the stable phase is reached.

These three mechanisms are consecutive or concomitant. The primary nucleation of the stable solid phase, or its seeding, is thus the trigger for a solid-mediated phase transition in a stirred crystallizer. Primary nucleation may occur on the surface of a substrate such as, for example, a homologous impurity or crystals of the metastable form.[16,17] Many authors have reported the heterogeneous nucleation of β-L-glutamic acid (stable phase) on the faces of α-L-glutamic acid (metastable phase) during the crystallization of α-L-glutamic acid.[18–20]

Once nucleation starts, the growth of the stable phase induces a decrease in concentration in solution. If this concentration becomes lower than the solubility of the metastable form, the latter dissolves, thus promoting the growth of the more stable form. This dissolution and growth process is often revealed through a concentration plateau positioned between the solubility of the metastable and stable forms denoted C^*_1 and C^*_2, respectively in Figure 9.3(a). The position of this plateau results from the competing kinetics of dissolution and growth. Two extreme cases are possible (Figure 9.3(b)). (1) The consumption of solute by growth is slower than the production of solute by dissolution, and the plateau is located in a "high" position, in the vicinity of the solubility of the metastable form C^*_1. Therefore, the growth mechanism of the stable form limits the transition and thus is the rate-controlling step. (2) The concentration plateau is just above the solubility of the stable form C^*_2. The dissolution mechanism of the metastable phase limits the transition and is thus the rate-controlling step.

9.3 Thermodynamic and Kinetic Aspects of a Co-crystallization Process

Two graphical representations of the isothermal phase diagram of the ternary system including the API, the co-crystallizing agent and the solvent exist, with

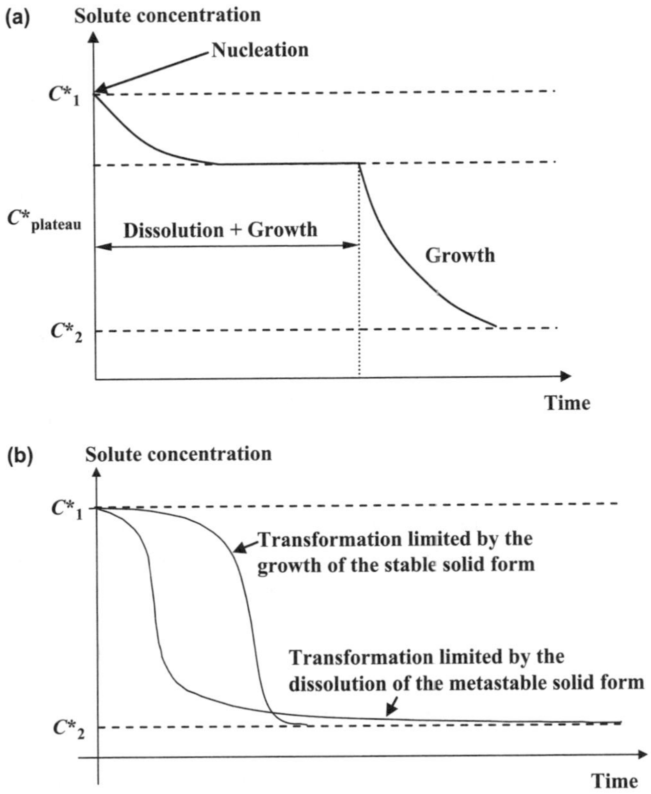

Figure 9.3 Evolution with time of the solute concentration during a solution-mediated phase transition, C^*_1 and C^*_2 are the solubilities of the metastable and stable phases, respectively. (a) General profile, (b) limit profiles.

orthogonal axis[21] and triangular axis,[22,23] respectively. For the sake of clarity, the representation with orthogonal axis will be used in this chapter.

9.3.1 Phase Diagram

Lets us consider a model system largely described in the literature[24–26]: an API molecule (carbamazepine (CBZ)) and one of its possible co-crystallizing agent (nicotinamide (NCT)). CBZ and NCT can give an equimolar (1:1) co-crystal (CBZ/NCT) involving a homosynthon in ethanol.[25] Figure 9.4 is a broad view of the phase diagram in isothermal condition ($T = 25\,°C$). The pink and green dashed lines stand respectively for the solubility of CBZ and NCT in the solvent (EtOH). The CBZ solubility curve was obtained at different NCT concentrations. The NCT solubility was measured only in pure ethanol and we plot a vertical line even though an inclined line is expected. The solubility of such binary co-crystal could be described by a solubility product of co-crystal

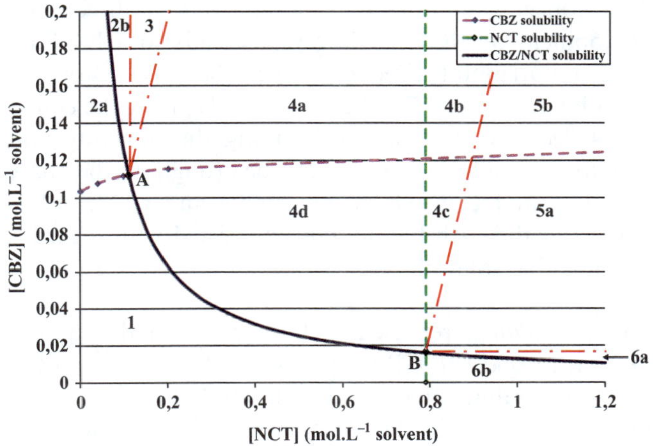

Figure 9.4 Phase diagram of CBZ/NCT system in ethanol at 25 °C. The solid lines are CBZ (horizontal) and NCT (vertical) solubility curves, the dashed line is the CBZ/NCT cocrystal solubility curve (reproduced with permission from Gagniere *et al.*, ref. 27).

components as previously described in the literature (see Chapter 11). The solubility curve of the CBZ/NCT co-crystal is plotted as a solid line. Two invariants-points A and B are respectively obtained at the intersection of the CBZ and NCT solubility curves with the co-crystal solubility curve. This diagram can be obtained at different temperatures (for instance see Gagniere *et al.*[27] for the CBZ/NCT co-crystal and Yu *et al.*[28] for the caffeine/acid glutaric co-crystal).

From the standpoint of thermodynamics, the phase diagram has to be analysed through the possible stable phases which may be present in the suspension. Four additional dotted-dashed lines are added in Figure 9.4. Then the phase diagram could be split in six domains:

- Domain 1: undersaturated clear solution.
- Domain 2: solution in equilibrium with CBZ crystals.
- Domain 3: solution in equilibrium with both CBZ crystals and CBZ/NCT co-crystals; the composition of the solution is given by point A.
- Domain 4: solution in equilibrium with CBZ/NCT co-crystals.
- Domain 5: solution in equilibrium with both NCT crystals and CBZ/NCT co-crystals; the composition of the solution is given by point B.
- Domain 6: solution in equilibrium with NCT crystals.

9.3.2 Relative Stability of the Solid Phases in Solution

During the crystallization in solution, several solid phases may develop separately or simultaneously. The development in solution of CBZ single crystals leads to a CBZ concentration decrease along a vertical direction. In the same way, the development of NCT single crystals leads to a NCT concentration decrease along a horizontal direction. Since there is a stoichiometric consumption of both CBZ and NCT when co-crystals grow, the concentrations of the two species decrease simultaneously along the bisecting line. From the standpoints of kinetics, subdivisions have to be brought to the phase diagram in domains 2, 4, 3, 5 and 6 in which metastable and stable solid forms may temporally exist together. All the domains are listed with regard to the development of the possible solid phases.

- Domain 1: a *solution* represented by a point in this domain is undersaturated with respect to CBZ crystals, NCT crystals and CBZ/NCT co-crystals. No crystalline phase is able to develop.
- Domain 2a: a *solution* in this domain is undersaturated with respect to CBZ/NCT co-crystals and supersaturated with respect to CBZ crystals. Only the CBZ solid phase is able to nucleate and grow.
- Domain 2b: a *solution* in this domain is supersaturated with respect to CBZ crystals and CBZ/NCT co-crystals. CBZ and CBZ/NCT solid phases are able to develop together. With regard to their respective solubilities, the CBZ/NCT co-crystal is the metastable phase whereas the CBZ crystal is the stable one.
- Domain 3: a *solution* in this domain is supersaturated with respect to CBZ crystals and CBZ/NCT co-crystals. CBZ and CBZ/NCT crystalline phases are able to nucleate and grow together. These two phases are both stable in this domain. Note that the solution is also supersaturated with respect to NCT crystals at very high concentration in solution. Thus domain 3 is theoretically divided in two sub-domains (not represented in Figure 9.4). This subdivision also exists in domain 5 (see below).
- Domain 4a: a *solution* in this domain is supersaturated with respect to CBZ crystals and CBZ/NCT co-crystals. These two crystalline phases are able to nucleate and grow together temporarily. With regard to the solubilities, CBZ is the metastable phase, whereas CBZ/NCT is the stable one.
- Domain 4b: a *solution* in this domain is supersaturated with respect to CBZ crystals, NCT crystals and CBZ/NCT co-crystals. These three crystalline phases are able to nucleate and grow together temporarily. With regard to the solubilities, CBZ and NCT crystals are the metastable phases, whereas CBZ/NCT is the stable one.
- Domain 4c: a *solution* in this domain is supersaturated with respect to NCT crystals and CBZ/NCT co-crystals. These two crystalline phases are able to nucleate and grow together temporarily. With regard to the solubilities, NCT is the metastable phase, whereas CBZ/NCT is the stable one.
- Domain 4d: a *solution* in this domain is undersaturated with respect to CBZ and NCT solid phases and supersaturated with respect to the CBZ/NCT solid phase. Only CBZ/NCT co-crystals are able to nucleate and grow.

- Domain 5a: a *solution* in this domain is supersaturated with respect to NCT crystals and CBZ/NCT co-crystals. NCT and CBZ/NCT crystalline phases are able to nucleate and grow together. These two phases are both stable in this domain.
- Domain 5b: a *solution* in this domain is supersaturated with respect to CBZ crystals, NCT crystals and CBZ/NCT co-crystals. CBZ, NCT and CBZ/NCT crystalline phases are able to nucleate and grow together. CBZ is the metastable phase whereas NCT and CBZ/NCT are both stable.
- Domain 6a: a *solution* in this domain is supersaturated with respect to NCT crystals and CBZ/NCT co-crystals. NCT and CBZ/NCT solid phases are able to develop together. But considering their respective solubilities, the CBZ/NCT co-crystal is the metastable phase whereas the NCT crystal is the stable one.
- Domain 6b: a *solution* in this domain is undersaturated with respect to CBZ/NCT co-crystals and supersaturated with respect to NCT crystals. Only the NCT solid phase is able to nucleate and grow.

9.3.3 Kinetic Pathways in a Phase Diagram

The previous section highlighted the importance of the kinetic aspects since the development of co-crystals may be in competition with the crystallization of single component crystals. Knowledge of the start and end points in the phase diagram of a co-crystallization process is not sufficient in order to operate and optimize a co-crystallization pathway. Information about the kinetic pathway followed in the phase diagram during the operation remains essential.

In situ probes are often used in order to provide information about solid phases and to measure the solute concentration during batch solution crystallization experiments.[29–31] For instance, Gagniere *et al.*[27] have qualitatively used an *in situ* video probe to monitor the evolution of single CBZ crystals and CBZ/NCT co-crystals in the slurry since the crystal habits of these two solid phases are largely different in ethanol. Measurement of the concentrations in solution of the two components of the co-crystal phase is very instructive about the competition between the metastable and stable phases. This monitoring is carried out most of the time using mid-infrared spectroscopy in ATR mode which provides estimates with time of the solute concentrations of the chemical components involved in the co-crystallization process. Three case studies taken from the literature are discussed hereinafter in order to present different situations.

9.3.3.1 First Case Study: Isothermal Co-crystallization in CBZ/NCT/Ethanol System (from Gagniere et al.[32]*)*

Experiments were performed in a restricted part of the CBZ/NCT phase diagram, more particularly in domains always undersaturated with respect to NCT. Batch runs were carried out in a thermostatted stirred vessel.

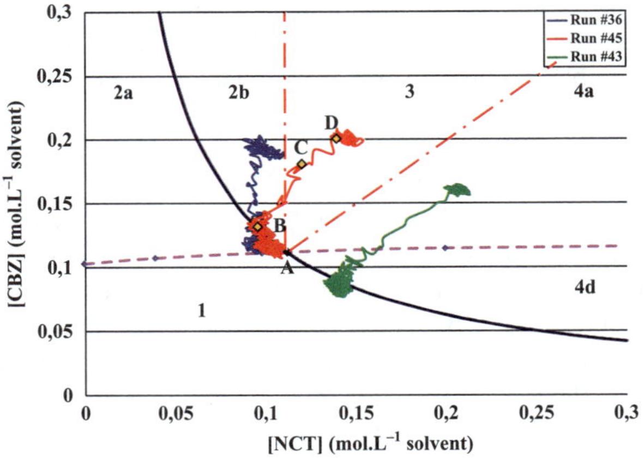

Figure 9.5 Phase diagram of CBZ/NCT system in ethanol at 25 °C. Solute concentration pathways depending on the initial starting point in the phase diagram (reproduced with permission from Gagniere *et al.*, ref. 32).

Undersaturated clear solutions of CBZ and NCT were obtained at 45 °C, and then cooled quickly to 25 °C. An isothermal plateau was then applied until equilibrium was reached. Crystal nucleation and growth occurred only during this isothermal step. Figure 9.5 displays three kinetic pathways in the phase diagram from representative experiments started in three domains where two solid phases may develop.

The first kinetic pathway considered (run #43) was started in domain 4a (Figure 9.5). It remained parallel to the bisecting line in domains 4a and 4d and ended on the co-crystal solubility curve in accordance with the thermodynamics. This pathway corresponds to the stoichiometric consumption of the two molecular species involved in the CBZ/NCT co-crystal growth. The absence of deviation from the bisecting line indicates that the metastable CBZ crystal population was not temporarily present in the slurry at a significant level to be detected by the technique used here. The time variation of the co-crystal supersaturation profile is plotted in Figure 9.6. After a first period for which supersaturation increased in the clear solution during cooling, the sharp decrease in supersaturation down to a value of approximately 1.1 was due to the quick growth of nuclei. In a last stage, the supersaturation decreased slowly since the system progressively reached equilibrium as the co-crystal growth rate declined. This profile is equivalent to the one obtained with a single component crystallized in a "classical" batch process.[33]

The second kinetic pathway considered (run #36) was started in domain 2b. It followed a vertical direction until equilibrium was reached. Only CBZ crystals have nucleated and grow. No significant deviation from the vertical line, which could indicate the temporary presence of co-crystals could be detected. Under similar conditions, the nucleation of CBZ/NCT co-crystals was reported but the kinetic pathway was not monitored.[32] The occurrence of a metastable CBZ/NCT crystal population by primary nucleation would have

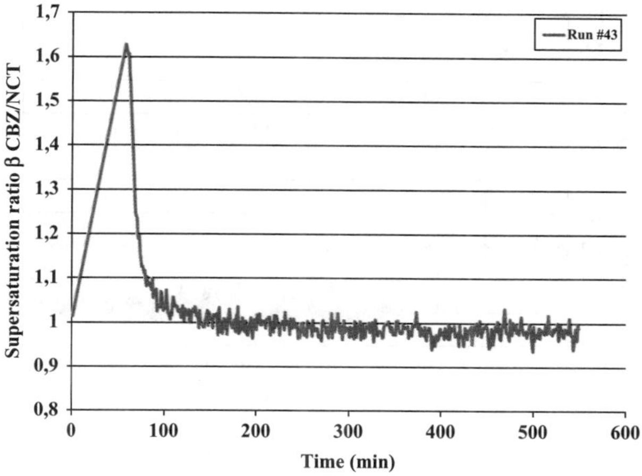

Figure 9.6 Time variations of the co-crystal CBZ/NCT supersaturation ratio for run #43 at 25 °C (reproduced with permission from Gagniere *et al.*, ref. 32).

deviated this pathway towards a lower NCT concentration in domain 2b (temporary growth of the co-crystals until the kinetic pathway crossed the solubility curve of the co-crystal), then towards a higher NCT concentration in domain 2a (dissolution of the CBZ/NCT co-crystals). With or without the occurrence of this metastable co-crystal population, the final point located on the CBZ solubility curve vertical to the starting point would have been the same.

The third kinetic pathway considered (run #45) was started in domain 3 and presented a more complex behaviour. The start and end points in this run were D and A, respectively. In the first period (from points D to C), the pathway remained parallel to the bisecting line. This corresponds to the stoichiometric consumption of CBZ and NCT. It turned out to be the primary nucleation and growth of CBZ/NCT co-crystals. From point C, the pathway started deviating from the bisector and deflected to the bottom of the phase solubility diagram, meaning that the CBZ crystal growth was consuming CBZ molecules. The occurrence of CBZ crystal primary nucleation was expected, since observations with an *in situ* video probe in the same conditions have shown later nucleation of prism-like CBZ crystals in the presence of needle-like CBZ/NCT co-crystals.[27] From C to B, both CBZ and CBZ/NCT solid phases grew together, and the kinetic pathway joined domain 2b. This stage ended when the co-crystal solubility curve was reached. In the final stage from B to A, the kinetic pathway deflected towards the invariant point A, crossed over the co-crystal solubility curve and entered in domain 2a. The NCT solute concentration increased due to a partial dissolution of the co-crystals. In the same time, the CBZ solute concentration kept decreasing. The CBZ molecules released by co-crystal dissolution were consumed by the CBZ crystal growth. This corresponds to a SMPT from CBZ/NCT co-crystals to CBZ crystals. Nevertheless this SMPT stopped after a short time when triple point A was reached.

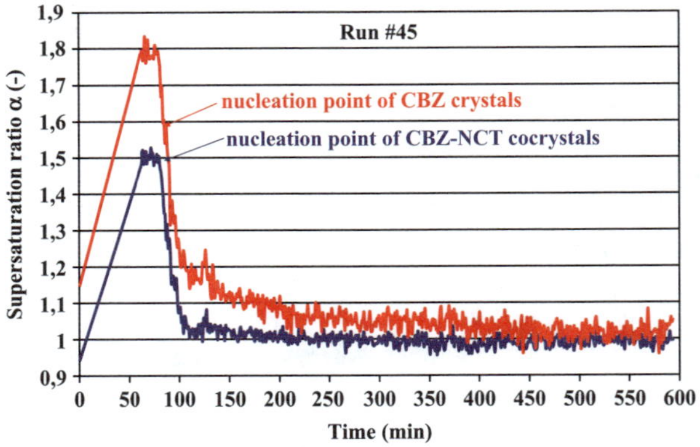

Figure 9.7 Time variations of the supersaturation ratio of CBZ/NCT co-crystals and of CBZ crystals for run #45 at 25 °C (reproduced with permission from Gagniere *et al.*, ref. 32).

This kinetic pathway confirms that a co-crystallization started in domain 3, converges to the invariant point A as predicted by the thermodynamics but could cross over several domains for kinetic reasons.

Figure 9.7 presents the time variations of the supersaturation profiles relatively for the CBZ and CBZ/NCT solid phases. The primary nucleation of CBZ/NCT first occurred at a supersaturation ratio 15% lower than the CBZ supersaturation ratio. A possible explanation lies in the fact that there may be a better affinity with the solvent of the CBZ/NCT nuclei than with that of the CBZ nuclei leading to a higher nucleation frequency of CBZ/NCT than CBZ. Ten minutes later, CBZ primary nucleation occurred. One could assume that the presence of the co-crystal form in suspension may favour a nucleation of CBZ crystals on the surface of CBZ/NCT solid form.

Finally, from the three situations presented in Figure 9.5, it should be noted that the kinetic pathway does not only depend on the initial conditions in terms of temperature and concentration, but also on the occurrence of the possible metastable and stable solid phases in the slurry which may interact and compete with each other.

9.3.3.2 Second Case Study: Heterothermal Co-crystallization in Caffeine/Glutaric Acid/Acetonitrile System (from Yu et al.[28]*)*

A complete phase diagram of the system caffeine, glutaric acid and acetonitrile was constructed from 10 to 35 °C (see Figure 9.8(a)). Batch runs were carried out in a thermostatted stirred vessel. A saturated solution with respect to the caffeine/glutaric acid co-crystal solid phase was prepared at 35 °C. The solution was cooled down to 34 °C to be slightly supersaturated. Seeding with co-crystals was then performed in order to avoid uncontrolled primary nucleation.

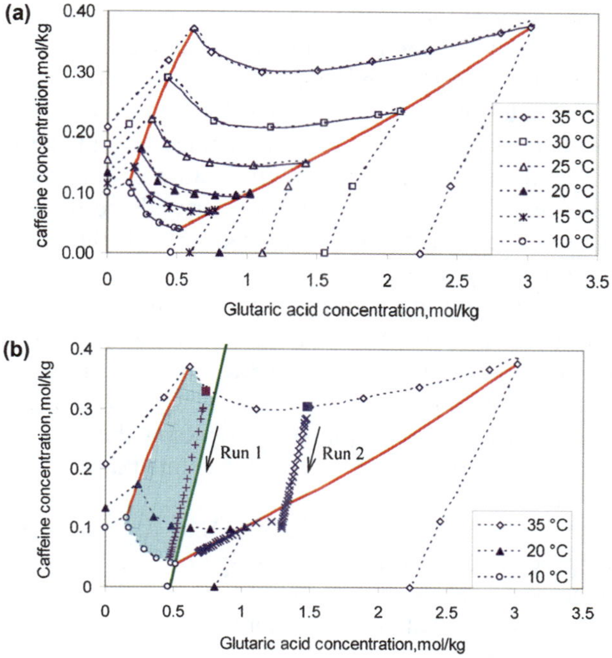

Figure 9.8 (a) Phase diagram of the caffeine–glutaric acid–acetonitrile system at different temperatures. Blue solid lines are the co-crystal solubility curve. Eutectic points for caffeine/co-crystal and for glutaric acid/co-crystal at different temperatures are connected by two bold red lines (reproduced with permission from Yu *et al.*, ref. 28). (b) The operating region (shaded) of co-crystallization between 35 and 10 °C. Solubility data at 20 °C are also presented. The trajectories of two crystallization runs are also shown. The starting positions of the two runs are each marked by a filled square (one blue, one red) (reproduced with permission from Yu *et al.*, ref. 28).

After a few minutes, co-crystal growth began. A constant cooling rate of 12 °C h^{-1} was then applied down to 10 °C. Figure 9.8(b) presents the kinetic pathways of two runs performed at two different initial compositions. The crystallization starting points were both in domain 4d at the starting temperature of 35 °C.

The kinetic pathway of the first run was parallel to the bisecting line which corresponds to the stoichiometric consumption of the two molecular species involved in the caffeine/glutaric acid co-crystal growth. Moreover the absence of deviation from the bisecting line indicates that the kinetic pathway remained in domain 4d throughout the cooling. In this condition, only the co-crystal phase is able to nucleate and grow since this domain is undersaturated with respect to the caffeine and glutaric acid solid phases. The pathway ended on the co-crystal solubility curve at the final temperature in accordance with the thermodynamics.

The kinetic pathway of the second run was also parallel to the bisecting line during the first part of the cooling; and co-crystal seeds were growing. Then the pathway deviated sharply and deflected to the left of the phase solubility diagram. This indicates that a population of glutaric acid crystals was growing in the slurry, which means that the solution became supersaturated with respect to glutaric acid and enabled the primary nucleation of the glutaric acid solid phase to take place. The two populations of crystals continued to grow on further cooling and the kinetic pathway tended towards the invariant point B. This situation could only be explained by the fact that the kinetic pathway at least entered domain 5 during cooling, crossing over the boundary between domains 4 and 5 (see Figure 9.8). Although the crystallization started in the stability zone of the co-crystal (domain 4d), the kinetic pathway joined the zone of common stability of co-crystals and glutaric acid crystals (domain 5a). In order to obtain a pure co-crystal phase by avoiding the occurrence of other solid phases (metastable or stable), Yu *et al.*[28] defined a safe operating region (see Section 9.4.4.1).

The lesson to be taken from this study is that the development of a co-crystallization process first requires knowledge of a large part of the phase diagram with temperature and second that the possible kinetic pathways which may occur should be evaluated. Considerations based on thermodynamic aspects are not sufficient.

9.3.3.3 Third Case Study: Solution-mediated Phase Transition in a Co-crystallization Process (from Gagniere et al.[34]*)*

Experiments were performed with the CBZ/NCT/ethanol system in domains of the phase diagram that were always undersaturated with respect to NCT (see Figure 9.9). Crystallizations were started in domains 2a or 3. Once the equilibrium was reached at 25 °C, a known amount of NCT in dry form was added in order to shift the overall composition of the slurry in domain 4a in the phase diagram. The previously stable CBZ solid phase became metastable. Henceforth the stable phase was the CBZ/NCT co-crystal. A SMPT started with the primary nucleation of the co-crystals if they were absent in the slurry and continued with the dissolution of the CBZ crystals and with co-crystal growth, until the new equilibrium was reached. Three kinetic pathways are displayed in Figure 9.9.

When NCT is added in run #52, both CBZ/NCT co-crystals and CBZ crystals co-exist in equilibrium with the solution at point A since the crystallization started in domain 3. NCT dissolved quickly. The pathway shifted towards domain 4 to a higher concentration of NCT in the phase diagram (from point A to point G in Figure 9.9(a)). The SMPT started readily since a primary nucleation was not necessary to trigger it. The pathway was parallel to the bisecting line from G to H. This corresponded to the stoichiometric consumption of CBZ and NCT due to the co-crystal growth. The amount of dissolved CBZ was not enough to deflect the pathway significantly, otherwise the

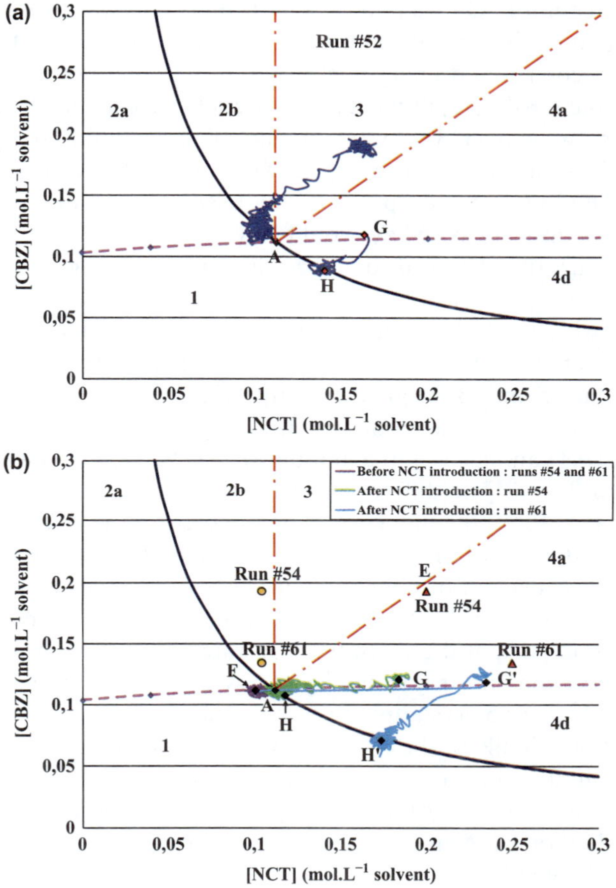

Figure 9.9 (a) Solute concentration pathway for run #52 in the phase solubility diagram at 25 °C (reproduced with permission from Gagniere *et al.*, ref. 34). (b) Solute concentration pathway for runs #54 and #61 in the phase solubility diagram at 25 °C. The orange circle symbols represent the global composition of the mixture before NCT introduction and the red triangle symbols represent the global composition of the mixture after NCT introduction (reproduced with permission from Gagniere *et al.*, ref. 34).

slope would have been lower. The passage from domains 3 to 4 was fast since the co-crystals were already present before introduction of NCT and they did not have to nucleate.

When NCT is added in run #54, the slurry is made up of only one population of CBZ crystals in equilibrium with the solution (see point F in Figure 9.9(b)). The amount of CBZ solid phase was 2.25 wt%. With quick NCT dissolution, the pathway moved towards domain 4 to point G in a few minutes. The pathway stopped around point G for 25 minutes, then followed the CBZ solubility curve (slightly under the curve), and finally ended at point H on the co-crystal solubility curve, which was the solution composition at equilibrium with respect to point E. Only a population of co-crystals was present in the final slurry. The standing period at point G corresponded to the induction time of the CBZ/NCT co-crystal primary nucleation. From point G to H, the pathway was close to the CBZ solubility curve. This could be explained by the kinetic competition between the CBZ crystal dissolution rate and the co-crystal growth rate: the CBZ consumed flow from the growth of the CBZ/NCT co-crystals was lower than the CBZ released flow from the CBZ crystals' dissolution. Otherwise, the crystallization pathway would have deviated from CBZ solubility curve. Consequently, the transition was limited by the growth of the co-crystals.

Compared with run #52, the passage from domain 2 to domain 4 in run #54 was much slower. This slowness was due to the induction time of the co-crystal primary nucleation and to the great quantity of co-crystals produced. Run #61 differed from run #54 by the amount of CBZ crystals in the initial slurry (0.63 wt% for run #61) and also by the amount of NCT used to trigger the SMPT from domains 2 to 4. The differences in the pathways could be translated in terms of kinetics. First, the induction time of co-crystal primary nucleation at point G' was reduced to 15 minutes due to a higher CBZ/NCT supersaturation ratio. Second, no plateau of CBZ dissolution–co-crystal growth was detected and the pathway was parallel to the bisecting line up until the final equilibrium (point H'). An important desupersaturation rate was also observed.

These observations are explained by a less numerous population of CBZ crystals and a higher co-crystal primary nucleation frequency. In this situation the consumption of CBZ by the co-crystal growth was not balanced by the CBZ crystals' dissolution. From a process point of view, the introduction of an important quantity of co-crystal agent brings a net gain of operating time due to a faster SMTP rate and an increased co-crystallization yield because of lower final co-crystal solubility.

Investigating such SMPTs is very fruitful since it gives some information about the kinetics of nucleation and growth of the different solid phases which may compete against each other in the co-crystallization process. Lastly, in the case of a final co-crystal phase which does not fulfil the purity requirements due to contamination by another crystalline phase, a retreatment process based on a SMPT may be set up in order to direct the process to the domain where the co-crystal is the only stable phase.

9.4 Development of a Co-crystallization Process

9.4.1 Prerequisite

Process development requires knowledge of the thermodynamics and kinetics of the system. From the earlier stages of development, an efficient co-crystal screening is essential. Of course this screening consists of finding out whether a co-crystal can be formed with one co-crystallizing agent or another, but it must also be able to detect the different possible stoichiometries and different polymorphic forms for each stoichiometry. The construction of the thermodynamic phase diagram permits determination of the stability domains of different stoichiometric co-crystals together with the relative stability of the different polymorphic forms. The co-crystal that will be developed is chosen for its end-use properties and for its bioavailability criteria. In terms of process development, it is preferable to select the stable polymorph of the co-crystal in case of polymorphism. In addition, the working domain has to be such that the chosen stoichiometry is stable.

9.4.2 Co-crystal Screening

Pharmaceutical co-crystals can be prepared by several methods, including mechanochemical, supercritical fluid-based technique, sonication, melt crystallization or traditional solution crystallization. In terms of screening strategy, mechanochemical methods (such as neat and liquid-assisted grinding) are among the most used (see Chapter 8 for more details). Although mechanochemical methods may appear ideal for co-crystal screening, the scale-up of the technique may be unrealistic. Supercritical fluid-based techniques recently received much attention from the scientific community, since this method may combine screening and production of co-crystals. In the manner of polymorphism, two main processes using supercritical fluid can be used in co-crystal screening: first, the RESS process (rapid expansion of a supercritical solution), wherein the supercritical fluid is used as a solvent, and second, the PCA process (particles compressed by anti-solvent) and SAS process (supercritical anti-solvent), in which the supercritical fluid is used as an anti-solvent.[35] These techniques allowed the formation of indomethacin–saccharin co-crystals[36] and co-crystals with caffeine.[37,38]

Sonication is also another method of generating co-crystals. It is applied to a solution or suspension.[39–43] Although this technique is little used in co-crystal screening, a recent study has shown that pure co-crystals could be obtained using ultrasound assisted solution co-crystallization, where conventional solution crystallization failed.[44]The use of melt crystallization methods have also been reported as efficient co-crystal screening methodologies. These techniques include thermal microscopy[45,46] and differential scanning calorimetry methods.[47]

Finally, solvent-based techniques such as cooling, evaporation, anti-solvent or drowning out are certainly the most commonly used methods for co-crystal screening. Indeed, they allow the variation of three important parameters in nucleation: the interfacial energy γ, the supersaturation ratio β, and the temperature T. Different solvents or solvent mixtures should be tested, since a change of solvent affects the solvent/solute interactions, and thus the interfacial energies. Table 9.1 presents solvents covering a wide variety of polarities and proticities that could be used. The solvents are generally chosen from those authorized by the pharmacopoeia.

Horst *et al.*,[48,49] recently developed an interesting co-crystal screening method in solution. The methodology is simple: the first step results in the determination of the two components solubility's (named xa^* and xb^*) at one

Table 9.1 Classification of organic solvents for screening co-crystals.

Type of Solvent	*Chemical Species*	*Low Boiling Point*	*High Boiling Point*
Protic–polar	Alcohol/Water	Methanol Ethanol	Hexane, butane Benzyl alcohol
Aprotic–polar	Ketone	Acetone	Methyl isobutyl ketone
	Nitrogen compound	Acetonitrile	Propionitrile, dimethyl formamide
	Amino compound		*N*-Methyl-2 pyrrolidone
	Ester	Ethyl acetate	Isopropyl acetate, butyl acetate
	Ether	Diethyl ether, methyl t-butyl ether	Dibutyl ether
	Chlorinated compound	Dichloromethane	
	Aromatic Chlorinated compound	Monochlorobenzene	*o*-Dichlorobenzene
Aprotic–non-polar	Alkane	Pentane, hexane	Decane
	Aromatics	Toluene, xylene	

temperature ($T1$). Second, a solution of concentration ($xa^*(T1)$, $xb^*(T1)$) has to be prepared. If the two components are able to co-crystallize, then a solution of concentration ($xa^*(T1)$, $xb^*(T1)$) will have a saturation temperature $T2$ such that $T2 > T1$. This method is quite simple and easy to set up: after the second step, the only thing is to verify if crystals have nucleated in suspension. However, the supersaturation ratio with respect to co-crystals can be very low with this technique, preventing the co-crystals' nucleation. A good compromise should be the cooling of the solution of concentration ($xa^*(T1)$, $xb^*(T1)$) in order to increase the supersaturation ratio with respect to co-crystals and so induce nucleation. Nevertheless, the temperature difference, between the initial and final point should be moderate in order to stay in the "safe" operating region, as defined below (see Section 9.4.4.1).

9.4.3 Determination of Operating Conditions in Batch Mode

The development of a co-crystallization in solution is similar to the development of the crystallization of the API itself since the mechanisms of nucleation and growth are similar in both cases. Control of the nature of the solid phase is essential during the operation and we have to keep in mind these two following points:

1. Start the co-crystallization with the desired co-crystal phase.
2. Prevent the occurrence of other solid phases due to inopportune primary nucleation events which may lead to the partial or total loss of the co-crystal population by SMPT.

The first point can be achieved by means of seeding. The second point can be obtained by defining an operating region in the phase diagram over the temperature range of the batch process. Finally in the case of a final desired co-crystal phase which does not fulfil the purity requirements due to the presence of another crystal phase, a retreatment process may be set up with the addition of the correct amount of one of the co-crystal components in order to provoke a SMPT which eliminates this undesirable solid phase and produces co-crystals.

9.4.3.1 Choice of "Safe" Operating Regions (Initial Conditions and Phase Diagram)

We have seen in the previous section that several solid phases may appear during the kinetic pathway. In order to start with only the desired co-crystal phase, the first recommendation is to locate the starting point in domain 4d in which only the co-crystal phase may nucleate. Since the initial temperature is generally high in order to solubilise the API and the co-crystallizing agent (for instance 20% weight) sufficiently, knowledge of the phase diagram at this initial temperature is required.

As soon as the co-crystals grow, the kinetic pathway may follow a line which slope corresponds to the stoichiometry of the co-crystal. On cooling, two situations exist: (i) the pathway remains in domain 4d and the final point is located on the co-crystal solubility curve; during the process, the co-crystal is the only stable phase developing in solution; (ii) the kinetic pathway crosses the boundary of the domain 4d and joins another domain in which another solid phase may also be stable (for instance the domain 5a where the co-crystallizing agent crystalline state is stable with the co-crystal – see the second case study above). The co-crystal purity may be poisoned by the co-crystallizing agent solid form. Consequently, knowledge of the phase diagram at this final temperature is also required.

Yu *et al.*,[28] made a proposal to prescribe an operating region for co-crystal production. The "safe" starting point is situated in the domain common to domain 4d at the initial and final temperatures. Depending on the evolution of the phase diagram with the temperature, this "safe" operating region may be only a part of domain 4d.

9.4.3.2 Seeding Strategy

For a few decades, seeding has been used in order to ensure reproducible production of particles in terms of crystal size distribution. Even though seeding could be treated as an art, seeding policy has been progressively established and we could consider that the effect of the main parameters of seeding (seed size, seed mass or seed concentration, seeding temperature, isothermal plateau of seeding, cooling profile) is largely described in the literature.[50–54] Seeding strategies were also investigated where polymorphism effects may compromise the crystalline purity of the final solid state form of the API.[55,56]

In the case of co-crystallization, a seeding strategy has two main goals: (i) to ensure the presence of the co-crystal phase in solution as soon as the crystalline growth starts, (ii) to avoid uncontrolled primary nucleation of another solid phase previous to the presence of the co-crystal seeds.

The conditions of co-crystal seed preparation are identical to those applied in the manufacture of an organic compound. We add the following comments:

- The introduction of the seeds must be done in a "safe" operating domain and it is preferable to expect the onset of co-crystal growth before starting the cooling. This is made practical by carrying out an isothermal plateau at the seeding temperature.
- The presence of co-crystals may act as a nucleation substrate for another solid phase by a heterogeneous nucleation mechanism. Nevertheless starting the crystallization by seeding in the safe operating region prevents the occurrence of such undesired heterogeneous nucleation.

9.4.3.3 Manipulation of Co-crystal Components

At the end of a co-crystallization process, the co-crystal crystalline purity may be evaluated by X-ray powder diffraction or by *in situ* sensing technologies (like Raman or infrared spectroscopy). In the case of a final co-crystal phase which does not fulfil the purity requirements, due to the presence of another solid phase, a retreatment process could be set up. One solution is to consider a SMPT in order to shift the overall composition of the slurry to domain 4d. This SMPT could be triggered off by adding a certain amount of one of the chemical species of the co-crystal system. The added chemical compound depends on the desired kinetic pathway. If the undesirable form is the API, an amount of the co-crystallizing agent has to be added to the slurry (see case study 3 in Section 9.3.3). Conversely dissolution of co-crystallizing agent particles can be done by addition of API crystals. This SMPT could be carried out in a stirred reactor at the final temperature of the co-crystallization, or at upper temperature in order to accelerate it.

9.5 Conclusions

Co-crystals represent a novel class of crystalline solids which possess scientific and regulatory advantages along with enormous intellectual property potential. In the earlier stages of pharmaceutical process development, an efficient co-crystal screening is essential in order to determine the different possible stoichiometries and the different polymorphic forms for each stoichiometry. When all the different phases are identified, the relative stability of the solid phases in solution has to be determined. This step requires the construction of the thermodynamic phase diagram of the system. However, this chapter has also highlighted the importance of kinetic aspects since the development of co-crystals may be in competition with the crystallization of single component crystals. Thus, depending on the operating conditions, several solid phases may appear during the kinetic pathways; some of them may disappear during the co-crystallization, whereas others continue to grow. This contribution points out the importance of a seeding strategy to ensure the development of pure co-crystals in suspension. We have also emphasized that the co-crystallization process has to be conducted in a "safe" operating region, defined thanks to knowledge of the phase diagram at the initial and final temperatures. Production of a pure co-crystal solid phase at the end of the process must be ensured in this region. However, in the case of a final co-crystal phase which does not fulfil the purity requirements, due to the presence of another crystalline phase, a retreatment process may be set up via an SMPT, by adding the correct amount of either co-crystallizing agent or API. Finally, from the standpoint of industrial manufacturing, producing particles with specific solid properties allows downstream operations to be performed more easily thanks to mastered filterability, low fines' concentration, low trend to cake on storage, good flow ability, and so on. This challenge, which is still relevant today in the case of monocomponent crystals, also concerns the case of co-crystals.

Acknowledgements

Financial support provided by Sanofi-Aventis for studies in co-crystallization at LAGEP is gratefully acknowledged.

References

1. N. Schultheiss and A. Newman, *Cryst. Growth Des.*, 2009, **9**, 2950.
2. J. Aaltonen, M. Allesø, S. Mirza, V. Koradia, K. C. Gordon and J. Rantanen, *Eur. J. Pharm. Biopharm.*, 2009, **71**, 23.
3. A. C. Zettlemoyer, *Nucleation*, Marcel Dekker, New York, 1969.
4. F. F. Abraham, *Homogeneous Nucleation Theory*, Academic Press, Amsterdam, 1974.
5. S. Toschev, Homogeneous nucleation, in *Crystal Growth: An Introduction*, ed. P. Hartman, North Holland, Amsterdam, 1973, p. 1.
6. D. Kashchiev, *Nucleation: Basic Theory with Applications*; Butterworth-Heinemann, Oxford, 2000.
7. J. P. Astier and S. Veesler, *Cryst. Growth Des.*, 2008, **8**, 4215.
8. A. E. Nielsen and J. M. Toft, *J. Cryst. Growth*, 1984, **67**, 278.
9. F. Rosenberger, Fundamentals of Crystal Growth, Springer Series in Solid-State Sciences, Vol. 1, Springer-Verlag, Berlin, 1979.
10. J. Zhang and G. H. Nancolias, *J. Phys. Chem.*, 1994, **98**, 1689.
11. P. G. Koutsoukos and E. Valsami-Jones, in *Phosphorous in Environmental Technologies*, ed. E. Valsami-Jones, IWA Publishing, London, UK, 2004.

12. L. Amathieu and R. Boistelle, *J. Cryst. Growth*, 1988, **88**, 183.
13. E. Garcia, C. Hoff and S. Veesler, *J. Crystal Growth*, 2002, **237–239**, 2233.
14. D. Mangin, E. Garcia, S. Gerard, C. Hoff, J. P. Klein and S. Veesler, *J. Cryst. Growth*, 2006, **286**, 121.
15. P. T. Cardew and R. J. Davey, *Proc. R. Soc. London, Ser. A*, 1985, **398**, 415.
16. J. Bauer, S. Spanton, R. Henry, J. Quick, W. Dziki, W. Porter and J. Morris, *Pharm. Res.*, 2001, **18**, 859.
17. E. S. Ferrari, R. J. Davey, W. I. Cross, A. L. Gillon and C. S. Towler, *Cryst. Growth Des.*, 2003, **3**, 53.
18. C. Cashell, D. Corcoran and B. K. Hodnett, *J. Cryst. Growth*, 2004, **273**, 258.
19. T. Ono, J. H. Ter Horst and P. J. Jansens, *Cryst. Growth Des.*, 2004, **4**, 465.
20. J. Scholl, D. Bonalumi, L. Vicum, M. Mazzotti and M. Muller, *Cryst. Growth Des.*, 2006, **6**, 881.
21. S. J. Nehm, B. Rodríguez-Spong and N. Rodríguez-Hornedo, *Cryst. Growth Des*, 2006, **6**, 592.
22. R. A. Chiarella, R. J. Davey and J. W. Peterson, *Cryst. Growth Des.*, 2007, **7**, 1223.
23. S. L. Childs, N. Rodríguez-Hornedo, L. S. Reddy, A. Jayansankar, C. Maheshwari, L. McCausland, R. Shipplett and B. C. Stahly, *Cryst. Eng. Comm.*, 2008, **10**, 856.
24. S. G. Fleischman, S. S. Kuduva, J. A. McMahon, B. Moulton, R. D. Bailey Walsh, N. Rodriguez-Hornedo and M. J. Zaworotko, *Crys. Growth Des.*, 2003, **3**, 909.
25. B. Rodriguez-Spong, Thesis Dissertation, Enhancing the pharmaceutical behaviour of poorly soluble drugs through the formation of co-crystals and mesophases, University of Michigan, 2005.
26. N. Rodriguez-Hornedo, S. J. Nehm, K. F. Seefeldt, Y. Pagan-Torres and C. J. Falkiewicz, *Mol. Pharmaceutics*, 2005, **3**, 362.
27. E. Gagniere, D. Mangin, F. Puel, O. Monnier, E. Garcia and J. P. Klein, *J. Cryst. Growth*, 2009, **311**, 2689.
28. Z. Q. Yu, P. S. Chow and R. B. H. Tan, *Cryst. Growth Des.*, 2010, **10**, 2382.
29. M. Boukerche, D. Mangin, A. Rivoire, O. Monnier, C. Hoff and J. P. Klein, *Proceedings of the 15th International Symposium on Industrial Crystallization*, ed. A. Chianese, 2002, Sorrento, Italy.
30. M. Henry, F. Puel, P. Perrichon, C. Wisniewski and G. Fevotte, *Int. J. Chem. React. Eng.*, 2008, **6**, 1.
31. L. Derdour, G. Fevotte, F. Puel and P. Carvin, *Powder Technol.*, 2003, **129**, 1.
32. E. Gagniere, D. Mangin, F. Puel, C. Bebon, J. P. Klein, O. Monnier and E. Garcia, *Cryst. Growth Des.*, 2009, **9**, 3376.
33. F. Lewiner, J. P. Klein, F. Puel and G. Fevotte, *Chem. Eng. Sci.*, 2001, **56**, 2059.
34. E. Gagniere, D. Mangin, F. Puel, J. P. Valour, J. P, Klein and O. Monnier, *J. Cryst. Growth*, 2010, **316**, 118.

35. D. Baltes, D. Mangin, O. Monnier, C. Hoff and J. P. Klein, *Proceedings of the 16th International Symposium on Industrial Crystallization*, ed. J. Ulrich, Dresden, 2005, Germany.
36. L. Padrela, M. A. Rodrigues, S. P. Velaga, H. A. Matos and E. G. De Azevedo, *Eur. J. Pharm. Sci.*, 2009, **38**, 9.
37. H. Kroeber, *Proceedings of the 17th International Symposium on Industrial Crystallization*, eds. J. P. Jansens and J. Ulrich, Maastrich, 2008, Holland.
38. M. Hermann, U. Förter-Barth, H. Kröber, P. B. Kempa, M. Del Mar Juez-Lorenzo and S. Doyle, *Part.Part.Syst.Charact.*, 2009, **26**, 151.
39. S. L. Childs, P. Mougin and B. Stahly, *WO*/2005/089375, 2005.
40. D. Bucar and L. R. MacGillivray, *J. Am. Chem. Soc.*, 2007, **129**, 32.
41. T. Friscic, S. L. Childs, S. A. A. Rizvic and W. Jones, *CrystEngComm*, 2009, **11**, 375.
42. L. McCausland, *US Pat.*, 2009/0179141A1, 2009.
43. S. L. Childs, P. M. Mougin-Andres and B. C Stahly, *US Pat.*, 2007287194 A1, 2007.
44. S. Aher, R. Dhumal, K. Mahadik, A. Paradkar and P. York, *Eur. J. Pharm. Sci*, 2010, **41**, 597.
45. D. J. Berry, C. C. Seaton, W. Clegg, R. W. Harrington, S. J. Coles, P. N. Horton, M. B. Hursthouse, R. Storey, W. Jones, T. Friscic and N. Blagden, *Cryst. Growth Des.*, 2008, **8**, 1697.
46. D. P. McNamara, S. L. Childs, J. Giordano, A. Iarriccio, J. Cassidy, M. S. Shet, R. Mannion, E. O'Donnell and A. Park, *Pharm. Res.*, 2006, **23**, 1888.
47. E. Lu, N. Rodriguez-Hornedo and R. Suryanarayanan, *CrystEngComm*, 2008, **10**, 665.
48. J. H. Ter Horst and P. W. Cains, *Cryst. Growth Des.*, 2008, **8**, 2537–2542.
49. J. H. Ter Horst, M. A. Deij and P. W. Cains, *Cryst. Growth Des.*, 2009, **9**, 1531.
50. M. Kohl, G. Fevotte, J. P. Klein, O. Monnier and C. Hoff, *Proceedings of the 4th International Workshop on Crystal Growth of Organic Materials (CGOM) (Special Event of the 6th BIWIC)*, ed. J. Ulrich, Bremen, 1997, p. 175.
51. N. Kubota, N. Doki, M. Yokota and A. Sato, *Powder Technol.*, 2001, **121**, 31.
52. A. Mersmann and F. W. Rennie, Operation of crystallizers, in *Crystallization Technology Handbook*, ed. A. Mersmann, Marcel Dekker, New York, 2001, Chap 9, p. 410.
53. J. S. Wey and P. H. Karpinski, Batch crystallization, in *Handbook of Industrial Crystallization*, ed. A. Myerson, Butterworth Heinemann, Woburn, MA, 2nd edn, 2001, Chap 10, p. 239.
54. E. Aamir, Z. K. Nagy and C. D. Rielly, *Chem. Eng. Sci*, 2010, **65**, 3602.
55. W. Beckman, *Org. Process Res. Dev.*, 2000, **4**, 372.
56. D. Mangin, F. Puel and S. Veesler, *Org. Process Res. Dev.*, 2009, **13**, 1241.

CHAPTER 10

Analytical Techniques and Strategies for Salt/Co-crystal Characterization

SUSAN M. REUTZEL-EDENS

Lilly Research Laboratories, Eli Lilly & Company, Indianapolis, IN 46285, USA

10.1 Introduction

Salts, and more recently co-crystals, have attracted much interest in the pharmaceutical industry for their promise in tailoring the physical properties of an active pharmaceutical ingredient (API) to meet the needs of the drug product and ultimately the patient.[1,2] *Salt forms*, produced by acid (A)–base (B) reactions in the solid state, are multi-component solids comprising minimally an A^-B^+ pair; they may be crystalline or amorphous. The term *co-crystal*, on the other hand, refers specifically to crystalline molecular complexes, which may include an A^-B^+ pair among the different components and by definition necessarily include solvates (hydrates), inclusion compounds, clathrates and solid solutions.

It is important to recognize that the terms 'salt' and 'co-crystal' are not meant to be mutually exclusive, except perhaps for simple acid–base pairs, where the only distinguishing feature of a salt and a co-crystal is the location of the acidic proton in the crystal structure. An equally important difference between salts and co-crystals in the context of crystal engineering is that while salts are necessarily made up of acids and bases, co-crystals are not.

RSC Drug Discovery Series No. 16
Pharmaceutical Salts and Co-crystals
Edited by Johan Wouters and Luc Quéré

Published by the Royal Society of Chemistry, www.rsc.org

Thus, co-crystals not only expand the options for altering the solid state properties of ionizable compounds (by not excluding acids or bases in the search for crystalline forms based on pK_a criteria), but they also create opportunities to improve the properties of non-ionizable drug substances. Fortunately, for drug compounds, which usually possess acidic or basic functionality and are always capable of hydrogen bonding, salts and co-crystals collectively provide numerous options beyond the parent molecule to engineer viable solid state forms. Crystal engineering, which entails designing crystalline molecular solids with the aim of tailoring specific physical or chemical properties, although only recently discussed in this context for pharmaceuticals, has in fact been the primary objective of salt and polymorph screening programs in the pharmaceutical industry for years.

The ability of a drug molecule to form salts and/or co-crystals, which themselves may adopt different structures (polymorphs, solvates), attests to the diversity of solid forms above and beyond those of the parent molecule that may be encountered in drug development. Effectively navigating the frequently complex solid form landscapes of drugs, their salts and co-crystals to identify a suitable form that can be incorporated by design into a quality drug product further underscores the need for reliable solid state characterization techniques. Crystallography, spectroscopy, microscopy, thermal analysis and other physical techniques are widely used, alone or in combination with one another, to examine the form and function of salts and co-crystals in the solid state. In this contribution, many of the most commonly used analytical techniques, along with a few of the emerging ones, for characterizing salts and co-crystals are explored. Brief descriptions of the methods themselves are provided and specific applications to the screening, structural characterization and assessment of properties of salts and co-crystals are presented.

10.2 Analytical Techniques

Solid state characterization techniques are used throughout drug development and pharmaceutical manufacturing from the earliest stages, when they serve to identify crystalline 'hits' coming from salt and co-crystal screens, all the way through to the final release (identity) testing of finished drug products. Many, but by no means all, of the techniques used to characterize salt and co-crystal forms during drug development are presented in this chapter. While emphasis is clearly placed on salts and co-crystals, the solid state techniques described herein are, for the most part, no different from those used to characterize the neat and solvated forms of the parent molecule. Indeed, any technique capable of distinguishing crystal forms can in principle be used to identify new solid state forms of salts and co-crystals. Of course, deciding which technique(s) to use will depend on a number of factors, including (1) what technique(s) are available, (2) what information is needed, (3) how much sample is available and required and (4) in what form is the sample (*e.g.*, single crystal *vs.* powder). For salts and co-crystals in particular, the latter consideration is becoming increasingly important as solid state and solvent drop grinding has gained popularity in the screening of salts[3] and co-crystals,[4] yielding materials which are not suitable for single crystal methods of structure determination.

10.2.1 X-ray Diffraction

X-ray diffraction is arguably the single most important analytical tool in solid state chemistry, with applications ranging from the detailed characterization of molecular and crystal structure (single crystal diffraction) to the detection of long-range order in crystalline materials (powder diffraction). Both single crystal and powder techniques require that the diffraction condition be satisfied as given by Bragg's Law (equation (10.1)):

$$n\lambda = 2d\sin\theta \tag{10.1}$$

where λ is the wavelength of X-ray radiation, d is the lattice spacing of parallel planes in a crystal and θ is the diffraction angle.[5] The diffraction condition will be satisfied when the angle (θ) between the incident X-ray radiation and the parallel planes of electron density in a crystal leads to constructive interference. In powder diffraction, the diffraction intensity is captured with a point or line detector and plotted in one dimension as a function of 2θ, while in single crystal diffraction, the diffraction intensity is captured in three dimensions with an area or CCD detector. Three-dimensional diffraction is then converted to a model of the electron density, which allows the reflections to be indexed to determine the unit cell parameters (lengths and angles), as well as the space group. From the diffraction intensities, which contain information about the distribution of electrons throughout the crystal, the atoms can be located in the unit cell.

Because the structural information derived from X-ray crystallography is unparalleled, single crystal X-ray diffraction in particular has become a cornerstone of most co-crystal research programs, validating in some cases crystal engineering strategies based on preferred hydrogen bonding motifs, while highlighting in others the unpredictable nature of molecular self assembly in complex systems. Single crystal X-ray structure analysis proved invaluable in identifying the mechanism by which the antibiotic loracarbef co-crystallized with paraben preservatives, for example. Discovered during the pediatric suspension development of loracarbef (methylparaben was used as a preservative to ensure room temperature (rt) shelf stability), the co-crystallization of [carba]cephalosporins and parabens (*para*-hydroxybenzoate esters) and their hydroxyacetophenone analogs was found to be surprisingly common, having been confirmed by X-ray structure analysis of several structurally related complexes.[6] Like many drug molecules, loracarbef and its cephalosporin analogs contain multiple functional groups, of which none appeared to be the single dominant *synthon* with which to predict hydrogen-bond directed co-crystallization. Indeed only through single crystal X-ray diffraction was the driving force for co-crystallization apparent. Based on five crystal structures, hydrogen bonding was observed between either the carbonyl O of the β-lactam

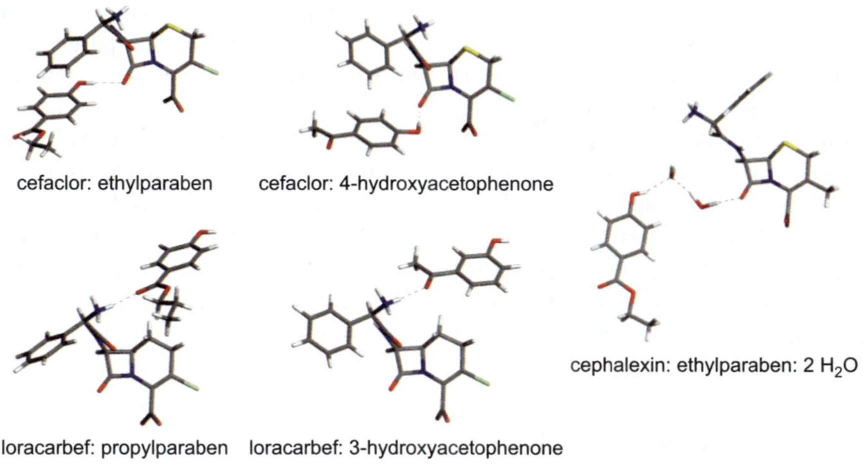

Figure 10.1 Cephalosporin and carbacephalosporin co-crystals.

host and the phenol -OH of the paraben guest or the ammonium NH of the host and the carbonyl O of the guest, Figure 10.1. The structural diversity seen in these five co-crystals alone attests to the lack of predictability of specific interactions being formed in the presence of competitive hydrogen bonding functionality. Certainly, had parabens been so evident as a co-crystal former with loracarbef, it is unlikely they would have been explored as preservatives in the pediatric suspension formulation of this antibiotic.

X-ray crystal structures will, in favorable cases, be solved from single crystals, allowing a variety of solid state properties, including stability, morphology, hygroscopicity, compressibility, and so on to be modeled and/or predicted and related structures to be engineered (crystal engineering). All too often, however, the structures of small molecule drug crystal forms are elusive because of the single crystal size/quality requirements of the X-ray methods or the methods of preparation (*e.g.*, solid state grinding, desolvation leading to crystal cracking) of the salt/co-crystal forms. Fortunately, tools for structure solution from powder diffraction data are becoming increasingly available for solving structures of moderately complex molecules not amenable to single crystal methods.[7,8] In direct-space structure solution methods, trial crystal structures are generated by movement of a structural model (chemical structure with defined intramolecular connectivity and typical bond lengths/angles) within a pre-defined unit cell. Trial structures are then searched by subjecting the structural models to a global optimization using Monte Carlo,[9,10] simulated annealing[11–13] or genetic algorithm techniques.[14–16] The correctness of each structure solution is assessed by comparing the calculated powder pattern for the trial structure to the experimental pattern, the fit being quantified by a crystallographic R-factor (R_{wp}) or figure-of-merit (χ^2). It must be understood that while single crystal and powder diffraction provide the same intrinsic information, when three-dimensional diffraction data is compressed into a one-dimensional powder pattern, information is lost (obscured by peak overlap) and the loss of symmetry independent reflections in a one-dimensional powder pattern will increase both the difficulty and uncertainty of structure solutions from powders.

X-ray powder diffraction (XRPD) provides considerably less structural information than single crystal X-ray diffraction; however, it is generally considered to be the 'gold standard' through which salt and co-crystal forms may be identified and polymorphism and solvate formation confirmed in polycrystalline solids.[17] Modern powder diffractometers, when equipped with temperature and/or humidity control accessories, greatly expand the range of applications of powder diffraction to include studies of solid state phase transitions, desolvation (dehydration), rehydration, melting and recrystallization.[18]

10.2.2 Thermal Analysis

Thermal analysis comprises a family of techniques in which a physical or chemical change is measured as a function of temperature while a material is subjected to a predefined heating or cooling program. For salts and co-crystals, melting, crystallization, sublimation, decomposition and solid state transitions may be observed and volatile components (residual solvent or gaseous by-products) may be quantified.[19] Among the most common thermal analysis techniques used for these applications in the pharmaceutical industry are differential thermal analysis (DTA), differential scanning calorimetry (DSC) and thermogravimetric analysis (TGA). In DTA, the temperature difference ($\mathrm{d}T/\mathrm{d}t$) between a sample and an inert reference material is measured, then plotted against time or temperature. Thermal events, such as melting or crystallization, will cause the temperature difference to change as heat is absorbed (endothermic) or released (exothermic) from the sample. Because of the inherent limitations of DTA for measuring transition enthalpies, this technique is typically used to determine the temperatures at which thermal events take place; quantitative thermal analysis is reserved for the related technique of DSC, which measures the energy (heat flow, $\mathrm{d}q/\mathrm{d}t$) necessary to establish a nearly-zero temperature difference between the sample and an inert reference. TGA, a technique wherein the mass change of a substance is measured as a function of temperature or time in a controlled atmosphere, is frequently conducted simultaneously with DSC or DTA, providing valuable information on the nature of desolvation processes and subsequent transformations. TGA instruments can also be interfaced with Fourier transform infrared (FTIR) or mass spectrometers (MS) to identify evolved gases as a sample is heated. Whereas samples for TGA are typically placed in open pans provided by the manufacturer, enclosed pans, which can be either hermetically sealed or crimped or rendered with a pinhole to allow for slower escape of evolved gases, are used for DSC and DTA.

The characterization of melting behavior is particularly important when selecting a salt or co-crystal for a drug product, since processing steps, such as milling or micronization, can generate considerable heat which not all salt and co-crystal forms are equally able to tolerate. Melting is a first order process that can be observed in the form of an endothermic peak in DSC (and DTA) curves. In favorable cases, the melting point of a salt or co-crystal form will be accurately identified as the onset temperature of the endothermic transition. Recall that unlike the peak temperature, the onset temperature of melting will be independent of the DSC heating rate. If heating induces a phase change or decomposition prior to melting, an endothermic peak may appear at a temperature well below the true melting point of the salt or co-crystal form, where it may be mistakenly interpreted as melting. In such cases, faster heating rates may be used to shift kinetic processes (*e.g.*, evaporation, crystallization and decomposition) to higher temperatures away from the melting event, although very fast heating rates can compromise the accuracy of the measurement (due to thermal lag). In fact, DSC experiments should be conducted at variable heating rates when decomposition is suspected, not only to ensure that true melting has in fact been measured, but also to understand the overall thermal stability of the salt or co-crystal form. Unfortunately, in some cases, kinetic events may simply not be shifted to high enough temperatures for even a reasonably accurate melting point to be measured.

Beyond establishing that a given salt or co-crystal has a sufficiently high melting point (or thermal stability) to withstand common pharmaceutical processing unit operations, melting point analysis by DSC has a special role to play in screening for salt and co-crystal forms that crystallize from the melt. Popularized by McCrone in the 1950s, mixed fusion methods specifically may be used to classify binary mixtures based on their melting point diagrams and, in the process, discover new phases that may have been missed by conventional solution crystallization methods.[20] To construct a binary phase diagram, mixtures of two components are usually prepared covering a range of compositions, then melted, cooled to allow for recrystallization and reheated, all in a DSC pan. If recrystallization does not occur upon cooling, the components may be ground together for an extended period of time and heated only once. For the pure components, compositions defined by the stoichiometry of the complex (salt or co-crystal) and eutectic minima, a single melting endotherm is typically observed by DSC. All other compositions will show two endotherms, one for the eutectic melt and the other which will define the liquidus curve(s).

Pure and eutectic melting data obtained for the binary mixtures are plotted as a function of the phase composition to construct the binary phase diagram, as shown for ephedrine and pimelic acid in Figure 10.2.[3] The melting 'peaks' in a binary phase diagram will point to the existence of potential salt or co-crystal forms at stoichiometries defined on the *x*-axis. For ephedrine and pimelic acid, 1:1 and 2:1 salts were proposed based on the observation of two peak melting temperatures at these compositions. To define the regions of thermodynamic stability for the relevant phases, vertical lines may be drawn from the peak melting temperatures and horizontal eutectic melting lines may be drawn to link the vertical lines either to one another or to the nearest pure component. To aid in the interpretation of binary phase diagrams, ideal liquidus curves

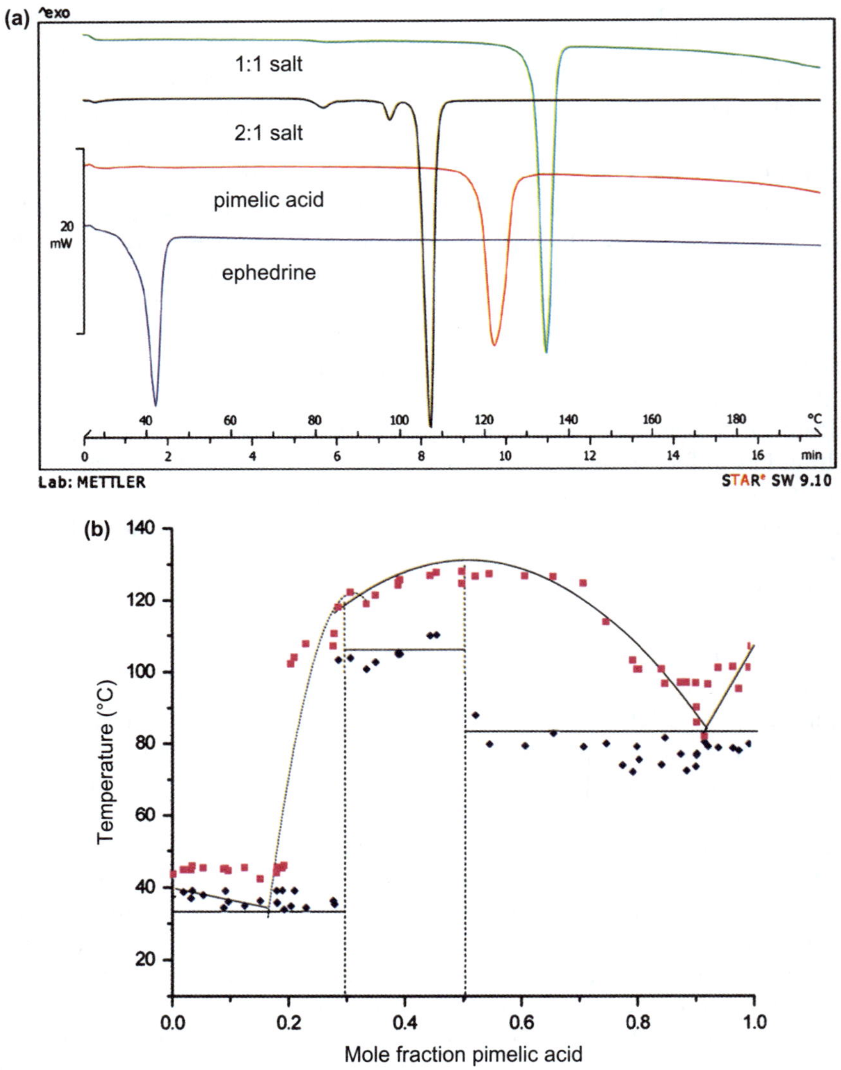

Figure 10.2 (a) DSC curves of ephedrine, pimelic acid and their 1:1 and 2:1 salts and (b) binary phase diagram, showing measured pure and eutectic melting points (adapted from Cooke *et al.*, ref. 3. Copyright © American Chemical Society).

may be calculated using the equations of Schroeder–van Laar and Prigogine–Defay.[21] Deviations from theory are typically interpreted in terms of non-ideality caused by ionization or intermolecular interactions in the liquid phase (melt).

Melting point analysis and binary phase diagrams are well suited for the characterization of neat salt and co-crystal forms; however, neat forms comprise only part of the solid form landscape of a given salt or co-crystal.

Solvates, hydrates included, are an equally important component, which must be factored into strategies to design and control crystal form.[22] TGA and DSC (or DTA) are essential tools for the identification of solvates and the characterization of desolvation phenomena altogether. Weight losses observed by TGA usually indicate that a solvent has been removed from a solvate crystal structure upon heating, although other potential sources of weight loss, such as decomposition or sublimation, must also be considered. The total weight loss, the onset temperature of desolvation and the shape of the weight loss curve itself all provide useful information relating to solvate stoichiometry, the stability of a solvate as a function of temperature and the relative binding of the solvent(s) of crystallization. Generally, the interpretation of TGA data is straightforward, especially for stoichiometric solvates. However, when desolvation of a stoichiometric solvate occurs with minimal heating (*e.g.*, above rt), the solvent can be lost during the initial equilibration step of the analysis as the solvate is effectively dried by the flowing N_2 purge gas, resulting in either a negligible or non-stoichiometric weight loss. Such weight losses are commonplace for non-stoichiometric hydrates, with the total weight loss reflecting the water content at a specific relative humidity, clearly a snapshot of the actual hydration state. Whereas different sample encapsulation methods, such as pinhole pans, may be used to minimize desolvation at the beginning of the experiment, TGA results should be reconciled with those obtained by moisture sorption analysis to confirm non-stoichiometric hydration or otherwise. TGA, as mentioned earlier, is frequently coupled with DSC (or DTA), allowing the endothermic transitions to be directly correlated to weight losses as a function of temperature or time, and can be interfaced with FTIR or mass spectrometers to deconvolute the weight losses in terms of specific solvents or decomposition/ sublimation products.

10.2.3 Microscopy

Crystals have fascinated scientists for centuries by their unique shapes (habits), many facets and sometimes spectacular colors. One example, where color alone was singularly responsible for the initial identification of different crystal forms is now one of the most widely-studied polymorphic systems, 5-methyl-2-[(2-nitrophenyl)amino]-3-thiophencarbonitrile, also known as ROY for its red, orange and yellow crystals.[23] As an intermediate to Eli Lilly and Company's schizophrenia drug olanzapine, polymorphism would have gone entirely unnoticed had early batches not been produced in different colors. A similarly striking example of a color change, in this case induced by co-crystallization, was recently reported for benzophenone and diphenylamine, Figure 10.3.[24] Interestingly, the co-crystallization of these components did not require grinding of any kind; simply bringing the solids into contact with one another was sufficient to drive co-crystallization and effect a color change.

Examples like these underscore the importance of careful and astute observation while working with crystal forms of all types. Of course, most drug

Figure 10.3 Photomicrographs showing color change induced by co-crystallization of benzophenone and diphenylamine brought into contact with one another by simply mixing the powders with a spatula. The bright yellow 1:1 co-crystal is observed at the interface of the powders (Chadwick *et al.*, ref. 24; reproduced by permission of The Royal Society of Chemistry).

crystals are simply too small to be easily viewed by the naked eye and therefore require magnification under a microscope for assessment of their optical properties, including shape, size, refractive index, birefringence and optical dispersion. In fact, chemical analysis has its roots in optical (light) microscopy based on the early recognition that different molecules produce crystals with different optical properties. In some respects, the role of optical microscopy has been reduced over the years, as crystallographic and spectroscopic techniques have to a large extent superseded microscopy for chemical analysis, and light microscopy must now share the spotlight with other forms of microscopy, such as electron microscopy and atomic force microscopy, which operate at considerably higher magnifications. Nonetheless, in the hands of an experienced microscopist, light microscopy can yield a wealth of information on the crystal structure and composition of materials, salts and co-crystals included.

Polarizing light microscopy is particularly useful for studying the optical properties of crystals. A polarizing light microscope is essentially a light microscope equipped with a circular rotatable stage and two polarizers, one positioned in the light path between the illuminator and the specimen and the other (the analyzer) placed on the other side of the specimen between the objective and the eyepiece. The vibration directions of the two polarizers are oriented 90° from one another, hence the term 'crossed polarizers'. When cross polarized light passes through an anisotropic crystal (all crystals in orthorhombic, monoclinic and triclinic space groups common to pharmaceuticals are included in this category), the crystal will show bright interference colors, as long as it is not in an extinction position or aligned on an optic axis. Rotating a crystal every 90° will at some point cause the vibration directions of the crystal to be aligned with those of the polarizer and analyzer, effectively causing the light to be extinguished; the crystals will be black against a black background. Whereas the observation of interference colors under a polarized light microscope is commonly taken as evidence of birefringence from crystals, that is, crystallinity, in a sample, the extinction positions in relation to the external crystal shape provide an indication of the crystal system. Viewing crystals along an optic axis under cross polarized light will produce interference figures comprising all the birefringence colors, extinction included, which can be characteristic of different crystal forms. For a detailed description of the applications of polarized light microscopy, the reader is directed to an excellent monograph by Nichols.[25]

To identify phase composition and uniformity unequivocally (through spectroscopic mapping[26]), both isothermally and with changes in temperature, if so desired, polarizing light microscopes may be equipped with a hot stage and/or accessorized with Raman, near-IR and mid-IR spectrometers. Hot stage microscopy is especially useful for screening salt and co-crystal forms that crystallize from the melt, with one of the more popular techniques being the Kofler contact method.[27] To screen for salts and co-crystals by the Kofler method, two compounds are sequentially melted on a microscope slide, the higher melting compound first, followed by the lower melting compound, so they are adjacent to one another and partially mixed by application of a coverslip.[28] Within the mixing zone, a composition gradient will exist. If a salt or co-crystal with different melting behavior is formed, it will be readily observed in the mixing zone by optical microscopy as the sample is heated on the hot stage. Under plane-polarized light, only the solid phases will be visible, so when the sample is heated to temperatures above the melting point minima (eutectics), but below the melting points of the salt (or co-crystal) and its components, black regions will appear under the microscope separating the higher melting crystal forms. A simple three phase system comprising two pure components and its complex (salt or co-crystal) will produce as many as two eutectic regions, as shown for 2-[4-(4-chloro-2-fluorophenoxy)phenol]pyridine-4-carboxamide (I) and glutaric acid, Figure 10.4.[29] Co-crystallization was, in

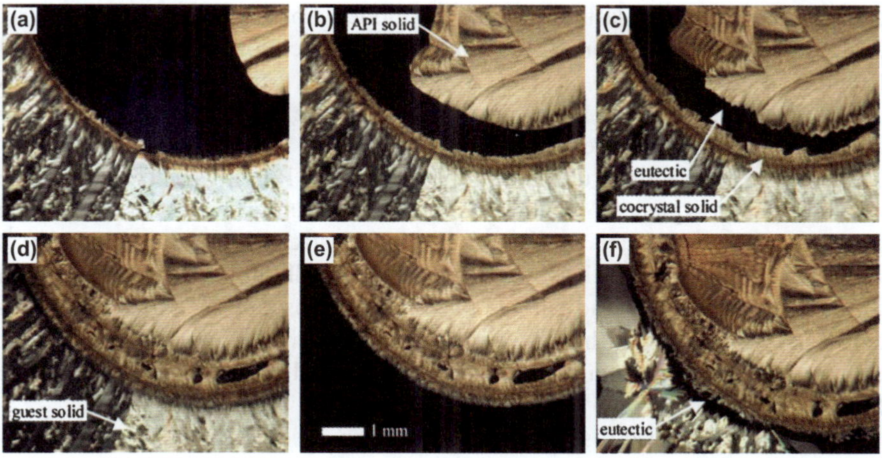

Figure 10.4 Photomicrographs showing a co-crystal of I and glutaric acid prepared using the Kofler technique, whereby the components are progressively heated (a–e), then cooled (f). At different temperatures, eutectics (dark areas viewed through cross polarizers) can be seen on either side of the co-crystal in frames (a–c) and (d and f) (reproduced with permission from McNamara *et al.*, ref. 29; copyright © Springer Science).

this particular case, subsequently confirmed by Raman spectroscopy. Thermal microscopy methods, such as the Kofler technique, allow the main features of the melting point diagram, such as the presence of eutectics and new mixed phases, to be determined in a single experiment using only milligrams of material. In comparison, determining a full phase diagram for a two component system by the mixed fusion method described earlier requires multiple measurements across a range of different component mixtures.

10.2.4 Vibrational (Raman, IR) Spectroscopy

Vibrational spectroscopy is typically used to characterize the fundamental vibrational modes of organic molecules either directly by their absorption of infrared (IR or mid-IR) radiation over the range of 4000–400 cm^{-1} or through inelastic scattering of radiation (producing Raman shifts from the excitation laser wavelength) of 3500–50 cm^{-1}. Other forms of infrared spectroscopy include near-IR spectroscopy, which probes molecular overtones and combination vibrations in the 14 000–4000 cm^{-1} range, and far-IR or terahertz spectroscopy, which probes lattice vibrations or phonon modes in the 133–1 cm^{-1} range. Vibrational (mid-IR and Raman) spectra, in providing definitive fingerprints of a molecule based on characteristic bond stretching and bending frequencies, have historically been used for chemical structure (functional group) identification. The sensitivity of mid-IR and Raman spectra to the solid state form has rendered them equally useful for phase identification and studies of hydrogen bonding in the solid state. Raman spectroscopy, when extended to lower wavenumbers (400–50 cm^{-1}), can also be used to probe intermolecular (lattice or phonon-mode) vibrations that are similarly characteristic of crystal forms. Although their applications are similar, IR and Raman spectroscopy are in fact complementary techniques, as vibrations require a change in dipole moment to be IR-active and a change in polarizability to be Raman-active.[30]

The different selection rules that apply to IR and Raman spectroscopy have interesting consequences when it comes to detecting water in crystalline hydrates. Whereas water produces strong OH stretching (3600–3100 cm^{-1}) vibrations in IR spectra, these vibrations are typically not observed in Raman spectra. Raman has the advantage over IR spectroscopy in that sample preparation is minimal, but the disadvantages of not being suitable for fluorescent compounds and potentially damaging to colored and thermally unstable compounds.

Raman and IR spectroscopy have both been used extensively in the characterization of salts and co-crystals to identify functional groups corresponding to the individual components, to characterize proton transfer in the solid state and to confirm the presence of new binary phases based on unique IR/Raman fingerprints. Confirmation of the latter is achieved simply through comparison with the pure component spectra, as shown for an indomethacin–saccharin co-crystal in Figure 10.5.[31] In this case, not only were the Raman carbonyl

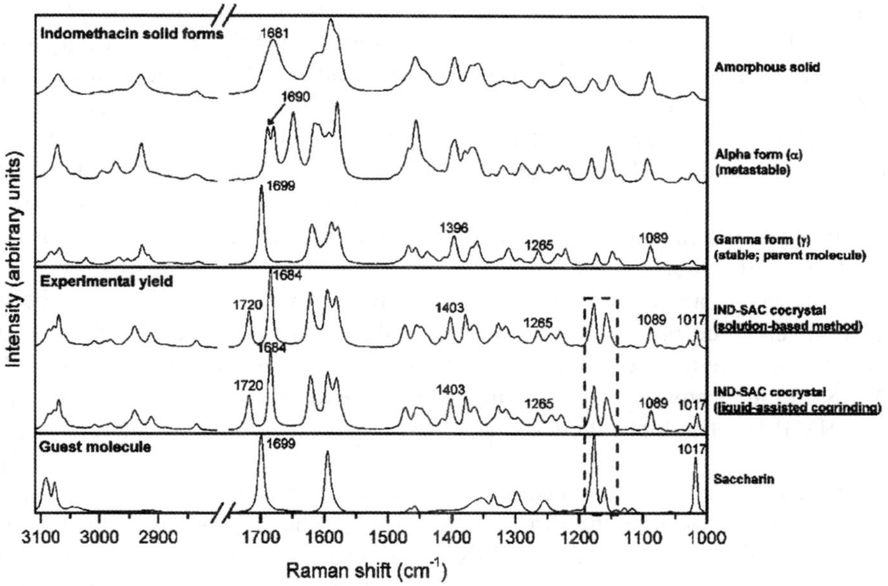

Figure 10.5 Baseline-corrected Raman spectra of an indomethacin–saccharin co-crystal prepared by solution-based or grinding methods (from Alessø *et al.*, ref. 31; copyright © American Chemical Society).

stretching bands of each component shifted as a result of homodimer formation in the co-crystal, but the co-crystal materials prepared by solution crystallization and liquid-assisted grinding methods were also confirmed to be one and the same form.

10.2.5 Solid State NMR Spectroscopy

Solid state nuclear magnetic resonance (SSNMR) spectroscopy, which in many respects complements the most powerful tool for structure determination, X-ray crystallography, has become an indispensable tool for the characterization of both structure and dynamics in molecular materials.[32] This non-destructive technique has the unique ability to probe electronic environments and the dipolar connectivity of NMR-active nuclei (*e.g.*, ^{13}C, ^{15}N, ^{19}F, ^{31}P, ^{23}Na and ^{1}H) in the solid state over a large timescale without the requirement for single crystal substrates or even homogeneous samples.[33] Thus, not only is SSNMR spectroscopy well suited for identifying polymorphs and solvates in bulk APIs, but this technique can also provide detailed structural information useful for both rationalizing material properties in terms of molecular and crystal structure and confirming structure solutions obtained by powder diffraction.[34]

The basic principles of NMR are the same for solution and solid state measurements; however, in order to obtain high-resolution solid state one-dimensional (1D) spectra, the standard solution methods have been modified to minimize two effects of the solid state on NMR: low sensitivity and severe line broadening. The sensitivity of the solid state experiment, which for ^{13}C is expected to be low based on its 1% natural abundance, is further diminished by long spin-lattice (T_1) relaxation times. Severe line broadening originates from both the orientation-dependent chemical shift, that is, chemical shift anisotropy (CSA), and dipole–dipole interactions, the latter of which occur when the magnetic moment of one nucleus influences the neighboring field of another nucleus.

While magic angle spinning (MAS) and high power decoupling have been used to enhance peak resolution, cross polarization (CP) has been used to enhance sensitivity. Schaefer and Stejska[35,36] were the first to combine CP, MAS and high power ^1H decoupling to produce high resolution solid state 1D ^{13}C NMR spectra. Since then, CP/MAS NMR spectroscopy has emerged as an essential, if not routine, tool for physicochemical analysis and is rapidly approaching the status (and acceptance) of X-ray diffraction as one of the most powerful tools for differentiating crystal forms.[37] The development of sophisticated two-dimensional (2D) SSNMR techniques that have been available to solution NMR spectroscopists for decades has in recent years significantly increased the amount of information that can be gained by SSNMR spectroscopy. For the interested reader, a number of excellent review articles have been written detailing the basic CP/MAS technique,[38] describing the hardware specific to SSNMR spectrometers[39] and highlighting practical aspects and pharmaceutical applications of high resolution SSNMR spectroscopy.[40–42]

SSNMR spectroscopy is unparalleled with respect to the diversity of techniques designed specifically to probe structure and dynamics with site selectivity, not to mention examine phase/component miscibility. Beyond the first-pass analysis of 1D spectra to differentiate potential salt and co-crystal forms from those of the individual components, both relaxometry and 2D correlation spectroscopy have been increasingly used to characterize salts and co-crystals. ^1H T_1 (or T_{1H}) relaxation time measurements can provide direct evidence of phase heterogeneity (to confirm the presence of phase impurities and/or rule out salt/co-crystal formation) based on the observation of multiple relaxation times characteristic of different component phases in a given material. ^1H $T_{1\rho}$ (or $T_{1\rho H}$) relaxation, which like T_{1H} relaxation, is strongly affected by efficient spin diffusion over the entire proton reservoir, is also frequently applied to study mixtures, and in favorable cases, both T_{1H} and $T_{1\rho H}$ measurements can allow domain sizes (hundreds of angstroms in the case of T_{1H}) to be calculated. In contrast to relaxometry, which provides direct evidence of component phase separation, dipolar correlation techniques, for example, ^1H–^{13}C CP-HETCOR (HETeronuclear CORrelation spectroscopy), can definitively prove molecular association between the different components in a salt or co-crystal. The ability of ^1H–^{13}C CP-HETCOR NMR spectroscopy to confirm co-crystal formation was recently demonstrated for a palmitic acid nicotinamide 1:1 co-crystal.[43]

As shown in Figure 10.6, close contacts (intra- or intermolecular) between ^1H and ^{13}C spins appear in the 2D spectrum as cross peaks. When correlations

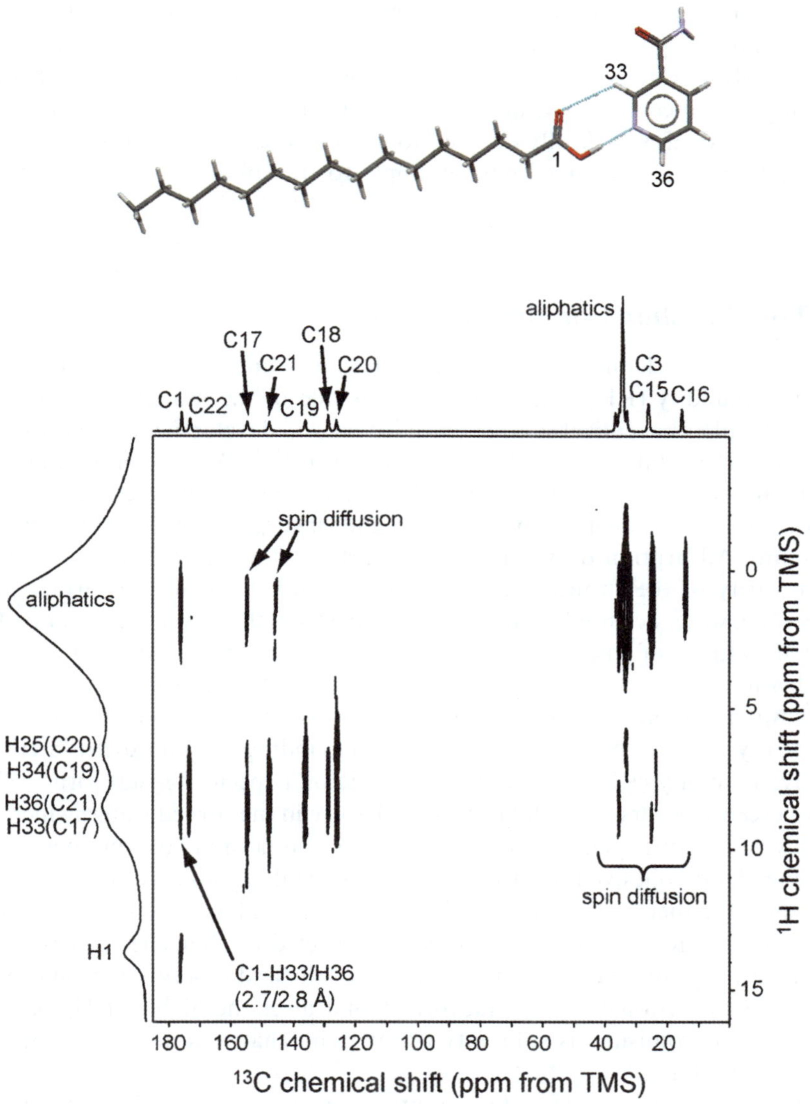

Figure 10.6 Top: Palmitic acid nicotinamide heterodimer formed in the 1:1 co-crystal (Cambridge Structural Database (CSD) ref code JEMDIP), bottom: ^{1}H–^{13}C CP-HETCOR spectrum of palmitic acid nicotinamide 1:1 co-crystal. 1D ^{1}H MAS ($\nu_r = 35$ kHz) and ^{13}C CP-TOSS (Cross Polarization-TOtal Sideband Suppression) ($\nu_r = 8$ kHz) spectra are shown as projections (adapted from Vogt *et al.*, ref. 43; copyright © 2008 American Chemical Society).

are observed for nuclei in different molecules, such as that seen between C1 of palmitic acid and both H33 and H36 of nicotinamide, molecular association, that is, co-crystal formation in this case, is confirmed. Spin diffusion over longer distances (>5 Å) will also produce correlations in CP-HETCOR spectra, providing additional proof of molecular association when the correlations are assigned to different constituents. Recently, both relaxometry and CP-HETCOR NMR spectroscopy were featured among many SSNMR techniques for co-crystal characterization in an excellent overview by Vogt *et al.*[43] Although specifically devoted to co-crystal characterization, the methods and strategies described equally apply to the solid state characterization of salts.

10.2.6 Moisture Sorption Analysis

Hygroscopicity, defined as the uptake of water vapor by a solid at a given relative humidity (RH) and temperature, is one of the most important attributes of a pharmaceutical solid to consider early in development when selecting a salt or co-crystal for a solid oral dosage form.[44–47] Water can interact with crystalline solids by adsorption onto the solid surface, deliquescence, capillary condensation (in microporous solids) and incorporation into the crystal structure. Adsorption of water onto the surface of a crystal depends on both the polarity of the surface functional groups and the available surface area. Generally, water accumulation at the surface occurs through hydrogen bonding to a maximum of just a few molecular layers. Deliquescence and capillary condensation, on the other hand, will lead to condensed water that may dissolve water-soluble compounds. Highly water-soluble salts and co-crystals will frequently deliquesce when the humidity surrounding a solid exceeds a critical relative humidity (RH_o) that is characteristic of a specific crystal form. When deliquescence occurs, the solid begins to dissolve in the sorbed water, which can have serious consequences, the least of which is a spontaneous increase in stickiness and cohesiveness of the powder. Since deliquescence has significant effects on the thermodynamic activity of water (a colligative effect), the counterion used to form a salt, for example, is expected to alter RH_o. Therefore, a useful strategy for dealing with deliquescent salts and co-crystals would be to choose a counterion or co-crystal former that gives the highest RH_o possible without compromising its solubility to the point that it is no longer pharmaceutically useful.

The dominant mechanism of water uptake in crystalline solids is absorption. With non-stoichiometric hydrates, water is continuously absorbed in channels of the crystal structure without a distinct phase change. Water can also displace organic solvents in crystalline solvates, often leading to isostructural hydrates and sometimes facilitating the removal of the organic solvents. In many cases, phase changes to more stable hydrated and even neat forms will occur. Here, it must be understood that crystals exposed to water vapor below their deliquescence point will not encounter liquid water. Therefore, nucleation and growth of a new phase will only occur at defect sites or in localized amorphous regions, wherein water accumulates, plasticizing the solid sufficiently to induce physical and/or chemical changes in an otherwise crystalline material.[48] As all of the thermodynamic properties of a drug crystal form are subject to change with any first order phase transition, it is best to identify a salt or co-crystal form that is stable under normal storage conditions, typically 10–75% RH at rt.

Water uptake in pharmaceutical solids may be measured gravimetrically by placing a pre-weighed sample in a constant RH environment, for example, a closed desiccator containing a saturated solution of an electrolyte, then periodically removing and weighing it. This process is repeated until the sample has reached equilibrium. The conventional desiccator method as described here has to a certain extent given way to automated techniques, however, it is still widely used to equilibrate samples when *in situ* analysis at a specified humidity is not possible. In automated moisture sorption analysis, also referred to as dynamic vapor sorption analysis, a sample is placed on a microbalance and exposed to a continuous flow of air or N_2 with a predetermined RH. After the sample has equilibrated for a specified period of time, the RH is automatically incremented to ascending or descending values, all the while continuously monitoring the weight *in situ*. Regardless of which water vapor sorption method is used, an isotherm may be calculated from the equilibrium moisture uptake at each RH increment. For an isotherm to be accurate, however, the mass must stabilize at each RH increment, so the duration of the experiment will depend in part on the nature of the moisture uptake. Whereas surface adsorption is typically relatively fast, bulk absorption via vapor diffusion is frequently quite slow. Finally, since only the sample weight is monitored during the course of generating a moisture sorption isotherm, physical changes must have an accompanying weight change to be observed (transformations between true polymorphs induced by water vapor sorption typically go undetected). As only the final material will be available for subsequent *ex situ* analysis of the phase composition, *in situ* analysis may be required if reversible form conversions are suspected based on observed weight gains/losses, for which environmental XRPD is perfectly suited. Other techniques, such as solid state NMR spectroscopy, may be used for *ex situ* analysis, provided that measures are taken to control the RH at which the sample is exposed.

10.3 Integrated Approaches to the Solid State Characterization of Salts and Co-crystals

When salts and co-crystals are [re]crystallized under varying conditions, the expectation is that energetically accessible polymorphs and solvates will appear as molecules or ions pack in different crystal structures at different rates to minimize their free energy. Although unique materials coming from solid form screens are typically flagged for further investigation based on initial screening data, such as XRPD patterns that do not correspond to one of the known forms, exactly how the new 'forms' should be characterized thereafter (assuming they can be reproduced) will depend on their chemical composition, purity and, of course, stability. Given the diversity of solid state forms that can exist and the many different solid state techniques that are typically available, the approaches that may be taken to identify forms and evaluate their properties are countless. Certainly any one technique may be sufficient to answer the question at hand, although most would agree that probing different aspects of solid state structure (and dynamics) through a combination of techniques is the best way to construct a complete, if not accurate, picture of the solid state. Some of the most important attributes to consider when encountering a potentially new salt or co-crystal form, along with common solid state techniques that can provide useful information, are summarized in Table 10.1. How different solid state techniques may be combined to answer important questions commonly asked more specifically during the initial selection of a salt or co-crystal, as well as later when problems are encountered during its development into a drug product, is described in the following sections.

10.3.1 Form Identification/Phase Purity

Owing to their ability to differentiate rapidly most solid state forms, powder X-ray diffraction and Raman spectroscopy have been utilized extensively, particularly in high-throughput salt/co-crystal form screening programs, to identify new solid phases. The appearance of new powder patterns or vibrational spectra is not sufficient to confirm salt or co-crystal formation between a host and guest, however, as polymorphs and solvates of either parent compound would similarly produce unique patterns/spectra. Therefore, to establish the existence of a new salt or co-crystal form unequivocally, two additional criteria must be satisfied: the chemical composition of the new 'form' *and* the molecular association of the components within the same physical form must be established. To meet the first of these criteria, common solution or 'chemical' methods for structural identification, such as solution-state NMR spectroscopy, mass spectrometry, elemental analysis, Karl Fischer titration, evolved gas analysis (thermogravimetric (TG)-FTIR spectroscopy and TG-MS) and liquid or gas chromatography, will almost always be employed alongside solid state techniques to identify novel polymorphs and solvates of salts and co-crystals.

The identification of two (or more) components in a material by chemical methods alone can infer salt, co-crystal or solvate formation when one or both of the components is a liquid or a gas under ambient conditions. For example, when elemental or evolved gas analysis reveals that HCl or HBr, both gases under STP conditions, are present, salts are very likely to be formed, although rare cases of co-crystals have been reported. This was the case for a molecular HCl complex found for the mono-HCl salt of *N*-{(1*S*)-1-[2-(1-{[(3S,4R)-1-*tert*-butyl-4-(2,4-difluorophenyl)pyrrolidin-3-yl]carbonyl}piperidin-4-yl)-5-chlorophenyl]ethyl}acetamide, wherein two equivalents of HCl were confirmed through elemental analysis, $AgNO_3$ titration and TG-MS.[49] Remarkably, not one, but four crystal forms of this most unusual 'bis-HCl' salt complex were identified, despite the presence of only one relatively basic site in the molecule. While HCl (or HBr) incorporation in a crystal usually signals salt formation and stoichiometric solvent retention generally points to solvate formation, the

Table 10.1 Solid state techniques for the characterization of salts and co-crystals.

Attribute	*Technique*	*Information*
Crystal structure	Single crystal neutron diffraction; single crystal X-ray diffraction; powder X-ray diffraction; solid-state NMR spectroscopy	Unit cell composition; atomic positions; anisotropic displacement parameters; bond lengths/angles; disorder; conformations; hydrogen bonding; crystal packing
Phase composition	Powder X-ray diffraction; solid-state NMR spectroscopy; Raman/IR spectroscopy; thermal analysis	Short-range order; long-range order; crystal form; phase purity
Chemical composition	Solution NMR spectroscopy; elemental analysis; Karl-Fisher titration; evolved gas analysis	Structure composition/degradation; evolved gases with increasing *T*; solvent identification/quantitation; stoichiometry
Habit/morphology	Light microscopy; electron microscopy	Crystallinity; crystal shape; particle size; phase purity
Thermal stability	DSC; TGA; environmental XRPD	Melting point; heat of fusion; heat of transition; desolvation temperature
Ionization state	Single crystal neutron diffraction; single crystal X-ray diffraction; solid-state NMR analysis; vibrational spectroscopy; X-ray photoelectron spectroscopy	Salt *vs.* co-crystal; disorder
Phase transformations	Environmental (VT, VRH)[a] XRPD; VT SSNMR spectroscopy; Raman/IR spectroscopy; thermal analysis; hot stage microscopy; moisture sorption analysis; slurry equilibration	Thermal/RH stability; thermodynamic transition temperature; critical water activity; transformation kinetics
Hygroscopicity	Moisture sorption analysis	RH stability; rate and mechanism of water uptake; critical water activity; deliquescence

[a]VT and VRH are variable temperature and variable relative humidity, respectively.

identification of multiple otherwise solid components in molecular materials that give unique powder patterns or solid state spectra does not in and of itself prove that the components are in the same physical form. In such cases, salt formation or co-crystallization can be confirmed by single crystal X-ray diffraction, although differentiating between a salt and co-crystal may not be possible (*vide infra*). 2D SSNMR correlation spectroscopy, as discussed earlier, may also be used to characterize directly molecular association between the ion pair in a salt or the host and guest in a co-crystal.[43] In most cases, however, salts and co-crystals are indirectly characterized by a combination of conventional physical and chemical methods.

Drug crystal forms may be differentiated by their chemical composition (number, identity and stoichiometry of components), the ionization state of acidic and basic components, if present (salts/co-crystals), and/or their solid state packing arrangements (polymorphs). Thus, aside from having a counterion or guest molecule to identify and quantify, and possibly the need to confirm protonation in the solid state, the approaches taken to characterize salt and co-crystal forms are no different from those used to characterize the neat and solvated forms of the parent molecule.[50] If a single crystal X-ray structure is available, the characterization of both chemical composition, stoichiometry included, and the crystal packing can, in most cases, be confirmed from these data alone. Otherwise, in order to establish that two crystal forms are polymorphic, their chemical composition must be proven to be identical and their crystal structures confirmed to be different using chemical and physical methods, respectively. Although TGA is usually used to identify solvates, simply observing that translucent crystals turn opaque or crack over time or with heating on a microscope hot stage is generally sufficient to establish solvate formation. Alternatively, crystals can be immersed in a transparent oil (*e.g.*, mineral oil) and as long as they do not dissolve in it, bubbling will be observed with heating on a microscope hot stage as solvent is lost from a solvate crystal structure. Spectroscopic techniques that can uniquely identify the solvent are also frequently used to confirm solvate formation. In this regard, SSNMR spectroscopy is particularly useful, since only bound (immobilized) solvent is detected under CP conditions. Moisture sorption analysis, of course, is invaluable for the characterization of hydrate stoichiometry, or its lack thereof. Barring cases of extreme chemical or physical instability, the solid state characterization of anhydrates and stoichiometric solvates is generally straightforward. Labile forms, isostructural forms[51] and non-stoichiometric solvates,[52] on the other hand, can present unique challenges to solid state characterization which frequently require multiple analytical techniques and sound experimental strategies.

Among the more important considerations in characterizing pharmaceutical materials generated during salt, co-crystal and polymorph screening is phase purity, for which many different techniques may apply. Optical microscopy can provide a first indication of multiple phases when unique and distinguishable morphologies are observed in a single batch of an API; however, other methods will be needed to confirm the presence of a phase impurity. One of the most definitive approaches to characterizing the phase purity of bulk samples is to predict (and refine, if needed) a powder pattern from the single crystal X-ray structure of the primary component, then compare it to the experimental pattern. When the positions and intensities of all of the reflections are accounted for in the experimental powder pattern, the material is presumed to be a pure phase. In the absence of single crystal diffraction data, measured XRPD patterns can be indexed to identify 'extra' peaks produced by crystalline phase impurities.

Preferred orientation effects can overwhelm the powder diffraction patterns of pharmaceutical solids, however, so small differences in relative peak intensities may initially be overlooked and many XRPD patterns may need to be collected before a trend that signals that a phase impurity is present becomes apparent. To minimize preferred orientation, transmission XRPD patterns may be collected for samples packed in spinning capillaries. Alternatively, SSNMR spectroscopy, which does not suffer from the effects of preferred orientation (crystals are tightly and randomly packed in ceramic rotors), can be used to readily assess phase purity. Mixtures of crystal forms generally give rise to additional isotropic peaks which are concentrated in spectral regions defined by the structures of the drug, counterion or co-crystal former in 1D SSNMR spectra. The appearance of 'extra' peaks will signal that another crystal form is present once other possible sources of peak splitting (*e.g.*, $Z' > 1$) are ruled out. In favorable cases, such as when the relative intensity of the split peaks is decidedly unequal, crystal form impurities may be identified from a *single* SSNMR spectrum. Finally, phase purity can be assessed using NMR relaxometry, although as discussed earlier, only phase heterogeneity, that is, the presence of phase impurities, can be unequivocally established.[43]

10.3.2 Proton Transfer in the Solid State

For acid–base complexes, the lone difference between a salt and a co-crystal is the location of the acidic H atom(s) in the crystal structure. While it is generally accepted that proton transfer will occur between an acid and base to form a salt in solution when the difference between their acid ionization constants (pK_a of base – pK_a of acid or ΔpK_a) is greater than two or three units,[53] crystallization may yield salts, co-crystals, or disordered solid forms that exhibit partial proton transfer when the ΔpK_a is less, with the exact location of the acidic proton being strongly dependent on the specific crystal packing environment. Here, it must be understood that the pK_a value is a solution property that is not specifically defined in crystals and as such, cannot be transferred to the solid state in a general way.[54]

While Aakeröy *et al.* have noted dramatically different structural behavior between carboxylic acid co-crystals and carboxylate salts, what effect, if any, the location of the proton specifically has on the physical properties relevant to delivering a drug in a controlled manner is unclear.[55] Indeed for all intents and purposes, whether the form is unionized, a salt or co-crystal should not matter as long as its physicochemical properties are suitable for the drug product. That said, from a regulatory perspective, the jury is still out. Today, the regulatory path for new salts requires more testing and clinical data when submitted to the U.S. Food and Drug Administration as a 505(b)(2) Abbreviated New Drug Application (ANDA), yet the testing requirements for the generic market entry of co-crystals have not been established.[2] Because of the potentially different regulatory pathways that may arise for salts and co-crystals in the future, it is important to look at the different analytical techniques used specifically to probe ionization in the solid state.

Single crystal diffraction remains the technique of choice for accurately locating atomic positions, as well as measuring anisotropic displacement parameters, in a crystal structure; however, Fourier difference maps derived from X-ray diffraction can only be used to approximate hydrogen atom positions. The definitive technique for precisely locating hydrogen atoms to characterize unequivocally proton transfer in the solid state is neutron diffraction. Unlike X-rays, which are scattered by the diffuse electrons of an atom, neutrons are scattered by its nucleus. This difference is especially significant for hydrogen atoms, because not only is the neutron scattering power of H comparable to that of C, O and N (the scattering intensity of X-ray diffraction varies monotonically with *Z*), but the center of gravity of the neutron also perfectly defines the atomic position, unlike that of the single electron of hydrogen, which is typically displaced towards the atom that it is bonded to.[56–58] Neutron diffraction, although not widely available, has been used to examine structural details associated with hydrogen atom positions in crystals with high accuracy. Steiner, for example, in compiling the distribution of hydrogen atom positions in O–H···O and N–H···O/O–H···N hydrogen bonds, found smooth correlations among many neutron crystal structures, where the hydrogen atom occupied sites associated with either O or N, as well as in between when hydrogen bonding was particularly strong and the proton was shared, Figure 10.7.[54] From these correlations and others, all hydrogen bonds were likened to an incipient proton transfer reaction, which, of course, is the key reaction transforming what might otherwise be a co-crystal into a salt. Indeed, a similar continuum does exist for salts and co-crystals, exemplified by structures wherein the acidic proton resides neither on the acid nor the base in the crystal structure, but instead is either shared between the two or disordered, with partial occupancy at both sites.[59] The concept of an incipient proton transfer reaction has been well demonstrated in studies of proton transfer mediated by temperature for a pentachlorophenol:pyridine 1:1 complex, as well as the ternary 4-dimethylaminobenzoic acid:3,5-dinitrobenzoic acid:4,4'-bipyridine complex recently reported by Thomas et al. and shown in Figure 10.8.[60]

Since hydrogen atom positions are more precisely and reliably located by neutron diffraction, it is a common practice to normalize O–H and N–H bond lengths in X-ray structures to match the average internuclear distances obtained for those bonds by neutron diffraction (*e.g.*, O–H = 0.983 Å, N–H = 1.009 Å). Normalizing the bond lengths, which entails shifting the H-atom along the bond vector, in some cases, by more than 0.1 Å, can lead to uncertainty in characterizing proton transfer in acid–base complexes. Fortunately, close inspection of structural parameters, such as bond lengths and angles, can also reveal whether or not ionization has occurred.

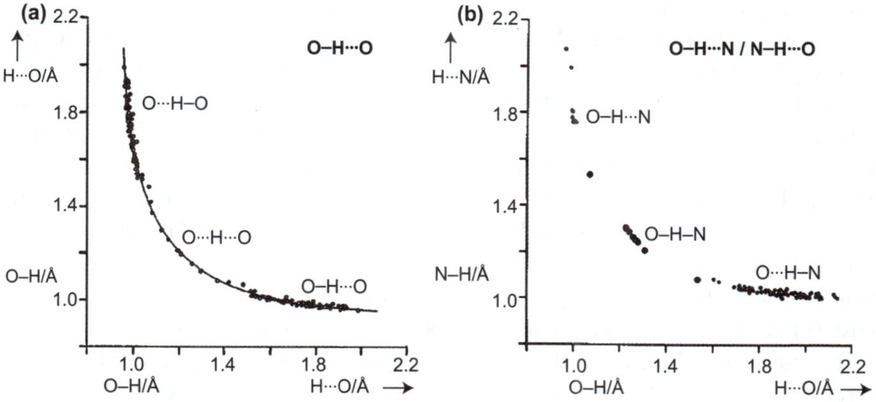

Figure 10.7 (a) Correlation of the O–H and H···O distances in O–H···O hydrogen bonds symmetrized with respect to the two O atoms and (b) correlation of N–H/O–H and H···O/H···N bond lengths in N–H···O and O–H···N hydrogen bonds based on neutron diffraction data (from T. Steiner, ref. 54, The hydrogen bond in the solid state, pp. 48–76; copyright Wiley-VCH Verlag GmbH & Co. KGaA, reproduced with permission).

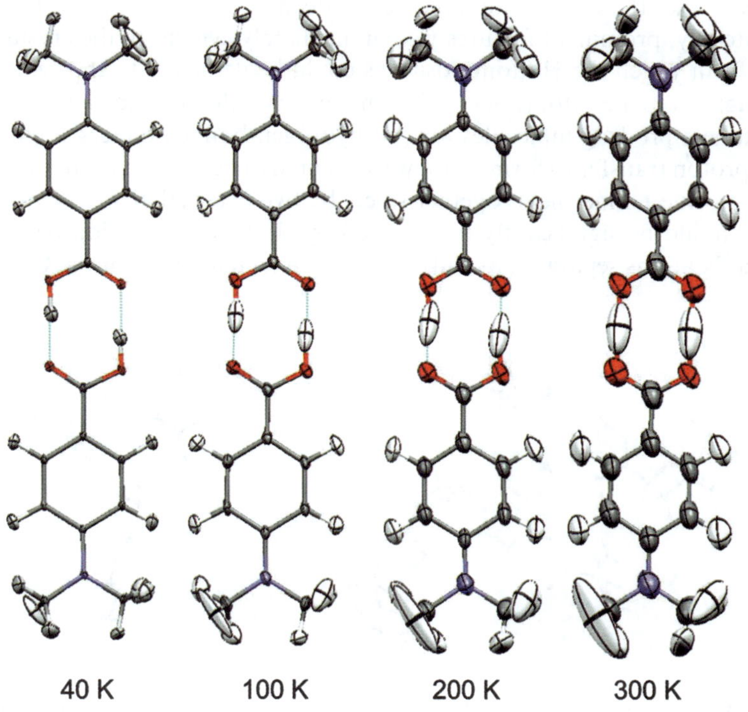

Figure 10.8 Anisotropic displacement parameters of 4-dimethylaminobenzoic acid in a ternary molecular complex determined by neutron diffraction, showing the evolution of the hydrogen atom from a well-localized position at 40 K to an elongated central position at 300 K (reproduced from Thomas *et al.*, ref. 60; copyright © 2010 American Chemical Society).

For example, the C–O bond lengths of carboxylic acids provide immediate evidence of proton transfer. Whereas neutral carboxylic acid C–O bond lengths that might be observed in co-crystals may differ by more than 0.08 Å (1.2 Å for C=O, 1.3 Å for C–OH), the C–O bond lengths in carboxylate salts have been shown to be within 0.03 Å of one another.[61,62] Deprotonation of a carboxylic acid to form a salt is not only expected to equalize the C–O bond lengths observed by X-ray diffraction, but has also been shown to have a dramatic effect on the C=O/C–O stretching vibrations observed by IR spectroscopy. Whereas neutral carboxylic acids in co-crystals have strong C=O absorption bands near 1700 cm^{-1}, along with weak C–O absorption bands near 1275 cm^{-1}, salts typically show single strong C–O stretches near 1650 cm^{-1}.

In addition to bond lengths and angles observed by X-ray crystallography, careful inspection of intermolecular contacts and torsion angles can indirectly reveal the ionization state of an acid or base in a crystal structure. This type of analysis was essential for the characterization of proton transfer in gemcitabine prodrug 1:1 and 2:1 *p*-toluenesulfonic acid complexes. As a *very* weak free base ($pK_a \sim 0.65$), gemcitabine prodrug was an oil, so strong acids were screened in the initial search of a crystalline salt form for the isolation, purification and storage of the drug substance.[63] The crystal structures of 1:1 and 2:1 *p*-toluenesulfonic acid complexes were solved; however, the H atom positions were not located with certainty and thus were not able to confirm salt formation or otherwise. Fortuitously, prominent features were immediately evident in the crystal structures, from which the H atom positions could be deduced. In the gemcitabine prodrug: *p*-toluenesulfonic acid 2:1 complex, crystallographically inequivalent gemcitabine prodrug molecules formed hydrogen-bonded dimers, Figure 10.9. Here, proton transfer to form a salt must occur, as a neutral aminopyrimidine-2-one ('cytosine') dimer held together solely by two external NH···O hydrogen bonds should be significantly destabilized by electronic repulsion between the internal N atoms separated by only 2.863 Å. In fact, a search of the Cambridge

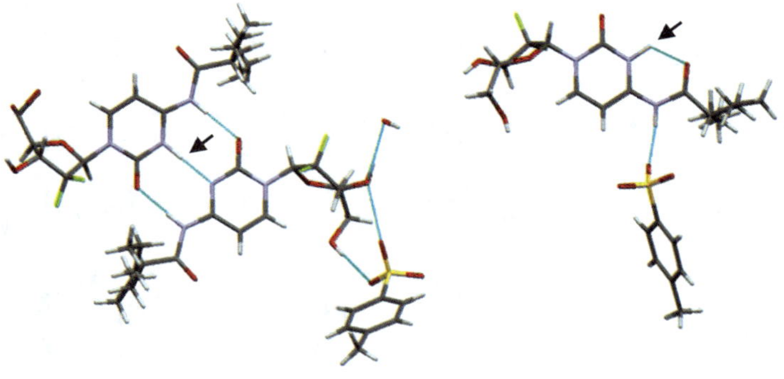

Figure 10.9 Gemcitabine prodrug:hemi-*p*-toluenesulfonic acid hemihydrate salt cocrystal (left) and gemcitabine prodrug:mono-*p*-toluenesulfonic acid neat Form I (right). The H atom positions (denoted by arrows) were inferred in each structure from intermolecular and intramolecular close contacts.

Structural Database (CSD) revealed that 0/153 structures containing neutral cytosine rings were dimerized in this way. With protonation of the cytosine ring, a complementary structure to the neutral molecule was created, hence the formation of a salt co-crystal, wherein a second equivalent of gemcitabine prodrug was bound in the crystal structure by triple hydrogen bonding, like that observed for the guanine–cytosine (GC) base pair in DNA. Perhaps not surprisingly, charge-assisted hydrogen bonding in the gemcitabine prodrug cytosine–cytosinium ion dimer appears to be stronger (based on shorter intermolecular contacts) than that in the GC base pair. Proton transfer to form the 1:1 *p*-toluenesulfonic acid salt of gemcitabine prodrug was similarly deduced from the orientation of the valproic acid side chain. As shown in Figure 10.9 (right), the valproic amide carbonyl is *syn*-coplanar with the *ortho*-cytosine ring nitrogen in this crystal form. Significant electrostatic repulsion would be expected to twist the valproic amide group out of the plane of the aminopyrimidine-2-one ring if proton transfer did not occur. Instead, the gemcitabine prodrug is protonated, allowing an intramolecular hydrogen bond to lock the amide carbonyl group of the valproic acid side chain into a coplanar orientation.

SSNMR spectroscopy has also been used to examine the ionization state of different functional groups in crystalline solids. One NMR approach to establishing the degree of proton transfer in the solid state is to measure chemical shift tensor values, which are known to be especially sensitive to changes in ionization state.[64] Li *et al.* examined proton transfer and hydrogen bonding in several acid–base complexes of an ErbB2 inhibitor using single crystal X-ray crystallography and ^{15}N SSNMR spectroscopy.[65] In this work, the crystal structures of the free base, a sesquisuccinic acid co-crystal, a dimalonic acid mixed ionic/zwitterionic complex and a dimaleate salt were determined and the H atoms attached to O and N were located from difference maps. Interestingly, among this series of acid–base complexes, the degree of proton transfer correlated with the ΔpK_a between the acid and base. Each form was then characterized by ^{15}N CP/MAS NMR spectroscopy at natural abundance, with spectral assignments made through short contact time spectral editing, a technique used to distinguish the protonated and non-protonated nitrogen atoms, as well as ^{15}N chemical shift calculations.

As shown in Figure 10.10, CP/MAS NMR spectra collected using short contact times produce sub-spectra of ^{15}N atoms directly attached to protons, owing to strong ^{1}H–^{15}N dipolar coupling of these centers; non-protonated N centers either do not appear in short contact time spectra or their signal(s) is strongly attenuated. Protonation of the free base to form the dimaleate salt was shown to significantly shift (>80 ppm) the ^{15}N resonances upfield in the CP/MAS NMR spectrum, while strong hydrogen bonding in the co-crystals shifted the same ^{15}N centers to a much smaller extent (20–40 ppm) relative to the free base. Proton transfer in acid–base complexes can also be established through dipolar (through space) and scalar (through bond) interactions. Emsley and co-workers showed that 2D solid state (^{1}H–^{15}N) MAS–*J*–HMQC NMR spectroscopy can be used to correlate isotropic chemical shifts between pairs of directly bonded ^{1}H and ^{15}N nuclei, allowing protonated ^{15}N sites to be

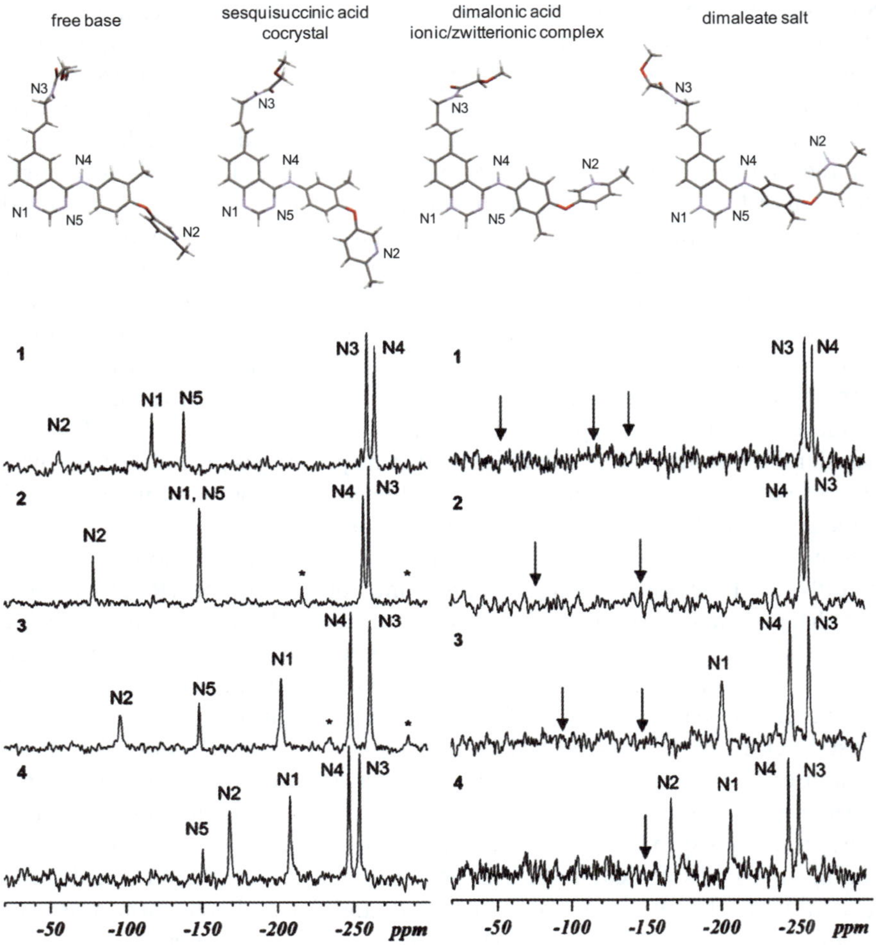

Figure 10.10 Normal (left) and short contact time (right) ^{15}N CP/MAS NMR spectra of free base (1), sesquisuccinic acid co-crystal (2), dimalonic acid mixed ionic/zwitterionic complex (3) and dimaleate salt (4) forms of an ErbB2 inhibitor. Only ^{15}N atoms attached to ^{1}H appear in the short contact time spectra (Adapted from Li *et al.*, ref. 65; Copyright © 2006 American Chemical Society).

readily identified.[66] Apperley *et al.* confirmed the site of proton transfer in sildenafil citrate, the active ingredient in Viagra®, using dipolar dephasing ^{15}N CP/MAS NMR spectroscopy to identify the protonated and non-protonated ^{15}N centers.[67] Peresypkin *et al.* differentiated protonated and non-protonated ^{15}N sites using low power decoupling to eliminate signals effectively from the strongly coupled NH groups.[49]

X-ray photoelectron spectroscopy (XPS), a surface technique for chemical analysis which probes the amount of energy required to emit an electron from an atomic core level, has also been shown to probe proton transfer in the solid state directly. Because XPS, like SSNMR spectroscopy, is sensitive to changes in the chemical environment of individual atoms, protonation of a N atom, for example, will cause its peak to shift in the N 1*s* emission spectrum. The sensitivity of XPS to the protonation state has been demonstrated for a 2:1 theophylline:oxalic acid co-crystal and theophyllinium salicylic-5-sulfonate monohydrate.[68] Whereas the N 1*s* emission spectrum of the 2:1 theophylline:oxalic acid co-crystal was virtually identical to that of pure theophylline, the N 1*s* peak of the salicylic-5-sulfonate salt was shifted by +2.3 eV. Here, the shift towards higher binding energy was interpreted in terms of a strong localized positive charge produced upon protonation, while the absence of peak shifting relative to pure theophylline was taken as evidence of co-crystal formation. The XPS characterization of a 1:1 theophylline:citric acid co-crystal prepared by dry grinding has also been reported.[69] In this case, not only was proton transfer ruled out based on the comparative N 1*s* emission spectra between the citric acid complex and pure theophylline, but C 1*s* and O 1*s* photoemission spectra were also used to confirm the identity and stoichiometry of the two components in the milled product. The ability of XPS to distinguish protonated and unprotonated N sites at the solid surface was recently put to good use in identifying the source of discoloration in bulk lots of a fumarate salt with compromised mechanical properties.[70] Adhesion of the N free base to the surface of the fumarate salt was in this case confirmed through analysis of HN^+/N peak ratios in the N 1*s* XPS emission spectra.

10.3.3 Thermodynamic Stability

Although in principle a salt or co-crystal form, be it anhydrous or a hydrate, need only be kinetically stable throughout the shelf-life of a drug product, the thermodynamically stable form, that with the lowest free energy, is generally preferred to mitigate the risk of undesired process- or storage-induced phase transformations. Establishing the relative thermodynamic stability of neat polymorphs and hydrates of a given salt or co-crystal is therefore paramount to ensuring that the stable form is selected. To this end, the phase stability must be at least qualitatively understood as a function of temperature and where hydrates are known to exist, as a function of RH. Fortunately, several techniques are available to measure the thermodynamic stability of drug crystal forms, both qualitatively and quantitatively, Table 10.2.[71] The most clear and direct way to assess the relative thermodynamic stability of two (or more) forms is to measure their equilibrium solubility in a common solvent system. As given by equation (10.2), the Gibbs free energy difference (ΔG) between two forms, A and B, at any given temperature is directly related to the ratio of their solubilities, S_A and S_B:

$$\Delta G_{B-A} = -RT\ln(S_B/S_A) \tag{10.2}$$

Equilibrium solubility measurements are not always technically feasible, however. For example, solution-mediated phase transformations may occur,

Table 10.2 Direct and indirect methods for evaluating the solid state stability of salt and co-crystal forms.

Method	*Measurement*	*Information*	*Temperature Range*	*Advantages*	*Potential Problems*
Slurry bridging	Suspension phase composition	Rank order of G	T_m to T_{bp} of solvent, up to T_m of drug	Simplicity	Solution-mediated phase transformations to other forms, labile solvates, slow transformation kinetics due to poor solubility or when near T_t
Equilibrium solubility	Solution concentration, suspension phase composition	ΔG; T_t by extrapolation	10–60 °C	Direct quantitative measurement of ΔG	Solution-mediated phase transformations, non-ambient temperature control throughout measurement
Kinetic solubility	Dissolution temperature with heating	ΔG; T_t by extrapolation	10–100 °C	Rapid analysis, automated measurement of solution clear points	Thermal lag with fast heating, solution-mediated phase transformations with slow heating
Intrinsic dissolution rate	Dissolution rate from constant surface area	ΔG; T_t by extrapolation	10–60 °C	Yields quantitative ΔG before form conversions occur in suspensions	Compression-induced phase transformations

DSC	Melting point; heat of fusion	ΔH (monotropy *vs.* enantiotropy via heat of fusion rule); T_t by extrapolation	near T_m	Small sample size, rapid analysis	Phase transformations below T_m, melting with decomposition, requires highly crystalline, phase pure samples
	Heat of transition	ΔH (monotropy *vs.* enantiotropy via heat of transition rule)	near T_t	Small sample size, rapid analysis	Reproducibility, not accessible for all polymorph pairs – only relates to forms involved in transition
Solution calorimetry	Heat of solution	ΔH (monotropy *vs.* enantiotropy via heat of solution rule)	0–40 °C	Yields ΔH for compounds which melt with decomposition	Larger sample size, solution-mediated phase transformations
Moisture sorption analysis	Moisture sorption isotherm	Rank order of G *vs.* RH or a_w, critical water activity (a_{crit})	RT to 60 °C (automated analysis); extended temperature range with RH chambers	Direct measurement of a_{crit} in absence of hysteresis, precise control of RH, measurement can be automated	Only sensitive to phase changes with accompanying weight gain/loss, hysteresis, slow transformation kinetics

possibly to an altogether different form, before the saturation solubility is attained. Depending on the stage of development, material may be limited as well. In any case, other methods may be needed or simply preferred. DSC is one of the most commonly used techniques to examine the relative stability of neat polymorphic modifications. Guided by the Burger–Ramberger rules, polymorph pairs may be characterized as monotropic (one form is more stable at all temperatures) or enantiotropic (the stability order reverses above/below a thermodynamic transition temperature) from the solid state transition enthalpies and/or relative melting temperatures and enthalpies observed by DSC.[72] Yu has shown that both pure and eutectic melting data obtained by DSC can also be used to quantitate the free energy differences (ΔG) between neat polymorphs.[73,74] Pure melting data, when coupled with eutectic melting data, equilibrium solubility, intrinsic dissolution rate and/or solution calorimetry data, allow polymorph stability to be quantitatively studied over a wide range of temperatures.

Evaluation of thermodynamic stability must also factor in the activity of the solvent when salts or co-crystals form hydrates (or organic solvates). Moisture sorption isotherms can provide insights into the thermodynamic stability relationships of neat and hydrated crystal forms as a function of RH (water activity) at a given temperature, provided that the rate at which the forms [inter]convert is sufficiently rapid *and*, as discussed earlier, a weight change accompanies the form conversion. Slurry equilibration, which accelerates the slow transformation kinetics of crystals surrounded by vapor, is extensively used to evaluate form stability qualitatively as a function of temperature and solvent (water) activity.[75] Whereas slurry conversions of true polymorphic forms will be independent of the solvent, solution-mediated phase transformations to and from hydrates may or may not depend on the water activity of the solvent system. Details of the practical aspects of assessing the thermodynamic stability of anhydrous and hydrated crystal forms through a combination of DSC, slurry conversion and solubility data have been given elsewhere.[76,77]

Selecting the most stable crystal form of a salt or co-crystal for a solid oral dosage form, of course, neither guarantees that it can be made in highly pure form and in high yield on a large scale nor does it assure one that over the shelf life of the drug product, conversion to an even lower energy form, such as a salt or co-crystal of different stoichiometry or free acid/base form, will not occur. When a salt form, for example, is thermodynamically less stable (more soluble) than its conjugate free acid (or base) in the pH microenvironment of a solid oral dosage form, it will be at risk for disproportionation. Here, the greater the difference in solubility of the salt versus its free form, the greater the thermodynamic driving force for disproportionation. The same pH-solubility dependence, and hence risks, exist for co-crystals made up of one or more ionizable components.[78] For this reason, it is generally advisable in the end to select a salt or co-crystal with *sufficient* solubility as opposed to the one with the greatest solubility.

To design a robust solution crystallization process that will consistently deliver the preferred salt or co-crystal form in high yield, equilibrium solubility measurements are typically made for a range of different compositions in a variety of solvents at several temperatures. The quantitative phase stability information is then plotted at each temperature in the form of an isothermal ternary phase diagram.[79] Figure 10.11 shows an isothermal ternary phase diagram that was constructed for a glutaric acid co-crystal of an undisclosed API, which in this case illustrates the effect that different solvents can have on the phase stability of a salt or co-crystal.[80] When mapping the phase diagram, it is imperative that the solid components in equilibrium with the liquid are characterized to ensure that any new forms are properly identified, especially given that different ratios of the components are unlikely to be extensively surveyed during polymorph and hydrate screening. Note that salts will give similar phase diagrams, although greater deviation from ideality may be expected because of differences in ionization.

From the ternary phase diagram in Figure 10.11, it can be seen that crystallizing even the stable form of a salt or co-crystal from solution can be complicated when the solubilities of the pure components are widely different.[81] This is because a much less soluble component will tend to crystallize from solution, a risk that has been widely discussed in recent years for co-crystals, but is equally important for salts. If required to navigate a phase diagram that is extremely asymmetric, such as that observed in methyl tertiary butyl ether, an excess of one of the components can be used to ensure that the co-crystal in this case is selected. In fact, Rodríguez-Hornedo *et al.* have shown that the common ion effect observed for salts, wherein an excess of a one of the components decreases the solubility of the salt below that of either component, can also be applied to co-crystals.[82] An alternative approach would be to identify a solvent, in this case, methyl isobutyl ketone or ethyl acetate (AcOEt), wherein the component solubilities are better matched. Recently, Rager and Hilfiker demonstrated that complex solvent mixtures (comprising as many as nine

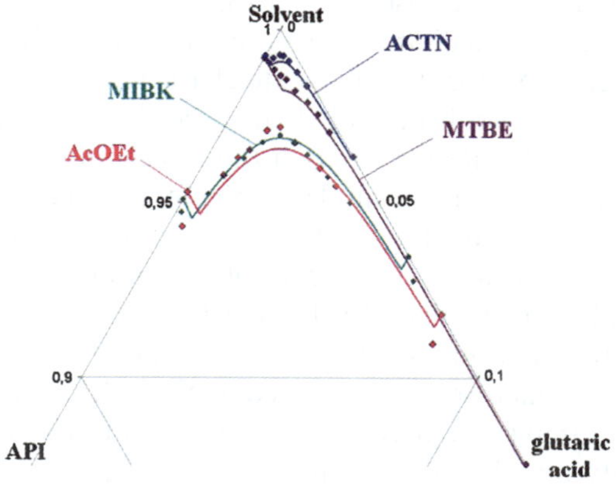

Figure 10.11 Ternary phase diagram of a glutaric acid co-crystal of an undisclosed API constructed from saturation solubility measurements in several solvents (reproduced with permission from Ainouz *et al.*, ref. 80; copyright © 2009 Elsevier BV).

solvents) can be used to level out the solubility differences between different compounds, an observation which should certainly increase the success rate in screening for salts and co-crystal forms by slurry and solution crystallization methods.[83]

10.3.4 Storage and Process-induced Phase Transformations

Identifying stable forms of salts and co-crystals and developing robust crystallization processes to select for them are critical, but not sufficient, to ensure the performance of drug products. For one thing, excipients routinely used to enhance drug product performance, when placed in intimate contact with a drug, may also compromise it either directly by reacting with the drug substance or indirectly by providing a source of water and/or pH modification. Additionally, common pharmaceutical processing steps, such as granulation, drying, milling and compaction, not only promote interactions between reactive components, but they may also temporarily place salt and co-crystal forms in regions of phase space (elevated temperatures, high pressures) where they become metastable and susceptible to conversions to other forms, crystalline and amorphous.[84] To ensure that physical transformations do not occur that have an adverse impact on the chemical stability, dissolution performance and bioavailability of a drug product, the influence of formulation excipients and the individual processing steps on the kinetics of phase transformations should be carefully examined. Storage conditions should also be specified that are within the region of the phase diagram where the thermodynamic stability of the selected salt or co-crystal form is assured (assuming, of course, that the thermodynamically stable form has been selected).

Characterizing the structure and stability of drug crystal forms in bulk APIs is relatively straightforward, with techniques such as X-ray powder diffraction, vibrational (Raman/IR) spectroscopy and solid state NMR spectroscopy now routinely employed for qualitative and quantitative analysis. The demands placed on analytical tools for the solid state characterization of drug products, on the other hand, can be substantially higher, depending on the complexity of the unit formula and the dosage strength (more accurately, the percentage of the drug substance in the unit formula). Therefore, to be useful for the solid state analysis of a drug product, a technique must not only have specificity, that is, the ability to distinguish the component phases of interest in the presence of overlapping signals, but it must also be sensitive enough to detect APIs and excipients of interest that may be present at low levels.

Confocal Raman microscopy, equipped with XYZ mapping to examine spatially resolved samples, is increasingly being used to assess the identity and content uniformity of solid state forms in low dose formulations. Solid state NMR spectroscopy, by virtue of its ability to detect specific nuclei with site selectivity, is also particularly well suited to characterizing component phases in bulk formulations. While in many cases the basic ^{13}C CP/MAS NMR technique is sufficiently sensitive and affords ample peak resolution for phase identification in formulations, greater selectivity and sensitivity can be achieved through observation of ^{19}F or ^{31}P nuclei, when applicable. Relaxation analysis and newer 2D NMR correlation techniques may also be used to explore component mixing. Despite many advances in instrumentation over the years, however, solid state NMR spectroscopy remains a research tool best used in a development setting to gain process understanding and to cross-validate XRPD, Raman or IR methods that will eventually be transferred to manufacturing and quality control laboratory settings.

10.4 Conclusions

In this chapter, several of the most commonly used solid state techniques, along with a few of the emerging ones, for characterizing salts and co-crystals have been described. The approaches taken to characterize salt and co-crystal forms, except for those specifically intended to probe ionization in the solid state, are the same as those used for neat and solvated forms of the parent molecule. Whether answering important questions commonly asked during the initial selection of a salt or co-crystal form or solving problems encountered later on during its development into a drug product, solid state techniques are best used in combination with one another to provide an accurate, if not complete, understanding of the solid state.

References

1. G. P. Stahly, *Cryst. Growth Des.*, 2007, **7**, 1007.
2. N. Schultheiss and A. Newman, *Cryst. Growth Des.*, 2009, **9**, 2950.
3. C. L. Cooke, R. J. Davey, S. Black, C. Muryn and R. G. Pritchard, *Cryst. Growth Des.*, 2010, **10**, 5270.
4. A. V. Trask and W. Jones, *Top. Curr. Chem.*, 2005, **254**, 41.
5. B. D. Cullity, *Elements of X-Ray Diffraction*, 2nd edn, Addison-Wesley Publishing Company, Reading, MA, 1978.
6. J. G. Amos, J. M. Indelicato, C. E. Pasini and S. M. Reutzel, *US Patents*, 5,412,094, May 2, 1995, and 6,001,996, December 14, 1999.
7. K. D. M. Harris, M. Tremayne and B. M. Kariuki, *Angew. Chem., Int. Ed.*, 2001, **40**, 1626.
8. M. Tremayne, in *Engineering of Crystalline Materials Properties*, ed. J. J. Novoa, D. Braga and L. Addadi, Springer, Dordrecht, 2008, p. 477.
9. K. D. M. Harris, M. Tremayne, P. Lightfoot, P. G. Bruce, *J. Am. Chem. Soc.*, 1994, **116**, 3543.
10. M. Tremayne, E. J. MacLean, C. C. Tang and C. Glidewell, *Acta Crystallogr., Sect. B*, 1999, **55**, 1068.
11. A. V. Trask, J. van de Streek, W. D. S. Motherwell and W. Jones, *Cryst. Growth Des.*, 2005, **5**, 2233.
12. S. H. Lapidus, P. W. Stephens, K. K. Arora, T. R. Shattock and M. J. Zaworotko, *Cryst. Growth Des.*, 2010, **10**, 4630.

13. A. J. Florence, N. Shankland, K. Shankland, W. I. F. David, E. Pidcock, X. Xu, A. Johnston, A. R. Kennedy, P. J. Cox, J. S. O. Evans, G. Steele, S. D. Cosgrove and C.S. Frampton, *J. Appl. Crystallogr.*, 2005, **38**, 249.
14. B. M. Kariuki, K. Psallidas, K. D. M. Harris, R. L. Johnston, R. W. Lancaster, S. E. Staniforth and S. M. Cooper, *Chem. Commun.*, 1999, 1677.
15. A. J. Hanson, E. Y. Cheung, S. Habershon and K. D. M. Harris, *Acta Crystallogr., Sect. A.*, 2005, **A61**, C162.
16. K. D. M. Harris, *Comput. Mat. Sci.*, 2009, **45**, 16.
17. R. Suryanarayanan, in *Physical Characterization of Pharmaceutical Solids*, ed. H. G. Brittain, Marcel Dekker, New York, 1995, p. 187.
18. S. X. Yin, R. P. Scaringe, M. F. Malley and J. Z. Gougoutas, *Amer. Pharm. Rev.*, 2005, **8**, 56.
19. D. Giron, *J. Therm. Anal. Calorim.*, 2002, **68**, 335.
20. W. C. McCrone, *Fusion Methods in Chemical Microscopy*, Interscience Publishers, New York, 1957.
21. J. Jacques, A. Collet and S. H. Wilen, *Enantiomers, Racemates, and Resolutions*, John Wiley & Sons, New York, 1981.
22. S. M. Reutzel-Edens, *Curr. Opin. Drug Discovery Dev.*, 2006, **9**, 806.
23. L. Yu, *Acc. Chem. Res.*, 2010, **43**, 1257.
24. K. Chadwick, R. Davey and W. Cross, *CrystEngComm*, 2007, **9**, 732.
25. G. Nichols, in *Polymorphism in the Pharmaceutical Industry*, ed. R. Hilfiker, Wiley VCH, Weinheim, 2006, p. 167.
26. A. Alkhalil, J. B. Nanubolu, C. J. Roberts, J. W. Aylott and J. C. Burley, *Cryst. Growth Des.*, 2011, **11**, 422.
27. L. Kofler and A. Kofler, *Thermal Micromethods for the Study of Organic Compounds and Their Mixtures*, ed. Wagner, Innsbruck, 1952, translated by W. C. McCrone, McCrone Research Institute, Chicago, 1980.
28. R. E. Davis, K. A. Lorimer, M. A. Wilkowski, J. H. Rivers, K. A. Wheeler and J. Bowers, *ACA Trans.*, 2004, **39**, 41.
29. D. P. McNamara, S. L. Childs, J. Giordono, A. Iarriccio, J. Cassidy, M. S. Shet, R. Mannion, E. O'Donnell and A. Park, *Pharm. Res.* 2006, **23**, 1888.
30. S. Sasic, O. Yukihiro, in *Pharmaceutical Applications of Raman Spectroscopy,* ed. S. Sasic, John Wiley & Sons, New York, 2008, p. 1.
31. M. Alessø, S. Velaga, A. Alhalaweh, C. Cornett, M. A. Rasmussen, F. van den Berg, H. Lopez de Diego and J. Rantanen, *Anal. Chem.*, 2008, **80**, 7755.
32. M. C. Etter, R. C. Hoye and G. M. Vojta, *Cryst. Rev.*, 1988, **1**, 281.
33. D. E. Bugay, *Pharm. Res.*, 1993, **10**, 317.
34. R. K. Harris, *Solid State Sci.* 2004, **6**, 1025.
35. J. Schaefer and E. O. Stejskal, *J. Am. Chem. Soc.*, 1976, **98**, 1031.
36. E. O. Stejskal, J. Schaefer and J. S. Waugh, *J. Magn. Reson.*, 1977, **28**, 105.
37. D. E. Bugay, *Adv. Drug Del. Rev.*, 2001, **48**, 43.
38. E. O. Stejskal and J. D. Memory, *High Resolution in the Solid State. Fundamentals of CP/MAS*, Oxford University Press, Oxford, UK, 1994.
39. D. L. Bryce, G. M. Bernard, M. Gee, M. D. Lumsden, K. Eichele and R. E. Wasylishen, *Can. J. Anal. Sci. Spectrosc.*, 2001, **46**, 46.

40. L. W. Jelinski and M. T. Melchior, *Appl. Spectrosc. Rev.*, 2000, **35**, 25.
41. R. E. Taylor, *Concepts Magn. Reson., Part A*, 2004, **22A**, 37.
42. M. Geppi, G. Mollica, S. Borsacchi and C. A. Veracini, *Appl. Spectroscopy Rev.*, 2008, **43**, 202.
43. F. G. Vogt, J. S. Clawson, M. Strohmeier, A. J. Edwards, T. N. Pham and S. A. Watson, *Cryst. Growth Des.*, 2009, **9**, 921.
44. K. Umprayn and R. W. Mendes, *Drug Dev. Ind. Pharm.,* 1987, **13**, 653.
45. G. Zografi, *Drug Dev. Ind. Pharm.,* 1988, **14**, 1905.
46. M. J. Kontny and G. Zografi, in *Physical Characterization of Pharmaceutical Solids*, ed. H. Brittain, Marcel-Dekker, New York, 1995, p. 387.
47. A. W. Newman, S. M. Reutzel-Edens and G. Zografi, *J. Pharm. Sci.*, 2008, **97***,* 1047.
48. C. Ahlneck and G. Zografi, *Int. J. Pharm.*, 1990, **62**, 87.
49. A. Peresypkin, N. Variankaval, R. Ferlita, R. Wenslow, J. Smitrovich, K. Thompson, J. Murry, L. Crocker, D. Mathre, J. Wang, P. Harmon, M. Ellison, S. Song, A. Makarov and R. Helmy, *J. Pharm. Sci.*, 2008, **97**, 3721.
50. T. L. Threlfall, *Analyst*, 1995, **120**, 2435.
51. S. M. Reutzel-Edens, R. L. Kleemann, P. L. Lewellen, A. L. Borghese and L. J. Antoine, *J. Pharm. Sci.*, 2003, **92**, 1196.
52. S. M. Reutzel and V. A. Russell, *J. Pharm. Sci.* 1998, **87**, 1568.
53. A. T. Serajuddin and M. Pudipeddi, in *Handbook of Pharmaceutical Salts: Properties, Selection, and Use*, ed. P. H. Stahl and C. G. Wermuth, Wiley-VCH, Weinheim, 2002, p. 137.
54. T. Steiner, *Angew. Chem. Int. Ed.*, 2002, **41**, 48.
55. C. Aakeröy, M. Fasulo and J. Desper, *Mol. Pharmaceutics,* 2007, **4**, 317.
56. R. Taylor and O. Kennard, *Acc Chem Res.*, 1984, **17**, 320.
57. C. L. Nygren, C. C. Wilson and J. F. C. Turner, *J. Phys. Chem. A* 2005, **109**, 1911.
58. Jeffrey, G. A. *An Introduction to Hydrogen Bonding*, Oxford University Press, Oxford, UK, 1997.
59. S. L. Childs, G. P. Stahly and A. Park, *Mol. Pharmaceutics*, 2007, **4**, 323.
60. L. H. Thomas, N. Blagden, M. J. Gutmann, A. A. Kallay, A. Parkin, C. C. Seaton and C. C. Wilson, *Cryst. Growth Des.*, 2010, **10**, 2770.
61. C. B. Aakeröy, I. Hussain and J. Desper, *Cryst. Growth Des.*, 2006, **6**, 474.
62. C. Aakeröy, A. Rajbanshi, Z. J. Li and J. Desper, *CrystEngComm*, 2010, **12**, 4231.
63. D. M. Bender, J. Bao, A. H Dantzig, W. D. Diseroad, K. L. Law, N. A. Magnus, J. A. Peterson, E. J. Perkins, Y. J. Pu, S. M. Reutzel-Edens, D. M. Remick, J. J. Starling, G. A. Stephenson, R. K. Vaid, D. Zhang and J. R. McCarthy, *J. Med. Chem.*, 2009, **52**, 6958.
64. C. Gardiennet-Doucet, B. Henry and P. Tekely, *Prog. Nucl. Magn. Reson. Spectrosc.,* 2006, **49**, 129.
65. Z. J. Li, Y. Abramov, J. Bordner, J. Leonard, A. Medek and A. V. Trask, *J. Am. Chem. Soc.*, 2006, **128**, 8199.
66. A. Lesage, P. Charmont, S. Steuernagel and L.Emsley, *J. Am. Chem. Soc.*, 2000, **122**, 9739.

67. D. C. Apperley, P. A. Basford, C. I. Dallman, R. K. Harris, M. Kinns, P. V. Marshall and A. G. Swanson, *J. Pharm. Sci.*, 2005, **94**, 516.
68. J. S. Stevens, S. J. Byard and S. L. M. Schroeder, *J. Pharm. Sci.* 2010, **99**, 4453.
69. J. S. Stevens, S. J. Byard and S. L. M. Schroeder, *Cryst. Growth Des.*, 2010, **10**, 1435.
70. J. S. Stevens, S. J. Byard, E. Zlotnikov and S. L. M. Schroeder, *J. Pharm. Sci.*, 2011, **100**, 942.
71. L. Yu, S. M. Reutzel and G. A. Stephenson, *Pharm. Sci. Tech. Today*, 1998, **1**, 18.
72. A. Burger and R. Ramberger, *Mikrochim. Acta II*, 1979, 259.
73. L. Yu, *J. Pharm. Sci.*, 1995, **84**, 966.
74. L. Yu, J. Huang and K. J. Jones, *J. Phys. Chem. B*, 2005, **109**, 19915.
75. M. D. Ticehurst, R. A. Storey and C. Watt, *Int. J. Pharm.*, 2002, **247**, 1.
76. T. L. Threlfall, *Org. Process Res. Dev.*, 2009, **13**, 1224.
77. S. M. Reutzel-Edens and G. A. Stephenson, in *Pharmaceutical Stress Testing, Predicting Drug Degradation, Second Edition*, ed. S. W. Baertschi, K. Alsante and R. A. Reed, Informa Healthcare USA, Inc., New York, 2011, pp. 254–285.
78. S. J. Bethune, N. Huang, A. Jayansankar and N. Rodríguez-Hornedo, *Cryst. Growth Des.* 2009, **9**, 3976.
79. T. Rager and R. Hilfiker, *Z. Phys. Chem.*, 2009, **223**, 793.
80. A. Ainouz, J.-R. Authelin, P. Billot and H. Lieberman, *Int. J. Pharm.*, 2009, **374**, 82.
81. S. N. Black, E. A. Collier, R. J. Davey and R. J. Roberts, *J. Pharm. Sci.*, 2007, **96**, 1053.
82. N. Rodríguez-Hornedo, S. J. Nehm, K. F. Seefeldt, Y. Pagán-Torres and C. J. Falkiewicz, *Mol. Pharmaceutics*, 2006, **3**, 362.
83. T. Rager and R. Hilfiker, *Cryst. Growth Des.*, 2010, **10**, 3237.
84. G. Z. Z. Zhang, D. Law, E. A. Schmitt and Y. Qiu, *Adv. Drug Delivery Rev.*, 2004, **56**, 371.

CHAPTER 11

Co-crystal Solubility and Thermodynamic Stability

L. ROY, M.P. LIPERT AND N. RODRÍGUEZ-HORNEDO*

University of Michigan, Department of Pharmaceutical Sciences, 428 Church Street, Ann Arbor, MI 48109-1065, USA

11.1 Introduction

Co-crystal-solution phase interactions have a major impact on our ability to understand and engineer co-crystal solubility. Recent advances in crystal engineering, in combination with the pioneering work of the 1950s and 1960s on molecular complexes in solution,[1–4] provide the opportunity to develop scientific concepts that link how molecules interact in the solid phase with how a collection of molecules behave in the (liquid) solution phase in order to tailor systems for specific purposes.

A molecular level understanding of co-crystals without the same depth of understanding of the solution phase leads to empirical approaches and technological developments guided by trial and error. A major goal of this chapter is to establish the link between key concepts and practical implications for fine tuning co-crystal thermodynamic solubility and stability. In writing this chapter, we are guided by the need to address the following questions:

- How the interplay between co-crystal lattice and solution phase interactions influences co-crystal solubility
- How solution chemistry determines co-crystal thermodynamic stability
- How the solubility of metastable co-crystal phases is measured
- What are key indicators of co-crystal thermodynamic stability?

RSC Drug Discovery Series No. 16
Pharmaceutical Salts and Co-crystals
Edited by Johan Wouters and Luc Quéré

Published by the Royal Society of Chemistry, www.rsc.org

This chapter focuses on the applications of fundamental concepts. The interested reader is referred to the original literature for development of the theoretical framework upon which this chapter is based.

11.2 Co-crystal Solubility

Solubility is of great importance in pharmaceutical systems since a drug must be dissolved before it is absorbed. When the drug absorption process is limited by dissolution, enhancing solubility can increase dissolution rate and in this way improve bioavailability.[5,6] In addition to the dissolution rate, the ability to dissolve the therapeutic dose of the drug and the selection of solubilization techniques are critical for product development.

Reduced aqueous solubility of many molecules is a major problem to overcome in order to achieve the desired therapeutic response.[7–11] The underlying causes of low solubility are captured by colloquial terms used to describe such molecules as "grease balls" when the resistance to solubility is hydrophobicity and as "brick dust" when solubility is limited by high lattice energies. This means that the strategies for improving solubility require a combination of technologies to influence solvent–solute and solute–solute interaction forces.

11.2.1 Key Concepts

Solubility is defined by the thermodynamic equilibrium of a solute between two phases, which in the context of this chapter are a solid phase and a liquid solution phase.[12,13] The criterion for equilibrium between coexisting phases is that the temperature, pressure and molar free energies or chemical potentials of each individual species in each phase are equal.[2,12,14] For a co-crystal, however, the sum of the molar free energies or chemical potentials of each co-crystal component plays a key role in determining phase equilibria. The molar Gibbs energy of the co-crystal A_mB_n in equilibrium with a solution phase is given by:

$$G_{A_mB_n} = G_{solution} = mG_A^L + nG_B^L = m\mu_A + n\mu_B \tag{11.1}$$

The activity of each component, a_i, is defined through the equation:

$$\mu_i = \mu_i^o + RT\ln a_i \tag{11.2}$$

where μ_i^o is the chemical potential of a reference state of i. Combining equations (11.1) and (11.2) gives:

$$\Delta G_{A_mB_n} = \Delta G_{solution} = RT(m\ln a_A + n\ln a_B) \tag{11.3}$$

and

$$\exp(\Delta G_{A_mB_n}/RT) = (a_A)^m(a_B)^n \tag{11.4}$$

The right hand side of equation (11.4) is often referred to as the solubility product, K_{sp}. Co-crystal solubility product behavior is well documented in the literature.[15–20] The activity of the solvent, which is the major component, is approximately unity and can be omitted from the equation. Co-crystal solubility means that a co-crystal is in equilibrium with a solution of a given thermodynamic activity product. Under these conditions, the solution is saturated with the given co-crystal phase.

Strategies to enhance solubility are based on the free energy of solution, which involves not only the crystal lattice interactions, but the interactions between the solutes and the solvent.[7,9,21–23] The two main factors associated with the overall free energy of solution are given by:

$$\Delta G_{\text{solution}} = \Delta G_{\text{lattice}} + \Delta G_{\text{solvation}} \quad (11.5)$$

where ΔG represents the Gibbs energies associated with the solubilization process, lattice, and solvation.

Figure 11.1 illustrates the steps involved in the solubilization or dissolution process: (1) breaking intermolecular bonds in the solid (solute–solute), (2) breaking intermolecular bonds in the solvent (solvent–solvent), and (3) forming solvent–solute bonds.[12] Thus, the solute–solute interaction forces must be overcome for the solute to dissolve. When the solvent–solute interactions match the solvent–solvent interactions, as in the case of ideal solutions, the energy of

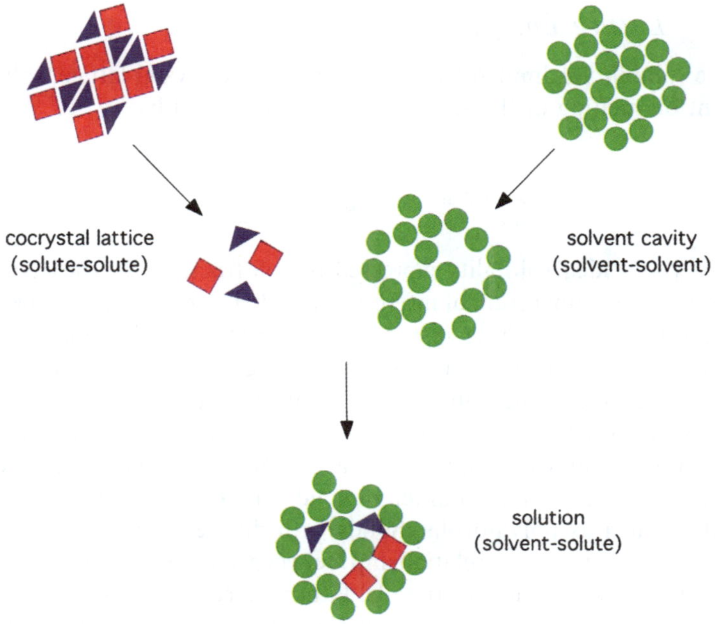

Figure 11.1 Schematic representation of the steps involved in the solution process of a co-crystal.

solubilization is given by the lattice energy ($\Delta G_{\text{lattice}}$) or melting of the solid, shown as the first step in Figure 11.1.

Reducing the lattice energy is often used to increase aqueous solubility.[21,24] Approaches that alter lattice energy include formation of amorphous phases, polymorphs, solvates, salts and co-crystals. On the other hand, solvation can play a decisive role in aqueous solubility as a result of the hydrophobicity of the drug.[10,21] Strategies that modulate solvent–drug interactions to increase solubility include ionization, complexation, micellar solubilization, prodrugs and co-solvents. Co-crystals and salts have the ability to influence both solid and solution phase chemistry and consequently the associated lattice and solvation energies.[7,20,25]

When attempting to explain the factors that contribute to co-crystal solubility, the focus is generally on solid state properties while solution phase contributions are not recognized. The situation becomes more complicated by attempting to assess co-crystal solubility by measurements under non-equilibrium conditions.

Solution interactions are ubiquitous in most pharmaceutical processes in both micro and macroscopic regimes. Water is a major component of the vapor and liquid phases that co-crystals encounter; therefore, it is natural to ask what the consequences of such solution phase processes are on co-crystal aqueous solution behavior and what errors can occur if they are ignored. The answers to these questions will be addressed in the following sections.

11.2.2 Factors that Influence Solubility

11.2.2.1 Lattice Energy

When the solubility is limited by the strength of the crystal lattice, solubility is dependent on melting of the solid solute as represented by:

$$\ln X_{\text{ideal}} = -\frac{\Delta H_{\text{m}}}{R}\left(\frac{T_{\text{m}} - T}{TT_{\text{m}}}\right) \qquad (11.6)$$

where X_{ideal} is the ideal solubility expressed as mole fraction, R is the gas constant, T is the reference temperature of interest, T_{m} and H_{m} are melting temperature in Kelvin and enthalpy of melting, respectively.[21] This form of the equation assumes that the heat capacity change upon melting is negligible and that solubility is a property of the solid solute only and is not solvent dependent.

The general expectation that lowering melting point increases solubility is based on this simplified solubility equation. Such behavior, however, only applies to ideal solutions or to different solid state forms of the same chemical composition such as polymorphs. Solids with different chemical components can also exhibit an inverse relationship between melting point and solubility when solvation energy changes parallel lattice energy changes.

The fact that melting point has often been shown to be a poor parameter to judge aqueous solubilities of co-crystals indicates that solubility is limited by solvation (hydrophobicity of the solute) and not by lattice energy.[20,26] Several

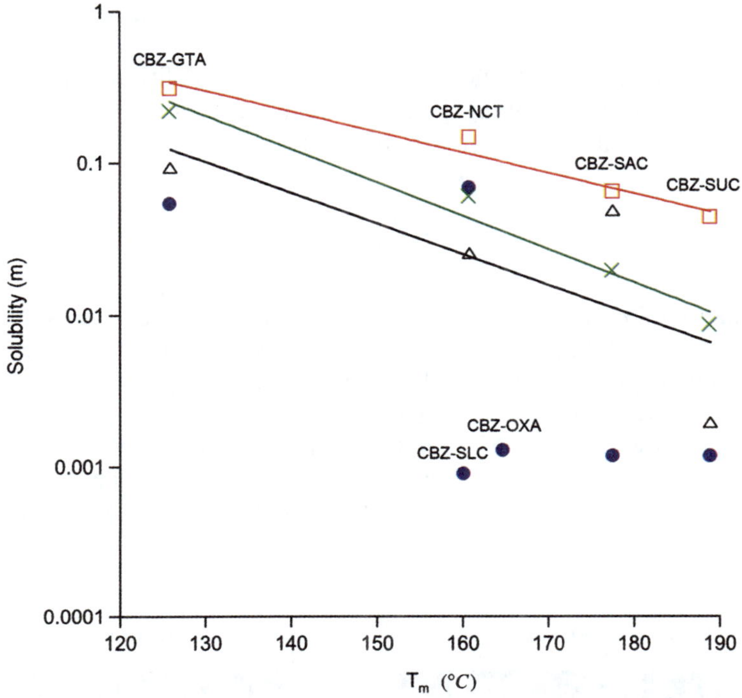

Figure 11.2 Co-crystal solubility of carbamazepine–glutaric acid (CBZ–GTA), carbamazepine–nicotinamide (CBZ–NCT), carbamazepine–salicylic acid (CBZ–SLC), carbamazepine–oxalic acid (CBZ–OXA), carbamazepine–saccharin (CBZ–SAC) and carbamazepine–succinic acid (CBZ–SUC) as a function of melting temperature in isopropyl alcohol (X), ethanol (open square), ethyl acetate (open triangle) and water (filled circle). Adapted with permission from Good and N. Rodríguez-Hornedo, ref. 20. Copyright 2009 American Chemical Society.

co-crystals of carbamazepine (CBZ) show different trends between melting point and solubility in organic and aqueous media. Figure 11.2 shows that while melting point is correlated with solubility in organic solvents it is not correlated with solubility in water. Solubility is expressed in moles per kilogram of solvent, (m). The main cause of this behavior is the limitation imposed by the solvent–solute interactions (hydrophobicity) which reduces the observed solubility below that predicted by equation (11.6).

11.2.2.2 Solution Energy

Solvent–solute interactions are often the main resistance to the observed solubility of hydrophobic molecules in aqueous media.[20,24] The deviation between ideal and observed solubility is quantitatively expressed by:

$$\ln X = \ln X_{\text{ideal}} - \ln \gamma \tag{11.7}$$

where the activity coefficient (γ) represents the contribution of solvent–solute interactions.

Lattice strength and solvation contributions to solubility can be calculated from measurement of melting point, enthalpy of melting, and equilibrium solubility in solvents of interest according to:

$$\ln X = -\frac{\Delta H_{\mathrm{m}}}{R}\left(\frac{T_{\mathrm{m}} - T}{TT_{\mathrm{m}}}\right) - \ln \gamma \tag{11.8}$$

The first term on the right hand side of equation (11.8) represents the lattice energy (ideal solubility) and the second term represents the solvation. A graphical method presented by Pinal and co-workers is applied to analyze the solubility of CBZ co-crystals in organic and aqueous solvents in Figure 11.3.[24] This graph shows that the lattice-induced changes in ideal solubilities of CBZ co-crystals range from 5-fold lower to 3-fold higher than the ideal solubility of CBZ. Solvation, represented by the activity coefficient, is shown to be the major

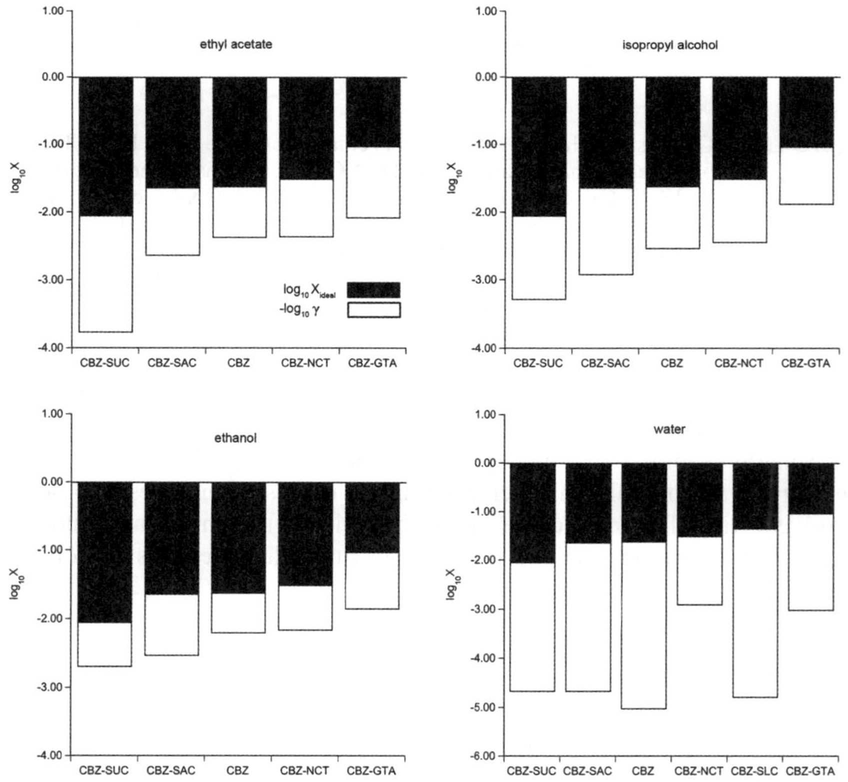

Figure 11.3 Lattice strength or ideal solubility (black bar) and solvation or activity coefficient (white bar) contributions to the observed solubility of carbamazepine (CBZ) and its co-crystals in organic and aqueous solvents.[20]

limitation to solubility in aqueous media. Observed solubility is lower than ideal solubility by as much as an order of magnitude in organic solvents and by as much as three orders of magnitude in water.

Since in organic solvents the magnitude of the solvent–solute interactions is proportional to the magnitude of the lattice strength, the observed solubility is proportional to ideal solubility. This explains the correlation of increasing co-crystal solubility with decreasing melting point in Figure 11.2. In water, however, the situation is quite different. Solvation (hydrophobicity) limits solubility to a greater extent than lattice strength and the magnitude of each factor is not proportional. Consequently, melting point is a poor indicator of aqueous solubility for these co-crystals.

Co-former solubility has also been correlated with co-crystal solubility for a series of co-crystals of the same drug. This behavior is a result of co-former decreasing the solvation barrier for a co-crystal to an extent proportional to that of the pure co-former. While universal indicators of co-crystal solubility would be useful, each case requires a careful analysis of the factors that influence solubility. Similar behavior has been reported for salt solubility.[7,25]

11.3 Tailoring Co-crystal Solubility via Solution Phase Chemistry

The versatility of co-crystals to fine tune solubility arises not only from their range of lattice properties but also from their solution phase interactions as a result of their diverse molecular properties. Since co-crystals are composed of multiple components, solution phase molecular associations are an important contribution to the solubility as illustrated in Figure 11.4. Some of these molecular processes include dissociation, complexation, ionization, and micellar solubilization in aqueous and biorelevant environments.

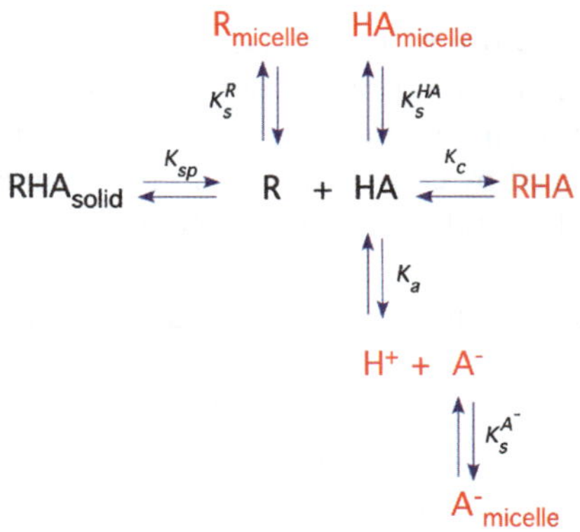

Figure 11.4 Co-crystal solution phase interactions and associated equilibria for a co-crystal RHA of a non-ionizable drug (R) and an ionizable co-former (HA) in micellar solution.

11.3.1 Ionization

Pharmaceutical co-crystals are often designed with acidic, basic, amphoteric, or zwitterionic molecules. These include a non-ionizable drug with acidic or amphoteric co-formers, (carbamazepine co-crystals with benzoic acid, saccharin, salicylic acid, 4-aminobenzoic acid);[27–29] basic drugs with acidic co-formers (caffeine with maleic acid, glutaric acid, oxalic acid;[30] itraconazole with succinic acid, fumaric acid, and malic acid);[31] and zwitterionic drugs with acidic co-formers (piroxicam with malonic acid, caprylic acid, fumaric acid;[32] gabapentin with 3-hydroxybenzoic acid),[33] among others. Given the wide range of drug and co-former ionization properties, one would expect that co-crystal solubility–pH dependence will vary greatly, even for a family of co-crystals of the same drug. The ability to predict such behavior is important in order to meet targeted aqueous solubilities and customize solubility–pH dependence of co-crystals.

Consider a 1:1 co-crystal RHA where R is a non-ionizable drug and HA is a monoprotic acidic co-former. If the co-crystal–solution equilibria are described by co-crystal dissociation and co-former ionization, the equilibrium expressions and associated constants are:

$$RHA_{solid} \overset{K_{sp}}{\rightleftharpoons} R_{soln} + HA_{soln}$$

$$K_{sp} = [R][HA] \tag{11.9}$$

$$HA_{soln} \overset{K_a}{\rightleftharpoons} A^-_{soln} + H^+_{soln}$$

$$K_a = \frac{[H^+][A^-]}{[HA]} \tag{11.10}$$

where K_{sp} is the solubility product of the co-crystal and K_a is the acid ionization constant. Species without subscripts refer to the solution phase. The analysis presented here assumes dilute solutions in which activities are relatively constant and can be replaced by concentrations in the equilibrium constants. This is an approximation to establish general trends. Non-idealities due to complexation, ionic interactions and solvent–solute interactions will need to be considered for a more rigorous analysis, particularly at high concentrations and ionic strengths.

The total co-former analytical concentration is the sum of the ionized and non-ionized species, which for a weak acid is given by:

$$[A]_T = [HA] + [A^-] \tag{11.11}$$

while total drug, which is non-ionizable, is given by:

$$[R]_T = [R] \tag{11.12}$$

Drug concentration in equilibrium with co-crystal can be expressed in terms of co-former equilibrium concentration, $[A]_T$, ionization constant, K_a, co-crystal K_{sp}, and solution $[H^+]$ by substituting [HA] and $[A^-]$ from equations (11.9), (11.10), and (11.11) into equation (11.12) and rearranging to give:

$$[R]_T = \frac{K_{sp}}{[A]_T}\left(\frac{1+K_a}{[H^+]}\right) \tag{11.13}$$

Total drug concentration at equilibrium, $[R]_T$, is shown as the dependent variable, since it is generally of interest to know how co-crystal solubility varies with co-former concentration and pH. Under these conditions:

$$S_{\text{co-crystal}} = [R]_T \tag{11.14}$$

The predicted co-former concentration effect on co-crystal solubility is consistent with previous models where ionization was not considered. That is, co-crystal solubility decreases with increasing co-former concentration.[16,17,29,33–35]

When co-crystal is in equilibrium with solutions of a stoichiometry equal to that of the co-crystal (*i.e.*, there is no excess co-former or drug in solution) then the co-crystal solubility is equal to the total concentration of the drug or co-former in solution:

$$S_{\text{co-crystal}} = [R]_T = [A]_T \tag{11.15}$$

and equation (11.13) is rewritten as:

$$S_{\text{co-crystal}} = \sqrt{K_{sp}\left(1+\frac{K_a}{[H^+]}\right)} \tag{11.16}$$

Under these conditions, solubility is referred to as stoichiometric solubility.

Equations (11.13) and (11.16) predict that solubility of co-crystal RHA will increase with decreasing $[H^+]$ (increasing pH). Co-crystal solubility is also dependent on the co-crystal K_{sp} and co-former K_a. At pH $\ll$ co-former pK_a, or $[H^+] \gg K_a$, co-crystal solubility approaches its intrinsic solubility $\sqrt{K_{sp}}$. At pH = pK_a, or $[H^+] = K_a$, the co-crystal solubility is $\sqrt{2K_{sp}}$ or 1.4 times the intrinsic co-crystal solubility. At pH $\gg$ co-former pK_a, or $[H^+] \ll K_a$, co-crystal solubility increases exponentially. The maximum concentration that can be experimentally achieved is, however, limited by the solubilities of the drug, co-former and co-former salts.

A reverse pH dependence is predicted for a co-crystal with a non-ionizable drug and a basic co-former:

$$S_{\text{co-crystal}} = \sqrt{K_{sp}\left(1+\frac{[H^+]}{K_a}\right)} \tag{11.17}$$

where K_a is the ionization constant of the conjugate acid of the base. In this case, the solubility increases with increasing $[H^+]$ (decreasing pH).

A valuable implication from this analysis is that co-crystals impart pH-dependent solubility to non-ionizable drugs when co-formers are ionizable. These predictions have been confirmed by studies with several co-crystals.[33,34]

11.3.1.1 Solubility of Co-crystals with Different Ionization Properties and Stoichiometries

Equations that describe co-crystal solubility dependence on $[H^+]$, K_{sp}, and K_a for several types of co-crystals have been derived and are presented in Table 11.1. The interested reader is directed to the original publication for the full derivations.[33,34] These equations describe the co-crystal stoichiometric solubility, when co-crystal is in equilibrium with solutions of stoichiometry equal to the co-crystal.

Table 11.1 shows that co-crystal solubility is governed by at least two parameters, K_{sp} and K_a, and one variable, solution pH. In most cases K_a values are known and K_{sp} can be calculated from experimentally measured co-crystal solubility at one pH. Alternatively, one can target solubility and pH values and calculate the required K_a and K_{sp}. Most frequently, several co-crystals of a given drug are discovered for which the solubility is not known. One can calculate the solubility–pH dependence from the K_a and K_{sp} values and streamline the co-crystal selection process.

Solubility–pH profiles in Figure 11.5 demonstrate the ability of co-crystals to modify solubility behavior relative to that of the drug. Solubility was calculated for four co-crystals with components that differ in their ionization properties using reported K_a values, experimental co-crystal solubilities and equations in Table 11.1. Figure 11.5(a) and (b) show that co-crystals of a non-ionizable drug can exhibit very different solubility–pH behavior, depending on the co-former ionization properties. A diprotic acid co-former will lead to increases in solubility with pH, as with a monoprotic acid. An amphoteric co-former will result in a U-shaped curve with a solubility minimum in a pH range between the two pK_a values. Similar behavior is predicted for a co-crystal of a basic drug and an acidic co-former where the ionizable groups reside in different molecules (Figure 11.5(c)). The pH range over which this minimum occurs is dependent on the difference between the two pK_a values; the greater the difference, the wider this minimum range will be, as has been shown for zwitterionic solutes.[21] This behavior is also predicted for a co-crystal of a zwitterionic drug and an acidic co-former (Figure 11.5(d)).

11.3.1.2 Co-crystal pH_{max} and Thermodynamic Stability

Co-crystal solubility is dependent on both solution composition and pH and so are the relative thermodynamic stabilities of the co-crystal and its components. Figure 11.6 shows this behavior for a hypothetical non-ionizable drug (R) and its 1:1 co-crystal (RHA) with an acidic co-former (HA). Co-crystal solubility was

Table 11.1 Co-crystal solubility dependence on $[H^+]$, K_a and K_{sp} for co-crystals with components that are acidic, basic, amphoteric and zwitterionic. Equations apply to co-crystal in equilibrium with solutions of same stoichiometry as the co-crystal (no excess component).

Co-crystal	*Solubility Equations*	*Equation*
RH_2A (1:1, neutral:diprotic acid)	$S = \sqrt{K_{sp}\left(1 + \frac{K_{a1,H_2A}}{[H^+]} + \frac{K_{a1,H_2A}K_{a2,H_2A}}{[H^+]^2}\right)}$	(11.18)
R_2H_2A (2:1, neutral:diprotic acid)	$S = \sqrt[3]{\frac{K_{sp}}{4}\left(1 + \frac{K_{a1,H_2A}}{[H^+]} + \frac{K_{a1,H_2A}K_{a2,H_2A}}{[H^+]^2}\right)}$	(11.19)
R_2HAB (2:1, neutral:amphoteric)	$S = \sqrt[3]{\frac{K_{sp}}{4}\left(1 + \frac{K_{a1,HAB}}{[H^+]} + \frac{[H^+]}{K_{a2,HAB}}\right)}$	(11.20)
BHA (1:1, basic:monoprotic acid)	$S = \sqrt{K_{sp}\left(1 + \frac{K_{a1,HA}}{[H^+]} + \frac{[H^+]}{K_{a,B}}\right)}$	(11.21)
B_2H_2A (2:1, basic:diprotic acid)	$S = \sqrt[3]{\frac{K_{sp}}{4}\left(1 + \frac{[H^+]}{K_{a,B}}\right)^2\left(1 + \frac{K_{a1,H_2A}}{[H^+]} + \frac{K_{a1,H_2A}K_{a2,H_2A}}{[H^+]^2}\right)}$	(11.22)
HAHX (1:1, monoprotic acid:monoprotic acid)	$S = \sqrt{K_{sp}\left(1 + \frac{K_{a,HX}}{[H^+]}\right)\left(1 + \frac{K_{a,HA}}{[H^+]}\right)}$	(11.23)
HABHX (1:1, amphoteric:monoprotic acid)	$S = \sqrt{K_{sp}\left(1 + \frac{K_{a,HX}}{[H^+]}\right)\left(1 + \frac{K_{a1,HAB}}{[H^+]} + \frac{[H^+]}{K_{a2,HAB}}\right)}$	(11.24)
$^-ABH^+H_2X$ (1:1, zwitterionic:diprotic acid)	$S = \sqrt{K_{sp}\left(1 + \frac{K_{a1,H_2A}}{[H^+]} + \frac{K_{a1,H_2A}K_{a2,H_2A}}{[H^+]^2}\right)} \times \sqrt{\left(1 + \frac{[H^+]}{K_{a1,^-ABH^+}} + \frac{K_{a12,^-ABH^+}}{[H^+]}\right)}$	(11.25)

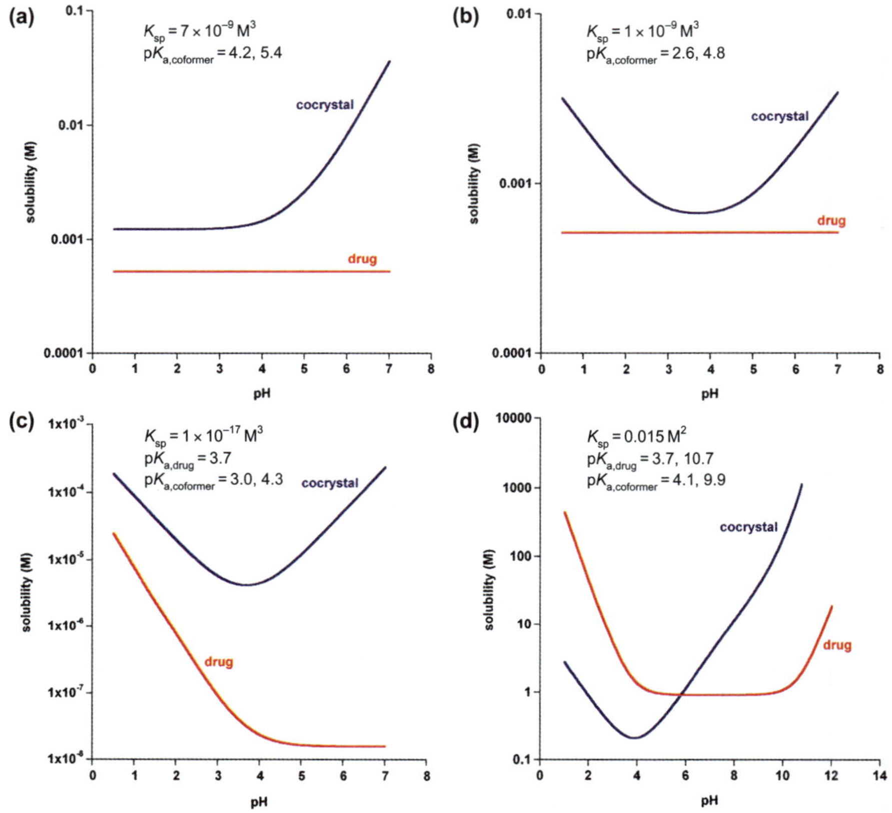

Figure 11.5 Solubility–pH profiles for (a) 2:1 R_2H_2A co-crystal calculated using equation (11.19), (b) 2:1 R_2HAB co-crystal calculated using equation (11.20), (c) 2:1 B_2H_2A calculated using equation (11.22), and (d) 1:1 $^{-}ABH^{+}H_2X$ co-crystal calculated using equation (11.25). pK_a values for drug and co-former and co-crystal K_{sp} are included in each graph. K_{sp} values were either experimentally determined or estimated from published work for the selected co-crystal in each graph (a) carbamazepine–succinic acid,[29] (b) carbamazepine–4-aminobenzoic acid hydrate,[34] (c) itraconazole–L-tartaric acid,[31] and (d) gabapentin–3-hydroxybenzoic acid.[33] Reprinted with permission from Bethune *et al.*, ref. 34. Copyright 2009 American Chemical Society.

calculated according to equation (11.13). The co-crystal solubility is predicted to increase with pH and to decrease as the co-former solution concentration increases. A reversal of thermodynamic stability can be clearly seen in Figure 11.6. The stable phase is co-crystal at low pH and drug at high pH values.

Co-crystal and drug solubility surfaces intersect at a given pH value and solution concentrations of R and A. Two important characteristics of the intersection points are that: (1) two solid phases (in this case, co-crystal and drug) coexist in equilibrium with solution, and (2) solution composition of co-crystal components ($[R]_{eu}$ and $[A]_{eu}$) is fixed at a given pH and temperature,

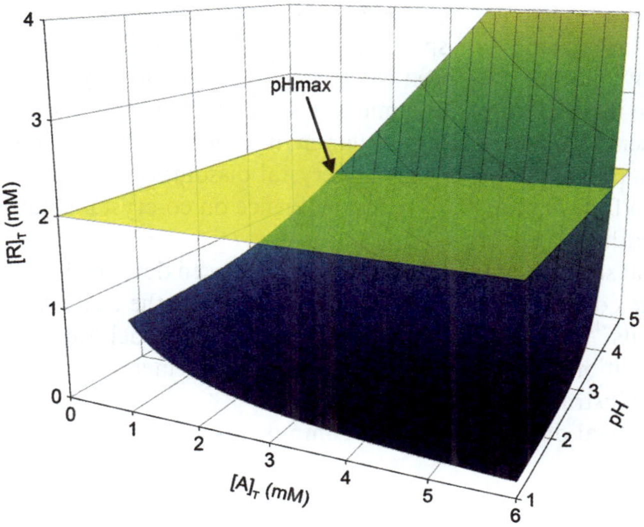

Figure 11.6 Dependence of co-crystal solubility or drug concentration, $[R]_T$, on co-former concentration, $[A]_T$, and pH for a 1:1 RHA co-crystal; calculated from equation (11.13) with $K_{sp}=1$ mM^2 and co-former $pK_a=3.0$. Solubility of drug, S_R, is represented by the yellow surface ($S_R=2$ mM) and co-crystal by the blue/green surface. The pH value at the intersection of the drug and co-crystal stoichiometric solubility curves corresponds to pH_{max} where a solution of stoichiometric composition is in equilibrium with two solid phases, co-crystal and drug. Reprinted with permission from Bethune *et al.*, ref. 34. Copyright 2009 American Chemical Society.

regardless of the ratio of the two solid phases. *This three-phase equilibrium is a eutectic or transition point.*[14] *When varying pH leads to the eutectic point, it is regarded as a* pH_{max}.[36] As with salts, the co-crystal pH_{max} is an important parameter that identifies co-crystal stability regions.

11.3.2 Micellar Solubilization

Why is co-crystal behavior in micellar surroundings relevant? Is it different from the behavior of single component crystals? The answer to the first question is that mastery of co-crystal properties depends on our ability to understand its interactions with the surrounding media. The answer to the second question is that every co-crystal component plays an important role on co-crystal behavior and this combination can be quite different from those of the individual component crystal phases.

Micelles are nano or micro structures with a hydrophobic core and a hydrophilic shell. They form above a critical concentration called the critical micellar concentration (CMC) as a result of the self-association of amphiphilic molecules to minimize hydrophobic interactions. Micellar surfactants and polymers are widely used in the pharmaceutical industry as formulation aids in solids, semi-solids, suspensions, emulsions, and solutions.[10,21,37] One of the most important attributes of micelles is that they can enhance the solubility of drugs that are insoluble or sparingly soluble in water. Theories of drug solubilization by micelles have been extensively studied and rely on a simple thermodynamic model that regards micellar solubilization as a distribution of the drug between the aqueous and micellar environments.[21,38–40] Although micellar surfactants are commonly used in co-crystal dissolution and solubility studies as well as in formulation,[26,41–44] their influence on co-crystal solution behavior has received little attention.[45,46]

Co-crystal solubility in micellar solutions has been described by the equilibria between the co-crystal phase and its components in the aqueous and micellar environments.[45–48] We first present a simplified model that assumes non-ionizing solution conditions and negligible co-former partitioning into the micelle, as illustrated in Figure 11.7.

The co-crystal dissociates into its components:

$$\mathrm{RHA_{solid}} \overset{K_{sp}}{\rightleftharpoons} \mathrm{R_{aq}} + \mathrm{HA_{aq}}$$

where

$$K_{sp} = [\mathrm{R}]_{aq}[\mathrm{HA}]_{aq} \tag{11.26}$$

and the hydrophobic drug (R) is solubilized by the micellar environment:

$$\mathrm{R_{aq}} + \mathrm{M} \overset{K_s^R}{\rightleftharpoons} \mathrm{R_m}$$

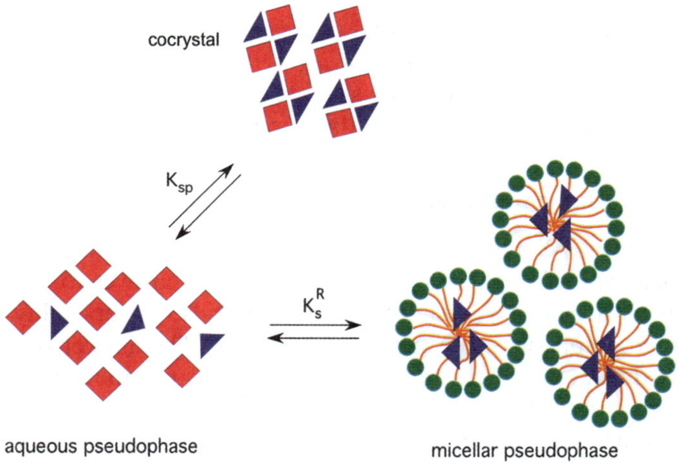

Figure 11.7 Schematic illustration of the equilibria between the co-crystal phase and its components in the aqueous and micellar solution pseudophases. This scheme represents micellar solubilization of one co-crystal component (drug), leading to excess co-former in the aqueous pseudophase. Adapted with permission from Huang and Rodríguez-Hornedo, ref. 45. Copyright 2010 American Chemical Society.

where

$$K_s^R = \frac{[R]_m}{[R]_{aq}[M]} \quad (11.27)$$

subscripts m and aq refer to micellar and to aqueous pseudophases, K_{sp} is the co-crystal solubility product, [M] is the micellar surfactant concentration, and K_s^R is the micellar solubilization constant for co-crystal component R.

The co-crystal solubility, $S_{RHA,T}$, under stoichiometric conditions, is equal to the total concentration of each co-crystal component in equilibrium with solution:

$$S_{RHA,T} = [R]_T = [A]_T \quad (11.28)$$

The contribution of micellar solubilization of the drug component R to the co-crystal solubility is given by:

$$S_{RHA,T} = [R]_{aq} + [R]_m = [HA]_{aq} \quad (11.29)$$

Co-crystal solubility can be expressed in terms of surfactant micellar concentration, co-crystal and micellar solution equilibrium constants as:

$$S_{RHA,T} = \sqrt{K_{sp}\left(1 + K_s^R[M]\right)} \quad (11.30)$$

The terms in square brackets refer to concentrations with the recognition that under dilute solution conditions they approximate activities. Equation (11.30) predicts that co-crystal solubility increases with co-crystal K_{sp}, drug micellar solubilization, K_s^R, and surfactant micellar concentration, [M]. The micellar surfactant concentration is the total surfactant concentration minus the CMC.

For a crystal of the non-ionizable drug (R_{solid}) the total solubility is given by:

$$S_{R,T} = [R]_{aq} + [R]_m = S_{R,aq}\left(1 + K_s^R[M]\right) \quad (11.31)$$

where $S_{R,aq}$ is the solubility of crystal R in the aqueous pseudophase. Equations (11.30) and (11.31) show that *drug co-crystal* and *drug crystal* solubilities exhibit different dependences on micellar concentration. Co-crystal solubility is dependent on $\sqrt{[M]}$ whereas drug crystal solubility is dependent on [M].

Figure 11.8 demonstrates that co-crystal and drug solubility curves intersect at a given surfactant concentration and identifies a critical stabilization concentration (CSC) where co-crystal and drug solid phases are in thermodynamic equilibrium with solution. Below the CSC, the co-crystal is thermodynamically unstable and can convert to the more stable drug solid phase. Above the CSC,

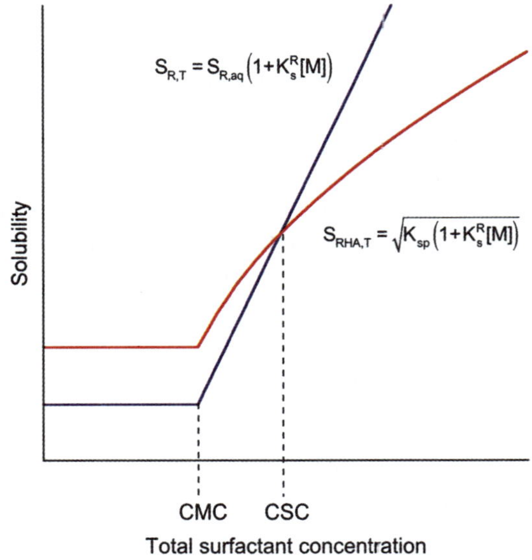

Figure 11.8 Solubility dependence of co-crystal (RHA) and drug (R) on surfactant concentration. Co-crystal solubility in micellar solutions is characterized by nonlinear behavior, which leads to intersection of the solubility curves of co-crystal and drug solid phases at the CSC, or critical stabilization concentration. Adapted with permission from Huang and Rodríguez-Hornedo, ref. 45. Copyright 2010 American Chemical Society.

the co-crystal becomes the stable phase relative to the drug phase and conversion to drug is not possible. At the CSC, the co-crystal and drug phases are in equilibrium with solution, which represents a eutectic point.

Differential solubilization, where micelles interact preferentially with one of the co-crystal components, is the underlying mechanism by which otherwise unstable co-crystals achieve thermodynamic stability in micellar solutions. This concept was recently reported and applied to describe the behavior of several co-crystals.[45,46] Pharmaceutical co-crystals generally comprise a hydrophobic drug and a relatively hydrophilic co-former, thus, differential solubilization may be widely encountered. This finding has important implications for the characterization of co-crystal solution behavior and can lead to huge errors in the evaluation of co-crystal solubility and dissolution if not recognized.

The effectiveness of a surfactant to stabilize a co-crystal or attain a CSC is determined by the relative magnitude of drug and co-former solubilization by micelles, K_s^R *and* K_s^{HA}. As illustrated in Figure 11.9, the greater the drug micellar solubilization, K_s^R, relative to that of the co-former, K_s^{HA}, the lower the CSC value. For co-crystals composed of drugs that are much more hydrophobic than their co-formers, $K_s^R \gg K_s^{HA}$.

The influence of micellar solubilization on the stability regions of the co-crystal, drug and co-former solid phases is illustrated by the triangular phase diagram in Figure 11.10. An incongruently saturating co-crystal can become

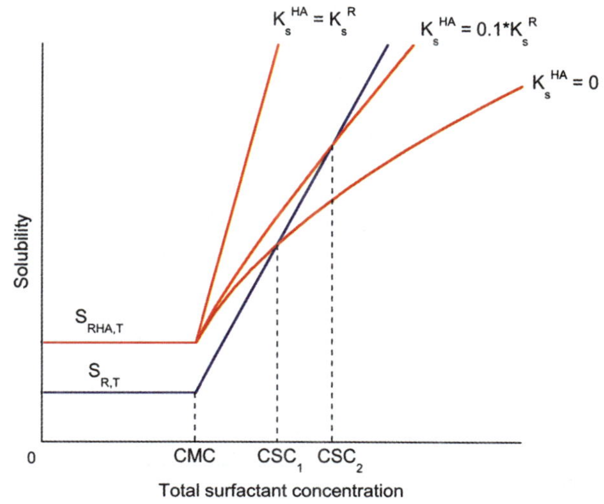

Figure 11.9 Differential solubilization of co-crystal components represented by the relative values of K_s^{HA} and K_s^{R} leads to nonlinear co-crystal solubility dependence and to intersection of the co-crystal and drug solubility curves. CSC refers to the critical stabilization concentration, at which both co-crystal and drug are thermodynamically stable. Adapted with permission from Huang and Rodríguez-Hornedo, ref. 45. Copyright 2010 American Chemical Society.

congruently saturating by differential solubilization of co-crystal components in micellar solutions. Congruently and incongruently saturating systems are defined in another chapter (see Chapter 12). The eutectic between co-crystal, solid drug and solution, E_1, is the more relevant eutectic for supersaturating co-crystals of hydrophobic drugs in aqueous solutions. The eutectic where co-crystal, solid co-former and solution are at equilibrium is designated as E_2 in Figure 11.10. Eutectic points and their importance in calculating co-crystal solution behavior are described more thoroughly in Section 11.4.

11.3.2.1 Ionization and Micellar Solubilization

The contributions of micellar solubilization and ionization to co-crystal solubility can be obtained by considering the relevant equilibrium expressions. For co-former micellar solubilization and ionization, the equilibrium expressions are:

$$\mathrm{HA_{aq}} + \mathrm{M} \overset{K_s^{HA}}{\rightleftharpoons} \mathrm{HA_m}$$

where

$$K_s^{HA} = \frac{[\mathrm{HA}]_m}{[\mathrm{HA}]_{aq}[\mathrm{M}]} \tag{11.32}$$

$$\mathrm{HA_{aq}} \overset{K_a}{\rightleftharpoons} \mathrm{A^-_{aq}} + \mathrm{H^+_{aq}}$$

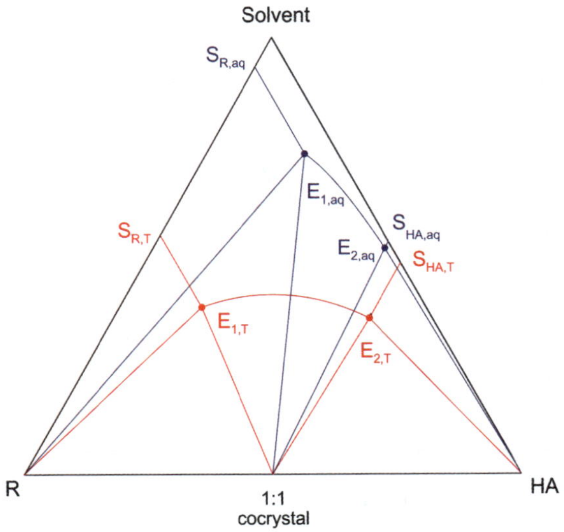

Figure 11.10 Hypothetical triangular phase diagram showing how differential solubilization of co-crystal components changes the eutectic points and co-crystal stability regions. E_1 (solid drug and co-crystal) and E_2 (co-former and co-crystal) are the co-crystal eutectic points. The subscript aq refers to aqueous and the subscript T refers to total (aqueous + micellar). Differential solubilization of the drug compared to the co-former shifts the co-crystal stability region, which results in a congruently saturating co-crystal. Adapted with permission from Huang and Rodríguez-Hornedo, ref. 45. Copyright 2010 American Chemical Society.

where

$$K_a = \frac{[H^+]_{aq}[A^-]_{aq}}{[HA]_{aq}} \quad (11.33)$$

The solubility of co-crystal RHA, $S_{RHA,T}$, in terms of co-crystal dissociation, drug and co-former micellar solubilization, and co-former ionization (described by equations (11.26), (11.27), (11.32), and (11.33)) is:

$$S_{RHA,T} = \sqrt{K_{sp}\left(1 + K_s^R[M]\right)\left(1 + \frac{K_a}{[H^+]} + K_s^{HA}[M]\right)} \quad (11.34)$$

This equation shows that co-crystal solubility increases with increasing co-crystal K_{sp}, component $K_s{}^R$, and $K_s{}^{HA}$, co-former ionization $K_a/[H^+]$, and surfactant micellar concentration, [M].

Equation (11.34) can serve as a guide for surfactant selection in order to meet a target co-crystal solubility. Values for co-crystal component thermodynamic parameters, K_s and K_a, are often available from the literature, or are experimentally accessible, and co-crystal solubility only requires a single measurement of co-crystal K_{sp} and solution pH. Equations (11.31) and (11.34) are shown graphically in Figure 11.11 for the case of a non-ionizable, hydrophobic drug and its co-crystal with an ionizable, hydrophilic co-former where $K_s^{HA} = 0$. This plot demonstrates that co-crystal and drug solubility surfaces intersect along a curve of given surfactant concentration and pH values and identifies stability regions for co-crystal or drug by two critical parameters. The first parameter is the CSC, or the surfactant concentration where co-crystal and drug solid phases are in equilibrium with solution. The second parameter is the pH_{max}, or the pH value at the CSC. Above the CSC or below pH_{max} the co-crystal becomes the stable phase relative to the drug phase. When one or more co-crystal components ionize, both CSC and pH_{max} are necessary to describe the solution conditions under which co-crystal and/or drug solid phases are thermodynamically stable. These critical parameters represent eutectic points.

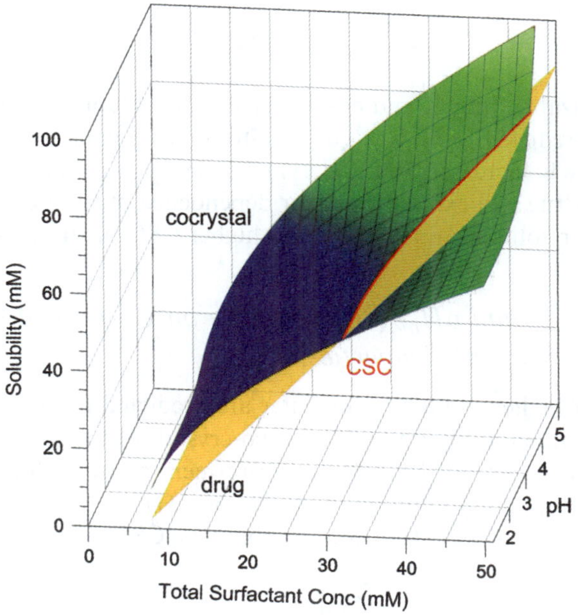

Figure 11.11 Co-crystal RHA (blue/green surface) and drug R (yellow surface) solubility dependence on surfactant concentration and pH. The intersection of the co-crystal and drug solubility surfaces represents the surfactant concentrations (CSC) and pH values (pH_{max}) where co-crystal and drug are in thermodynamic equilibrium with solution. Solubilities were calculated from equations (11.31) and (11.34) with $K_{sp} = 1$ mM^2 ($S_{RHA,aq}/S_{R,aq} = 5$), $S_{R,aq} = 0.2$ mM, p$K_a = 4$, $K_s^R = 1$ mM^{-1}, $K_s^{HA} = 0$ mM^{-1}, and CMC = 8 mM.[46]

11.3.2.2 Estimation of Co-crystal Solubilization from Drug Solubilization

A useful estimate of the surfactant influence on co-crystal solubilization can be calculated from knowledge of the drug solubilization according to:

$$\frac{S_{\mathrm{RHA,T}}}{S_{\mathrm{RHA,aq}}} = \sqrt{\frac{S_{\mathrm{R,T}}}{S_{\mathrm{R,aq}}}} \tag{11.35}$$

This expression is obtained by combining equations (11.31) and (11.34) for a non-ionizable drug R when $K_s{}^{R}$ is unaffected by co-former and $K_s{}^{HA} = 0$. A surfactant concentration that increases drug solubility 100-fold is predicted to increase co-crystal solubility 10-fold. Equation (11.35) implies that a surfactant will increase the solubility of all 1:1 co-crystals of a drug by the same ratio as long as the stated assumptions are justified.

Equation (11.35) can be rewritten for a general co-crystal stoichiometry, R_nX_m, as:

$$\frac{S_{\mathrm{R}_n\mathrm{X}m,\mathrm{T}}}{S_{\mathrm{R}_n\mathrm{X}m,\mathrm{aq}}} = \left(\frac{S_{\mathrm{R,T}}}{S_{\mathrm{R,aq}}}\right)^{\frac{n}{n+m}} \tag{11.36}$$

for a non-ionizable drug R and co-former X. The solubility increase for a 2:1 co-crystal is predicted to be $10^{2/3}$, or 21.5-fold its aqueous solubility when the drug solubility is increased 100-fold. Thus co-crystal stoichiometries richer in hydrophobic drug will exhibit a weaker dependence of total co-crystal solubility on micellar solubilization, leading to higher CSC or pH_{max} values.

11.3.2.3 CSC and pH_{max} Dependence on Co-crystal and Surfactant Properties

Co-crystals of higher solubilities in water are predicted to exhibit higher CSC values as illustrated in Figure 11.12. For co-crystals of the same drug, aqueous solubilities can be altered by different co-formers or by co-former ionization behavior in solution (by adjusting the solution pH).

The influence of co-crystal aqueous solubility on CSC may be calculated from:

$$CSC = \frac{\left(\frac{S_{\mathrm{RHA,aq}}}{S_{\mathrm{R,aq}}}\right) - 1}{K_s^{\mathrm{R}}} + CMC \tag{11.37}$$

by solving for the surfactant concentration at which $S_{\mathrm{RHA,T}} = S_{\mathrm{R,T}}$ from equations (11.30) and (11.31). This expression applies to a 1:1 co-crystal with no micellar solubilization of co-former and negligible solution complexation of co-crystal components. CSC is shown to be inversely proportional to drug micellar

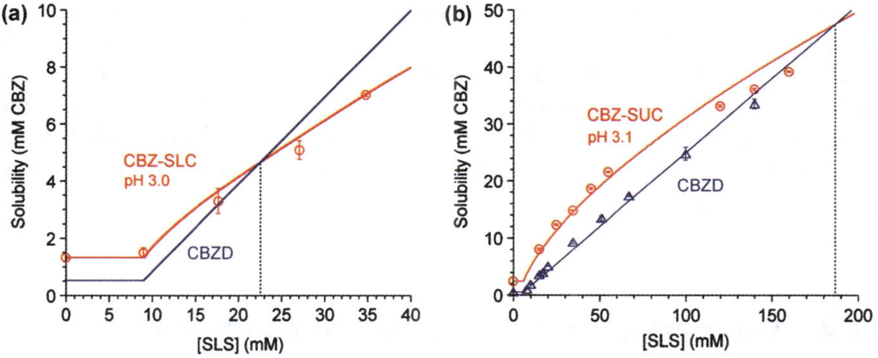

Figure 11.12 Influence of micellar solubilization on co-crystal solubility and CSC for two co-crystals of carbamazepine with different stoichiometries and aqueous solubilities at 25 °C. The dashed line shows the SLS concentration at the CSC, where co-crystal and drug are thermodynamically stable. The solid lines represent predicted co-crystal and drug solubilities according to equations (11.30) and (11.34) for a 1:1 co-crystal, carbamazepine–salicylic acid (CBZ–SLC). The equation for a 2:1 co-crystal, carbamazepine–succinic acid (CBZ–SUC) has been presented in the literature.[34,46]

solubilization and directly proportional to co-crystal aqueous solubility. This equation allows estimation of the required K_s^R to achieve the CSC for co-crystal and its drug component, and in this way provide guidance for the rational selection of surfactant and its concentration.

Considering the contributions of co-former solubilization and ionization, in addition to drug solubilization, leads to a more general form of equation (11.37) expressed by:

$$CSC = \frac{\frac{K_{sp}}{S_{R,aq}^2}\left(1 + \frac{K_a}{[H^+]}\right) - 1}{K_S^R - \frac{K_{sp}}{S_{R,aq}^2} K_s^{HA}} + CMC \tag{11.38}$$

where $[H^+]$ is the $[H^+]_{max}$ associated with the pH_{max}. According to equation (11.38), the CSC for a 1:1 co-crystal RHA is dependent on several critical parameters: $K_{sp}/S_{R,aq}^2$, K_a, $[H^+]$, K_s^R, and K_s^{HA}. This equation can also be solved for $[H^+]$ to predict the pH_{max} dependence on micellar surfactant concentration and other co-crystal and surfactant properties.

Equation (11.38) shows that if a CSC exists, there also is a pH_{max} value associated with that CSC and vice versa. CSC is predicted to increase as ionization increases. Higher levels of ionization increase co-crystal solubility and thus more surfactant is required to achieve the CSC. Equation (11.38) can also be used to engineer the pH_{max} of a co-crystal based on selection of an appropriate surfactant and concentration. Equations that consider co-crystals with other ionization properties and stoichiometries have been presented elsewhere.[46–48]

In summary, the theoretical treatment of co-crystal micellar solubilization predicts that the CSC, where co-crystal and drug phases are in thermodynamic equilibrium, is most readily achieved by (1) preferential drug solubilization ($K_s^R \gg K_s^{HA}$), (2) co-crystals of lower aqueous solubility relative to drug, and (3) co-crystal stoichiometries that are higher in co-former. These predictions have been confirmed by the experimental behavior of co-crystals in several surfactant solutions.[45–48]

11.4 Co-crystal Thermodynamic Stability and Solubility Evaluation

Since co-crystal solubilities can exceed the solubility of a crystal form of the drug, they generate supersaturation and are predisposed to phase transformations. The concept of supersaturating drug delivery systems (SDDS) has been widely applied to improve bioavailability by selection of high energy forms such as amorphous solid phases,[49,50] salts,[7,36] self-emulsifying drug delivery systems (SEDDS),[10,51,52] and more recently, co-crystals.[44] Development of SDDS requires application of thermodynamic and kinetic concepts to manage conversions to more stable forms. While these concepts can be generally applied to stabilization of SDDS, specific strategies will depend on the properties of each form.

11.4.1 Co-crystal Thermodynamic Stability

The criteria for co-crystal thermodynamic stability were presented in Section 11.3. An important feature of co-crystals is that they coexist in equilibrium with solution. This occurs when their molar free energy or chemical potential is equal to the sum of the chemical potentials of each co-crystal component in solution. Thus, the individual component chemical potentials in a solution saturated with co-crystal can vary as long as their sum is constant. In terms of activities, it is the activity product that is constant.

Thermodynamic stability relationships between a co-crystal and its components depend on the component activities in solution (a_i). The co-crystal is the stable phase relative to its components in a solution where activities do not exceed the component solubilities ($a_i \leq S_i$). A solution can be at equilibrium, saturated with co-crystal phase only ($a_i < S_i$), or doubly saturated with co-crystal and one of its components ($a_A = S_A$), for example when co-crystal and solid A are in equilibrium with solution. This three-phase equilibrium describes the eutectic or transition point and plays an important role in virtually all co-crystal processes as it identifies the co-crystal stability domains.

Experiments identify the eutectic point as the activities of co-crystal components, $a_{drug,eu}$ and $a_{co\text{-}former,eu}$, or concentrations, $[drug]_{eu}$ and $[co\text{-}former]_{eu}$, at which a solution is saturated with co-crystal and another solid phase. Since the solution is saturated with two phases, its concentration is independent of the ratio of the mass of the phases at the eutectic. Co-crystal-solution phase equilibria are therefore characterized by eutectic points that:

1. are sensitive to solution chemistry, for example ionization (Figure 11.15) and micellar solubilization (Table 11.2)
2. identify a pH_{max} for co-crystals whose components can ionize in solution
3. identify a CSC when an additive exhibits differential solubilization for one of the co-crystal components
4. serve as a measure of co-crystal solubility.

Eutectic points are not unique to the co-crystal domain. The pH_{max} for pharmaceutical salts and the critical water activity for hydrates are widely used eutectic points to establish stability conditions in the development of dosage forms.[36,53]

11.4.1.1 Relationship Between Eutectic Points and Phase Diagrams

Phase stability regions can be identified without the need to measure a full phase diagram experimentally. Eutectic solution compositions can be measured to identify stability regions and if desired to calculate a full phase diagram by considering the solution phase equilibria and associated equations.[17,34,35]

Eutectic points and stability regions for co-crystals of carbamazepine–4-aminobenzoic acid in ethanol are shown in Figure 11.13(a) and (b). This system has two co-crystals with different stoichiometries (1:1) and (2:1); thus, there are four eutectic points. Variations in $[\mathrm{drug}]_{eu}$ relative to [drug] at saturation in the absence of co-former provide a measure of co-crystal and drug solubility dependence on solution phase interactions. Co-crystal thermodynamic stability can be quantitatively predicted by combining eutectic point measurements with equations that describe solution behavior.[20,46]

Eutectic points can also establish stability regions for co-crystal hydrates, where a co-former modulates both co-crystal solubility and water activity.[35] The theophylline–citric acid co-crystal (THP–CTA) has a hydrated and an anhydrous co-crystal. This system exhibits three eutectic points (E_1, E_2, and E_3). The solid phases in equilibrium at the eutectic points are drug hydrate/co-crystal hydrate at E_1, co-crystal hydrate/anhydrous co-crystal at E_2, and anhydrous co-crystal/co-former hydrate at E_3. The region where co-crystal hydrate is thermodynamically stable is given by the concentrations at E_1 and E_2. Co-former solution composition is controlling co-crystal solubility as well as its hydrated state. Phase diagrams in Figure 11.14(a) and (b) were generated from eutectic point measurements according to co-crystal solubility product behavior.

11.4.1.2 Influence of Solution Chemistry on Eutectic Points

Figure 11.15 shows the influence of pH on a co-crystal of a non-ionizable drug with an ionizable co-former. The greater the co-former ionization, the higher

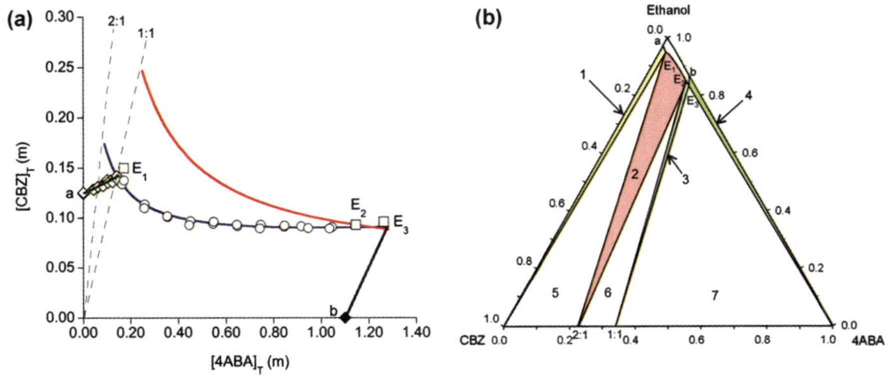

Figure 11.13 Carbamazepine–4-aminobenzoic acid (CBZ–4ABA) in ethanol at 25 °C. (a) Phase solubility diagram showing solution concentrations ($[CBZ]_T$ and $[4ABA]_T$) at equilibrium with anhydrous carbamazepine, CBZ(III), (open diamond), 2:1 co-crystal (open circle), or 4ABA (filled diamond). Eutectic points E_1, E_2, and E_3 are represented by open squares. The dashed lines represent reactant stoichiometry in solution corresponding to that of co-crystals. Predicted solubility dependence on co-former concentration for CBZ(III), 2:1 co-crystal and 1:1 co-crystal have been presented in the literature.[17] (b) Triangular phase diagram generated from experimental and calculated solubilities. Compositions are shown on a mass fraction basis. Stability regions are: 1, CBZ (III), 2, 2:1 co-crystal, 3, 1:1 co-crystal, 4, 4ABA, 5, CBZ(III)/2:1 co-crystal, 6, 2:1 co-crystal/1:1 co-crystal, and 7, 1:1 co-crystal/4ABA; solubilities of CBZ (III) and 4ABA in neat ethanol are indicated by a and b in each plot. Adapted with permission from Jayasankar *et al.*, ref. 17. Copyright 2009 American Chemical Society.

the eutectic co-former concentrations. This behavior is explained by the increase in co-crystal solubility with pH and by the equations presented in Section 11.3.

The eutectic constant K_{eu} has been introduced for chiral systems containing racemic compounds and was recently applied to describe the stability of co-crystals.[54–57] K_{eu} is defined as the activity ratio of co-former to drug for the co-crystal/drug eutectic point:

$$K_{eu} \equiv \frac{a_{coformer,eu}}{a_{drug,eu}} \approx \frac{[coformer]_{eu}}{[drug]_{eu}} \tag{11.39}$$

K_{eu} dependence on external conditions is critical to establish co-crystal thermodynamic stability. For instance, $K_{eu} > 1$ for a 1:1 co-crystal or $K_{eu} > 0.5$ for a 2:1 co-crystal indicates that the co-crystal is thermodynamically unstable with respect to the drug. A decrease of K_{eu} below the co-crystal stoichiometric ratio, for example, < 1 for a 1:1 co-crystal or < 0.5 for a 2:1 co-crystal indicates a reversal in the thermodynamic stability, where the co-crystal is more stable

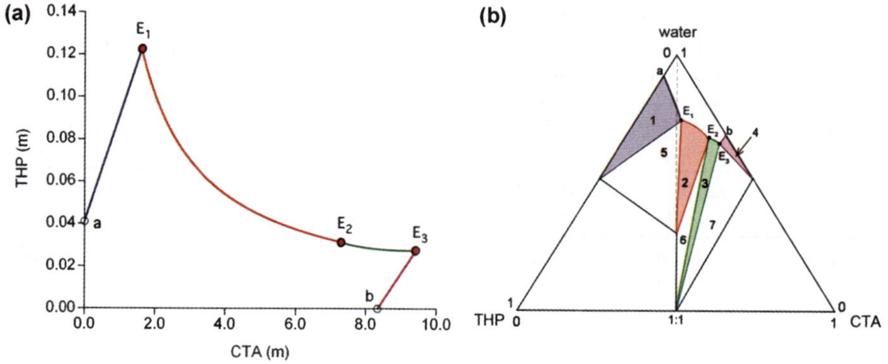

Figure 11.14 Phase diagrams for theophylline–citric acid (THP–CTA) anhydrous and co-crystal hydrate in water at 25 °C. Eutectic points are indicated by E_1 (THP hydrate/THP–CTA hydrate), E_2 (THP–CTA hydrate/THP–CTA anhydrous) and E_3 (THP–CTA anhydrous and CTA hydrate). Solubilities of THP hydrate and CTA hydrate are indicated by a and b in each plot.[35] (a) Phase solubility diagram generated from measured eutectic points and models that describe co-crystal solubility behavior. (b) Schematic triangular phase diagram showing the stability domains for anhydrous and hydrated co-crystals with co-formers that modulate the water activity. Stability regions for the crystalline phases are: 1, crystalline drug hydrate; 2, co-crystal hydrate; 3 anhydrous co-crystal; 4, co-former hydrate; 5, crystalline drug hydrate/co-crystal hydrate; 6, anhydrous/hydrated co-crystals; 7, anhydrous co-crystal/hydrated co-former.

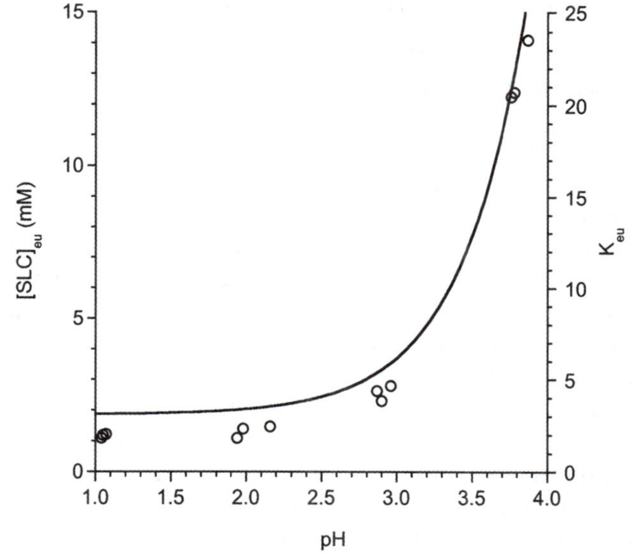

Figure 11.15 Co-former eutectic concentration ($[SLC]_{eu}$) and eutectic constant ($K_{eu} = [SLC]_{eu}/[CBZ]_{eu}$) dependence on pH for the CBZ–SLC co-crystal. Curve represents theoretical calculations from equation (11.13). Drug eutectic concentration $[CBZ]_{eu} = 6.1$ (± 0.3) mM in this pH range. Adapted with permission from Bethune *et al.*, ref. 34. Copyright 2009 American Chemical Society.

Table 11.2 Eutectic concentrations and eutectic constant (± standard error) for the equilibrium of CBZ–SLC and CBZ dihydrate at 25 °C in water with and without sodium lauryl sulfate ($n = 3$).

[SLS] (mM)	*pH*	*$[CBZ]_{eu}$ (mM)*	*$[SLC]_{eu}$ (mM)*	*K_{eu} (exp.)*	*K_{eu} (pred.)*[a]
0	3.0 ± 0.1	0.61 ± 0.03	2.88 ± 0.04	4.75 ± 0.17	–
35	3.2 ± 0.1	9.38 ± 0.07	5.25 ± 0.03	0.56 ± 0.01	0.59

[a]Calculated from theoretical solubility dependence of co-crystal and drug on surfactant concentration. $K_s^{CBZ} = 0.6\ mM^{-1}$, $K_s^{SLC} = 0.06\ mM^{-1}$.[45]

than the drug. In instances where activities are not well approximated by concentrations, K_{eu} may deviate from the stoichiometric ratio.

The influence of micellar solubilization on co-crystal eutectic points is presented in Table 11.2 for carbamazepine–salicylic acid (CBZ–SLC) in aqueous media. Addition of surfactant reverses the co-former to drug molar concentration ratios at the eutectic. In the absence of surfactant, $[SLC]_{eu} > [CBZ]_{eu}$, indicating that the co-crystal requires excess co-former to be at equilibrium with pure drug. This situation is reversed in 35 mM (1%) SLS solution, where $[CBZ]_{eu} > [SLC]_{eu}$ demonstrating that there is a CSC at SLS concentrations below 1%.

11.5 Eutectic Point Measurement

The eutectic point can be measured by a number of methods. Two simple and generally applicable methods and their shortcomings are discussed.

1. *Solid phase analysis*: This method is generally useful for estimating eutectic points in terms of the initial concentration of one of the co-crystal components. In the case of a co-crystal that converts to drug solid when exposed to solvent, it involves determination of the $[\text{co-former}]_{eu}$ at which the co-crystal and the drug phases are at equilibrium with solution. It simply requires solid phase analysis as a function of co-former solution concentration or another variable of interest, such as pH, surfactant concentration or other solution conditions. The approach is similar to that used for stability assessment of hydrate/anhydrous phases. It is useful as a first approximation but has several limitations: (a) it neglects solution composition changes during equilibration and the values may be far from the equilibrium concentrations, and (b) it is highly sensitive to the material purity in relation to the amount of co-crystal phase. Trace levels of a co-crystal component may be undetected by XRPD and DSC leading to incorrect stability assessment in suspension or by moisture uptake studies when eutectic concentrations are not measured.[58–61]
2. *Solution concentration and solid phase analysis*: Measurement of the solution concentration of *each* co-crystal component when the solution is

in equilibrium (saturated) with the desired solid phases at the eutectic provides the most accurate assessment of co-crystal stability. This approach requires quantitative measurement of solution composition and qualitative analysis of solid phase composition. Knowledge of both $[\text{co-former}]_{eu}$ and $[\text{drug}]_{eu}$ greatly facilitates our understanding of co-crystal solution interactions and our ability to make quantitative predictions.

11.6 Co-crystal Solubility Measurement

Methods to determine co-crystal solubility are based on thermodynamic and kinetic approaches. Thermodynamic equilibrium experiments provide a measure of co-crystal solubilization processes, while kinetic studies provide insight about the time scales of dynamic processes and concentration fluctuations during co-crystal dissolution. From equilibrium measurements, we can extract information about the origin of co-crystal solution phase behavior and fine tune solution processes by controlling thermodynamic solubility. We can use the knowledge gained from equilibrium studies to design kinetic studies and separate the thermodynamic and kinetic contributions to the co-crystal dissolution and transformation behavior.

11.6.1 Thermodynamic Equilibrium Methods

If co-crystals are to solve solubility problems one must assess their true or thermodynamic solubility so that development strategies are guided by the fundamental properties of co-crystals. Measuring the solubility of co-crystals that generate supersaturation of the parent drug is often experimentally impossible due to conversion. Eutectic points, described in Section 11.4, provide a measure of co-crystal solubility under thermodynamic equilibrium conditions. The solution at the eutectic point is saturated with co-crystal and solution concentrations represent experimentally accessible thermodynamic solubility values. Once co-crystal solubility is determined at the eutectic, the solubility under different solution conditions (pH, co-former, micelle concentration) can be obtained from solubility models that consider the appropriate solution phase equilibrium expressions.

How eutectic point measurement leads to information about co-crystal solubility under stoichiometric solution conditions includes:

1. $[\text{co-former}]_{eu}/[\text{drug}]_{eu}$ ratios or K_{eu} values above the co-crystal stoichiometric ratio are associated with increased co-crystal thermodynamic activity of the drug (co-crystal solubility higher than drug solubility on a molar drug basis).
2. A higher $[\text{co-former}]_{eu}/[\text{drug}]_{eu}$ ratio, for a series of co-crystals of the same drug with different co-formers, translates to higher co-crystal solubility. This simple observation will establish a rank order of co-crystal solubilities.

3. How solution conditions such as pH, polymers, surfactants or formulation aids influence $[\text{co-former}]_{eu}/[\text{drug}]_{eu}$ ratios yields information about the co-crystal solubility relative to that of the drug. Increasing $[\text{co-former}]_{eu}/[\text{drug}]_{eu}$ ratios are associated with increased co-crystal to drug solubility.

To assess the eutectic point variation with solution phase chemistry quantitatively, solubility equations that consider the appropriate contributions are considered. K_{eu} in terms of ionization for a (1:1) co-crystal of a non-ionized drug and an acidic co-former is:

$$K_{eu} = \frac{[\text{coformer}]_{eu,T}}{[\text{drug}]_{eu,T}} = \frac{[\text{coformer}]_{eu,ionized} + [\text{coformer}]_{eu,unionized}}{[\text{drug}]_{eu,T}} \tag{11.40}$$

Substitution of the relevant expressions (equations (11.11) and (11.13) into (11.40)) gives:

$$K_{eu} = \frac{\frac{K_{sp}}{S_{drug}}\left(1 + \frac{K_a}{[H^+]}\right)}{S_{drug}} \tag{11.41}$$

Co-crystal K_{sp} and drug solubility represent the eutectic concentrations of free components in this example. K_{eu} is predicted to increase with decreasing $[H^+]$ (increasing pH) leading to an increase in co-crystal solubility relative to the drug. For co-crystals with other acidic or basic components, the reader is referred to recent publications on co-crystal ionization.[33,34,62]

K_{eu} can also be expressed in terms of the co-crystal to drug solubility ratio (α) under stoichiometric solution conditions from:

$$K_{eu} = \left(\frac{S_{cocrystal}}{S_{drug}}\right)^2 = \alpha^2 \tag{11.42}$$

The relationship between K_{eu} and α for carbamazepine co-crystals in aqueous solutions with different pH values and in several organic solvents is shown in Figure 11.16. The data points represent K_{eu} values and solubilities obtained from measured eutectic solution concentrations of co-crystal components according to equations (11.40) and (11.42). Co-crystal stoichiometric solubilities increase relative to drug solubility as the $[\text{co-former}]_{eu}$ increases relative to the $[\text{drug}]_{eu}$. Other co-crystal stoichiometries and solution equilibria are presented in the original publication.[57]

Eutectic concentration measurements are essential for understanding *why and under what conditions co-crystal solubility and thermodynamic stability will change*. The experiments have important advantages as they: (a) require little material and time, and (b) guide the rational selection of formulation aids and dissolution media that will best capture the co-crystal drug release behavior.

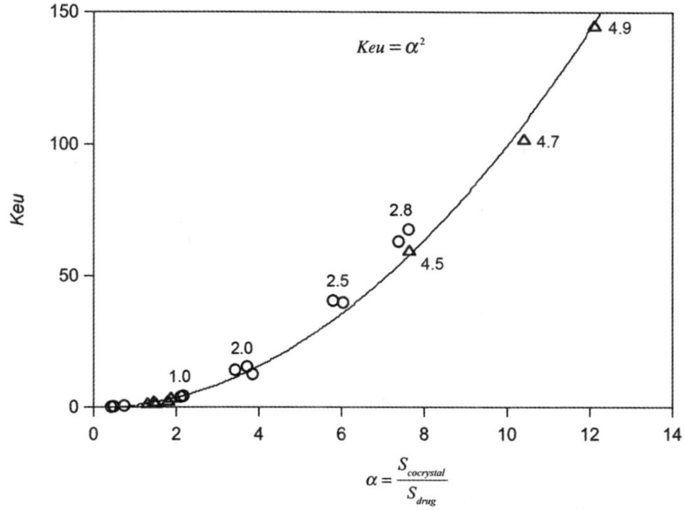

Figure 11.16 K_{eu} dependence on co-crystal to drug solubility ratio (α). Curve corresponds to $K_{eu}=\alpha^2$. The points correspond to carbamazepine co-crystals of salicylic acid (triangles) and saccharin (circles) in different solvents. The pH is indicated for aqueous samples. Adapted with permission from Good and Rodríguez-Hornedo, ref. 57. Copyright 2010 American Chemical Society.

11.6.2 Kinetic Methods

As described in previous sections, solution phase interactions play an important role in co-crystal solubility. The influence is greater than for single component crystals (or their hydrates) since each co-crystal component will modify solution behavior to different extents depending on their interactions with the environment. Kinetic studies are useful when informed by co-crystal thermodynamic solubilities and their solution phase dependence. Simply adding a co-crystal to a solution and measuring drug concentration as a function of time may fail to capture important properties of the co-crystal and lead to inaccurate assessment of its performance.

11.6.2.1 When and Why Kinetic Measurements Fail

The more soluble a co-crystal is, the greater the risk of failure of kinetic measurements.

Figure 11.17 shows a schematic representation of the concentration *versus* time profiles during kinetic studies. The maximum concentration achieved during these experiments is determined by the co-crystal dissolution rate and crystallization rate of a less soluble form. This maximum concentration is therefore not correlated with solubility under conditions of rapid conversion and highly soluble co-crystals may even elude detection.

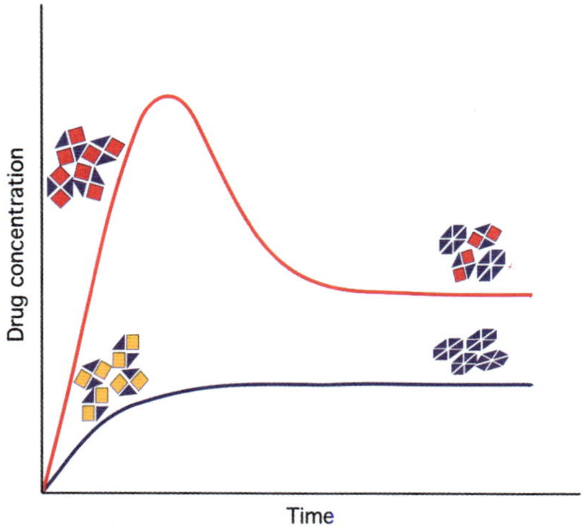

Figure 11.17 Pitfalls of kinetic measurements in assessing co-crystal solubility. The highest maximum concentration is not always associated with the highest solubility. High solubility co-crystals may even elude detection due to a fast transformation rate to drug.

11.7 Conclusions

Co-crystals are of great importance in engineering properties of pharmaceuticals and other materials. An important attribute of co-crystals is that their components can be selected in such a way that properties can be tailored to meet required specifications. The practical use of co-crystals can be realized through understanding how co-crystals of acceptable purity are produced and which thermodynamic properties dictate co-crystal behavior. Pharmaceutical co-crystals have the ability to enhance and modulate the delivery of poorly water-soluble drugs, but the development of co-crystal is often limited by their poor thermodynamic stability in aqueous environments. This chapter presents a conceptual framework along with quantitative models to fine tune co-crystal thermodynamic stability and solubility. The secret of controlling co-crystal stability lies in understanding its sensitivity to the solution environment. Key thermodynamic parameters that define co-crystal stability and solubility in pure solvents and in the presence of additives (eutectic points, pH_{max}, and CSC) are presented. These parameters are critical in selecting, developing, and using co-crystals.

References

1. T. Higuchi and I. H. Pitman, *J. Pharm. Sci.*, 1973, **62**, 55–58.
2. T. Higuchi and K. A. Connors, in *Advances in Analytical Chemistry and Instrumentation*, ed. C. Reilley, 1965, vol. 4, pp. 117–212.
3. T. Higuchi and S. Bolton, *J. Am. Pharm. Assoc.*, 1959, **48**, 557–564.

4. T. Higuchi and D. A. Zuck, *J. Am. Pharm. Assoc., Sci.Ed.*, 1952, **41**, 10–13.
5. G. Amidon, H. Lennernas, V. Shah and J. Crison, *Pharm. Res.*, 1995, **12**, 413–420.
6. J. Dressman, M. Vertzoni, K. Goumas and C. Reppas, *Adv. Drug Delivery Rev.*, 2007, **59**, 591.
7. A. T. M. Serajuddin, *Adv. Drug Delivery Rev.*, 2007, **59**, 603–616.
8. A. T. M. Serajuddin, P. C. Sheen, D. Mufson, D. F. Bernstein and M. A. Augustine, *J. Pharm. Sci.*, 1988, **77**, 325–329.
9. S. Rane and B. Anderson, *Adv. Drug Delivery Rev.*, 2008, **60**, 638.
10. R. Liu, *Water-insoluble Drug Formulation*, CRC Press, Boca Raton, Florida, 2008.
11. M. Kennedy, J. Hu, P. Gao, L. Li, A. Ali-Reynolds, B. Chal, V. Gupta, C. Ma, N. Mahajan, A. Akrami and S. Surapaneni, *Mol. Pharmaceutics*, 2008, **5**, 981–993.
12. J. H. Hildebrand, *Solubility*, The Chemical Catalog Company, Inc., New York, 1924.
13. J. H. Hildebrand and R. L. Scott, *The Solubility of Nonelectrolytes*, Reinhold, New York, 1950.
14. A. Findlay and A. N. Campbell, *The Phase Rule and its Applications*, Longmans, Green and Co., New York, 1938.
15. N. Rodríguez-Hornedo, *Mol. Pharmaceutics*, 2006, **3**, 362.
16. S. J. Nehm, B. Rodriguez-Spong and N. Rodríguez-Hornedo, *Cryst. Growth Des.*, 2006, **6**, 592–600.
17. A. Jayasankar, L. S. Reddy, S. J. Bethune and N. Rodríguez-Hornedo, *Cryst. Growth Des.*, 2009, **9**, 889–897.
18. K. Ito and K. Sekiguchi, *Chem. Pharm. Bull.*, 1966, **14**, 255–262.
19. T. Rager and R. Hilfiker, *Z. Phys. Chem.*, 2009, **223**, 793–813.
20. D. J. Good and N. Rodríguez-Hornedo, *Cryst. Growth Des.*, 2009, **9**, 2252–2264.
21. S. H. Yalkowsky, *Solubility and Solubilization in Aqueous Media*, American Chemical Society, Oxford University Press, New York, 1999.
22. Y. Miyako, H. Tai, K. Ikeda, R. Kume and R. Pinal, *Drug Dev. Ind. Pharm.*, 2008, **34**, 499–505.
23. A. Avdeef, *Adv. Drug Delivery Rev.*, 2007, **59**, 568–590.
24. Y. Miyako, H. Tai, K. Ikeda, R. Kume and R. Pinal, *Drug Dev. Ind. Pharm.*, 2008, **34**, 499.
25. B. D. Anderson and R. A. Conradi, *J. Pharm. Sci.*, 1985, **74**, 815–820.
26. M. K. Stanton and A. Bak, *Cryst. Growth Des.*, 2008, **8**, 3856–3862.
27. S. G. Fleischman, S. S. Kuduva, J. A. McMahon, B. Moulton, R. D. B. Walsh, N. Rodríguez-Hornedo and M. J. Zaworotko, *Cryst. Growth Des.*, 2003, **3**, 909–919.
28. J. A. McMahon, J. A. Bis, P. Visweshwar, T. R. Shattock, O. L. McLaughlin and M. J. Zaworotko, *Z. Kristallogr.*, 2005, **220**, 340–350.
29. S. L. Childs, N. Rodríguez-Hornedo, L. S. Reddy, A. Jayasankar, C. Maheshwari, L. McCausland, R. Shipplett and B. C. Stahly, *CrystEngComm*, 2008, **10**, 856–864.

30. A. V. Trask, W. D. S. Motherwell and W. Jones, *Cryst. Growth Des.*, 2005, **5**, 1013–1021.
31. J. F. Remenar, S. L. Morissette, M. L. Peterson, B. Moulton, J. M. MacPhee, H. R. Guzman and O. Almarsson, *J. Am. Chem. Soc.*, 2003, **125**, 8456–8457.
32. S. L. Childs and K. I. Hardcastle, *Cryst. Growth Des.*, 2007, **7**, 1291–1304.
33. L. S. Reddy, S. J. Bethune, J. W. Kampf and N. Rodríguez-Hornedo, *Cryst. Growth Des.*, 2009, **9**, 378–385.
34. S. J. Bethune, N. Huang, A. Jayasankar and N. Rodríguez-Hornedo, *Cryst. Growth Des.*, 2009, **9**, 3976–3988.
35. A. Jayasankar, L. Roy and N. Rodríguez-Hornedo, *J. Pharm. Sci.*, 2010, **99**, 3977–3985.
36. P. H. Stahland and C. G. Wermuth and P., *Handbook of Pharmaceutical Salts: Properties, Selection, and Use*, Wiley-VCH, New York, 2002.
37. R. G. Strickley, *Pharm. Res.*, 2004, **21**, 201–230.
38. K. S. Birdi, *Handbook of Surface and Colloid Chemistry*, CRC Press, Boca Raton, Florida, 2003.
39. D. Myers, *Surfactant Science and Technology*, Wiley-VCH, New York, 1992.
40. S. D. Christian and J. F. Scamehorn, *Solubilization in Surfactant Aggregates*, Marcel Dekker, New York, 1995.
41. J. F. Remenar, M. L. Peterson, P. W. Stephens, Z. Zhang, Y. Zimenkov and M. B. Hickey, *Mol. Pharmaceutics*, 2007, **4**, 386–400.
42. M. S. Jung, J. S. Kim, M. S. Kim, A. Alhalaweh, W. Cho, S. J. Hwang and S. P. Velaga, *J. Pharm. Pharmacol.*, **62**, 1560–1568.
43. A. V. Yadav, A. P. Dabke and A. S. Shete, *Drug Dev. Ind. Pharm.*, **36**, 1036–1045.
44. D. P. McNamara, S. L. Childs, J. Giordano, A. Iarriccio, J. Cassidy, M. S. Shet, R. Mannion, E. O'Donnell and A. Park, *Pharm. Res.*, 2006, **23**, 1888–1897.
45. N. Huang and N. Rodríguez-Hornedo, *Cryst. Growth Des.*, 2010, **10**, 2050–2053.
46. N. Huang and N. Rodríguez-Hornedo, *CrystEngComm*, 2011, DOI: 10.1039/C1CE05381G.
47. L. Roy, M. Lipert, N. Huang and N. Rodríguez-Hornedo, *AAPS Journal*, 2010, **12**(S2)Abstract T3112.
48. L. Roy and N. Rodríguez-Hornedo, *AAPS J.*, 2010, **12**(S2)Abstract R6072.
49. C. Leuner and J. Dressman, *Eur. J. Pharm. Biopharm.*, 2000, **50**, 47–60.
50. J. Brouwers, M. E. Brewster and P. Augustijns, *J. Pharm. Sci.*, 2009, **98**, 2549–2572.
51. C. Pouton, *Adv. Drug Delivery Rev.*, 1997, **25**, 47.
52. P. Gao, B. D. Rush, W. P. Pfund, T. H. Huang, J. M. Bauer, W. Morozowich, M. S. Kuo and M. J. Hageman, *J. Pharm. Sci.*, 2003, **92**, 2386–2398.

53. D. J. W. Grant and T. Higuchi, *Solubility Behavior of Organic Compounds*, Wiley, New York, 1990.
54. Y. L. Wang, R. LoBrutto, R. W. Wenslow and I. Santos, *Org. Process Res. Dev.*, 2005, **9**, 670–676.
55. M. Klussmann, A. J. P. White, A. Armstrong and D. G. Blackmond, *Angew. Chem., Int. Ed.*, 2006, **45**, 7985–7989.
56. J. Jacques, A. Collet and S. H. Wilen, *Enantiomers, Racemates, and Resolutions*, Wiley, New York, 1981.
57. D. J. Good and N. Rodríguez-Hornedo, *Cryst. Growth Des.*, 2010, **10**, 1028–1032.
58. A. V. Trask, W. D. S. Motherwell and W. Jones, *Cryst. Growth Des.*, 2005, **5**, 1013–1021.
59. A. V. Trask, W. D. S. Motherwell and W. Jones, *Int. J. Pharm.*, 2006, **320**, 114–123.
60. S. Karki, T. Friščić, W. Jones and W. D. S. Motherwell, *Mol. Pharmaceutics*, 2007, **4**, 347–354.
61. T. Friščić, L. Fabian, J. C. Burley, D. G. Reid, M. J. Duer and W. Jones, *Chem. Commun.*, 2008, 1644–1646.
62. C. L. Cooke and R. J. Davey, *Cryst. Growth Des.*, 2008, **8**, 3483–3485.

CHAPTER 12

Application of Phase Diagrams in Co-crystal Search and Preparation

TIMO RAGER AND ROLF HILFIKER

Solvias AG, Department for Solid-State Development, Römerpark 2, 4303 Kaiseraugst, Switzerland

12.1 Introduction

Typical selection criteria for co-crystal formers include toxicological considerations, physico-chemical properties such as water solubility, and the functional groups that are available for interaction with the substance to be co-crystallized (*e.g.*, a drug substance). Some degree of predictability for the formation of a crystalline multi-component compound is often claimed for certain combinations of functional groups. The expression crystal engineering has been coined for this strategy and combined with the concept of synthons from organic chemistry. Computer modeling may further improve the predictive potential based on interaction energies and steric aspects.

Thermodynamic considerations offer a complementary approach for targeted co-crystal search and development. They can provide selection criteria for suitable co-crystal formers and can also help to predict the most promising preparation conditions. This includes the potential to reduce the experimental effort in co-crystal search and crystallization optimization significantly. Phase

RSC Drug Discovery Series No. 16
Pharmaceutical Salts and Co-crystals
Edited by Johan Wouters and Luc Quéré

Published by the Royal Society of Chemistry, www.rsc.org

diagrams are the method of choice to visualize the thermodynamic relationships and to simplify their application.

12.2 Binary Phase Diagrams

The binary phase diagram of two components A and B contains information on whether the interaction of the two components is strong enough to generate a thermodynamically stable co-crystal. This is illustrated in Figure 12.1 with the aid of simulated phase diagrams for a co-crystal with a fixed stoichiometry of 1:1. The Gibbs free energy functions of the individual components and the co-crystal are shown on the left-hand side of this figure and the resulting phase diagrams are on the right-hand side. For the sake of convenience, ideal behavior of the two components in the liquid phase was assumed for the simulation of the phase diagrams, that is, an ideal mixing entropy and no mixing enthalpy. In addition, complete immiscibility in the solid state was assumed, and the enthalpies and entropies of crystallization of all solid phases were defined as temperature independent. Although these assumptions are a severe over-simplification of the reality, they facilitate the analysis considerably while still permitting qualitatively correct conclusions to be drawn.

With the above assumptions the condition of thermodynamic equilibrium between the solid and the liquid phase leads to the following equations for the liquidus of each component or compound:

$$T = \frac{\Delta_m H_A}{\Delta_m S_A - R \cdot \ln x_A} \quad \text{(for component A)} \tag{12.1}$$

$$T = \frac{\Delta_m H_B}{\Delta_m S_B - R \cdot \ln x_B} \quad \text{(for component B)} \tag{12.2}$$

$$T = \frac{\Delta_m H_{AB}}{\Delta_m S_{AB} - \Delta_{mix} S_{AB} - R \cdot \ln(x_A \, x_B)} \quad \text{(for the 1:1 compound AB)} \tag{12.3}$$

In rearranged form, these equations are also known as the Schröder–van Laar relation. $\Delta_m H_X$ is the melting enthalpy of one mole of each component or compound X, $\Delta_m S_X = \Delta_m H_X / T_X$ its melting entropy (with T_X being the melting temperature of the pure component or compound), x_A and x_B the mole fractions of A and B, and T the melting temperature at the given composition. $\Delta_{mix} S_{AB} = -2R \ln(0.5)$ is the mixing entropy of one mole A and one mole B in the liquid phase.

Component A was assumed to be a given substance with fixed thermodynamic properties, that is, melting enthalpy, melting entropy, and (along with this) melting point. The enthalpy and entropy of crystallization of the co-crystal were set to a constant value as well, whereas the thermodynamic parameters of the co-crystal former B have been varied.

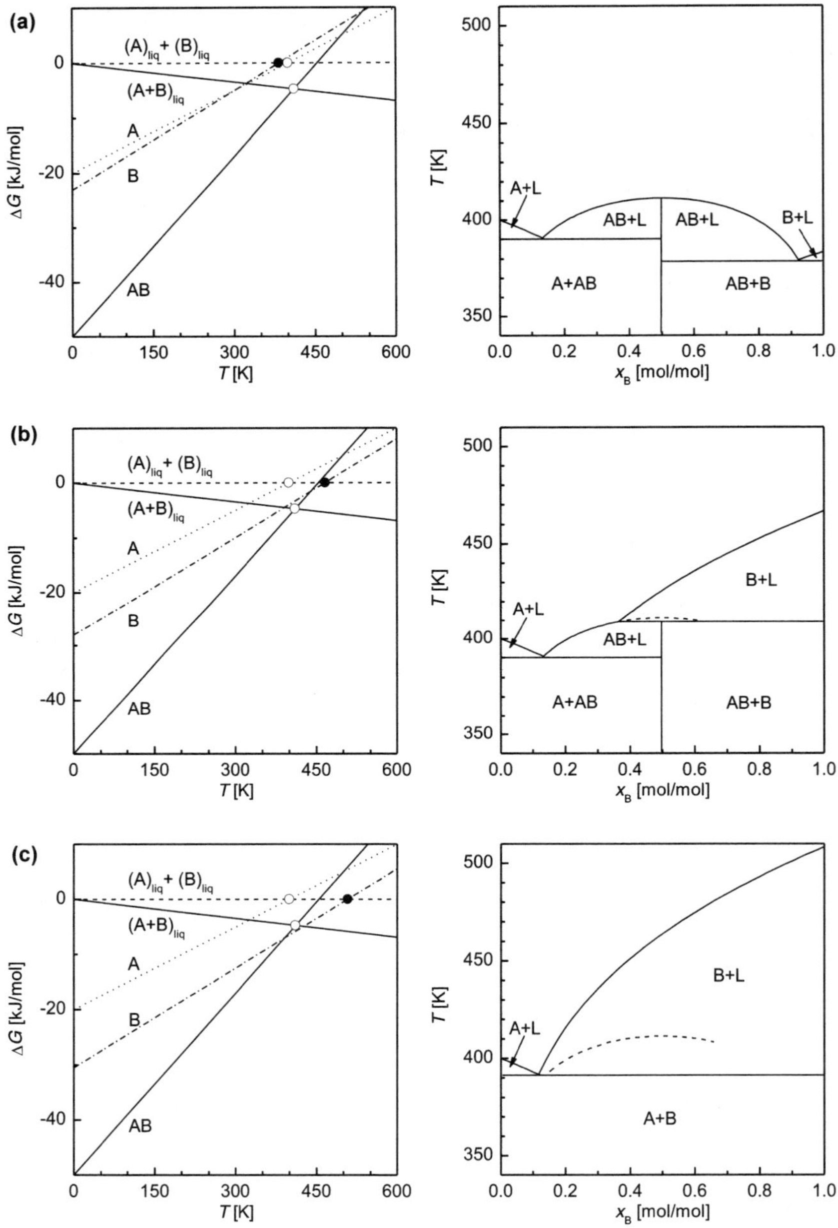

Figure 12.1 Gibbs free energy functions (left column) and phase diagrams (right column) for binary systems with different thermodynamic properties of the co-crystal former B. The free energy functions are plotted for the separate liquid phases of A and B ($(A)_{liq} + (B)_{liq}$) as the reference state ($\Delta G = 0$). Relative to this, the 1:1 mixture of liquid A and B ($(A + B)_{liq}$) is stabilized by $2RT\ \ln(0.5)$ as a result of the mixing entropy. The melting points of A and AB are fixed and indicated by open circles, whereas the variable melting point of B is indicated by a solid circle. The liquidus curves in the phase diagrams are calculated based on the assumption of ideal behavior (Equations (12.1) to (12.3)).

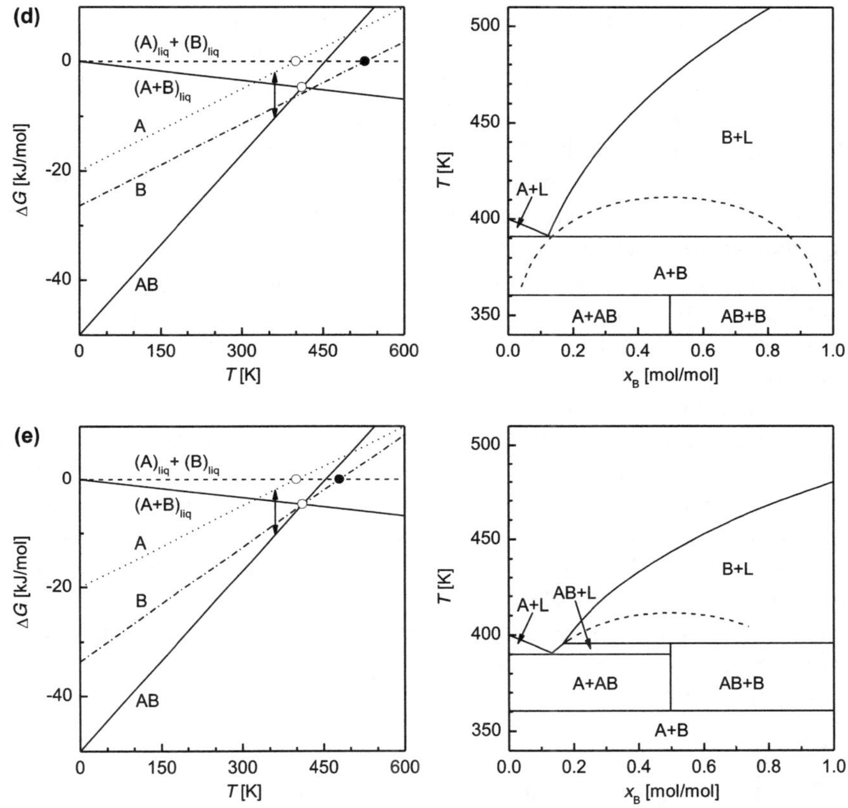

Figure 12.1 Continued

In Figures 12.1(a) to 12.1(c), the melting enthalpy of form B (crossing point of the ΔG function with the ordinate) was changed, whereas its melting entropy (slope of the ΔG function) was kept constant. In addition, equal entropies for the sum of one mole A and B and for one mole of the co-crystal were assumed. Thus, the difference in Gibbs free energy at any temperature (which determines the relative thermodynamic stability) is constant and identical to the difference in enthalpy.

When the Gibbs free energies of A and B are comparable in size and add up to a more positive value than the free energy of the co-crystal (Figure 12.1(a)), the melting curve exhibits a maximum at the composition of the co-crystal (*dystectic*) and the co-crystal melts *congruently*. The maximum is accompanied by two temperature minima (*eutectics*) on both sides. When the molar Gibbs free energy of B at the melting temperature of the co-crystal becomes more negative than half the free energy of the co-crystal (Figure 12.1(b)), the dystectic and the eutectic on the B-rich side of the phase diagram disappear and a *peritectic* emerges instead. At the peritectic temperature, the co-crystal melts *incongruently* with re-crystallization of solid B. As soon as the sum of the Gibbs free energies of the components falls below the free energy of the co-crystal (Figure 12.1(c)), the co-crystal becomes thermodynamically unstable and will not appear in the phase diagram at all.

As an additional variation, in Figure 12.1(d), a smaller melting entropy of B compared to the preceding figures was presumed in combination with a small melting enthalpy. The formation of the co-crystal is therefore favored by enthalpy but disfavored by entropy. As a result of this, the co-crystal is only stable at low temperature and disproportionates into the individual components above a certain temperature (360 K), which is located below the appearance of any liquid phase. As the other alternative, a large melting enthalpy and a large melting entropy of component B were presumed in Figure 12.1(e). The co-crystal will be thermodynamically stable at elevated temperature, but may disproportionate over time at a lower temperature.

Attractive interactions between the two co-crystal components in the liquid phase may significantly change the picture in Figure 12.1 in that the eutectic and dystectic temperatures are further decreased and the temperature differences between the eutectics and the melting points of the components will increase. However, the stabilities of a mixture of the individual solid components and the co-crystal relative to the molten state are affected by these attractive interactions to the same extent, and the relative thermodynamic stability of the co-crystal and the individual components will still be reflected correctly in the phase diagram.

12.3 Thermodynamic Selection Criteria for Co-crystal Formers

Inspection of the binary phase diagrams indicates that, from a thermodynamic point of view, the best chances of finding a stable co-crystal are given for co-crystal formers with a low melting point and a low melting enthalpy. A high melting point or a high melting enthalpy are unfavorable, in particular a combination of both. The latter characteristics of a co-crystal former indicate that it already forms a very stable crystal by itself. Co-crystal formation with such a substance is unlikely to provide any energetic advantage and is therefore less likely to occur than with a substance whose own crystal structure is weak. Based on this consideration, it is expected that molecular compounds with a liquid component (*i.e.*, solvates) will be encountered more frequently than molecular compounds with a second crystalline component (*i.e.*, co-crystals).

Of course, the disadvantages of a high melting point and melting enthalpy of the pure co-crystal former may be overcome by particularly favorable interactions with the drug molecule. However, this represents an additional challenge. If two molecules with a similar structure and the same functional groups are envisaged for co-crystal formation, consideration of the thermodynamics will indicate that the one with the lower melting point and melting enthalpy be chosen first.

12.4 Ternary Phase Diagrams

Despite the existence of some large-scale technical processes for crystallization from the melt or from the gas phase, solvent-based crystallizations are by far the most common ones for organic molecules. This is first of all due to the fact that crystallizations from solution represent a convenient method of removing impurities from the re-crystallized material. In addition to this, the temperature for crystallization from the melt or from the gas phase may not be accessible because of the thermal instability of the substances. Also, a high viscosity or a high glass transition temperature of the melt may prevent complete crystallization and, finally, a lump of crystalline material instead of a powder would be obtained, which could involve problems during further handling.

The presence of a solvent in the crystallization process alleviates these problems, mainly by lowering the activation energy barrier for the rearrangement of the solute molecules into a new crystalline structure. Thus, the solvent has the classical function of a catalyst and will only affect the kinetics but not the thermodynamics of the transformation. This is valid as long as the solvent does not become part of the crystal structure, that is, is not forming a solvate with the solute. With respect to co-crystals this signifies that the existence of a thermodynamically stable binary compound is not a function of the solvent.

Regardless of this, the addition of a solvent as a third component provides new opportunities to minimize the overall Gibbs free energy of the *ternary* system by mixing in the liquid phase. In the case of ideal mixing, this entropy-driven gain in Gibbs free energy for one mole of components is given by:

$$\Delta_{\text{mix}}G = RT(x_A \ln(x_A) + x_B \ln(x_B) + (1 - x_A - x_B)\ln(1 - x_A - x_B)) \quad (12.4)$$

where x_A and x_B are the mole fractions of A and B, and $1{-}x_A{-}x_B$ the mole fraction of S. Enthalpic terms will add to this mixing free energy in the case of real systems.

A comparison of the free energies of different states of the system at each composition leads to a lowest Gibbs free energy surface (Figure 12.2(a)) and a corresponding phase diagram (Figure 12.2(b)). Both of them are preferably represented in a mole fraction-based equilateral triangle. The area of such a triangle covers all possible combinations of the three components and has the useful property that the coordinates of any point (*i.e.*, its distances from the sides) sum up to a constant value, which is set to one mole.

Figure 12.2(b) illustrates that an increasing solvent content of the ternary mixture brings about a narrowing of the domains of coexistence of two mechanically mixed crystalline phases. In turn, the single crystalline phases in equilibrium with the solution expand. Ultimately, the domains of coexistence of two solid phases at a given temperature and pressure reduce to an invariant point, which represents the composition of the solution in equilibrium with any mixture of the two neighboring solid phases. On the other hand, pure solid phases of the components and of the co-crystal exist over a certain range of composition of the liquid phase.

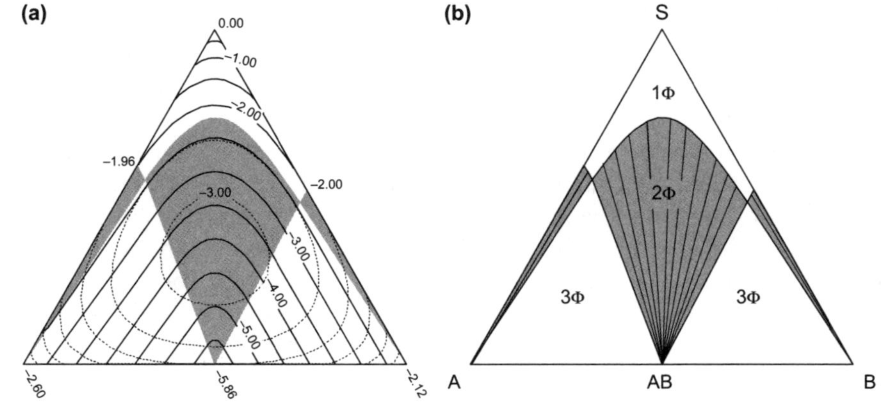

Figure 12.2 Properties of the system A–B–S as a function of composition in mole fractions based on the data in Figure 12.1(a) at 75 °C under the assumption of ideal miscibility in the liquid phase. (a) Surface of lowest Gibbs free energy with contour lines of constant free energy in steps of 0.5 kJ mol^{-1}. The Gibbs free energy of each pure liquid component was set at 0.0 kJ mol^{-1}. The metastable part of the Gibbs free energy surface of the homogeneous liquid phase is indicated by dashed lines. (b) The corresponding ternary phase diagram with tie-lines and indication of the number of coexisting phases.

The phase boundaries between the two-phase regions and the liquid phase at a given temperature can be approximated by the following polynomial equations (see Rager and Hilfiker (2009), Further Reading, ref. 13):

$$x_A = s_A\left(1 + \sum_i a_i x_B^i\right) \qquad \text{(for component A)} \qquad (12.5)$$

$$x_B = s_B\left(1 + \sum_j b_j x_A^j\right) \qquad \text{(for component B)} \qquad (12.6)$$

$$x_A \cdot x_B = K_{AB}\left(1 + \sum_i a_i x_B^i\right)\left(1 + \sum_j b_j x_A^j\right) \quad \text{(for the 1:1 component AB)} \quad (12.7)$$

In these equations, x_A and x_B are again the mole fractions of components A and B, s_A and s_B are their solubilities in the solvent S, K_{AB} is the solubility product of the compound AB, and a_i and b_j are virial coefficients describing deviations from ideality in the ternary mixture. The equilibrium constants s_A, s_B, and K_{AB} are correlated with the Gibbs free energy of each crystalline phase. The Gibbs free energy of formation of the co-crystal from the components at temperature T can be expressed as:

$$\Delta_r G^\circ_{AB} = RT \ln\left(\frac{K_{AB}}{s_A s_B}\right) \tag{12.8}$$

It follows from this equation that the solubility product of a thermodynamically stable co-crystal is lower than the product of the solubilities of the individual components.

The aspect of the ternary phase diagram can change significantly, depending on the thermodynamic properties of the components and the co-crystal as well as their interactions with the solvent. As a consequence, suitable conditions for the formation of a pure co-crystal solid phase may be found at very different compositions and will also vary with the solvent (Figure 12.3). For example, the domain of existence of the co-crystal will be narrow when the equilibrium constant for co-crystallization is small (Figure 12.3(a)), and the domain of existence will be broad when the equilibrium constant is large (Figure 12.3(b)). The pure co-crystal solid phase will be found in the center of the phase diagram when the solubilities of both components are comparable (Figure 12.3(c)), but it will be located nearer to the more soluble component in the case of highly asymmetric solubilities (Figure 12.3(d)). Attractive interactions in solution (*i.e.*, a co-solvating effect of the second component) will increase the concentrations of the components necessary to reach the supersaturation limit of the co-crystal (Figure 12.3(e)), whereas repulsive interactions in solution (*i.e.*, decreasing solubility of each component in the presence of the other) will shift the co-crystal domain upwards (Figure 12.3(f)).

Obviously, the aspect of the ternary phase diagram also depends on the stoichiometry of the co-crystal. As an example, the domains of existence of a 1:1 and a 1:2 co-crystal are shown in Figure 12.4. The comparison demonstrates that despite the differences in the phase diagrams the location of the solubility curve of the co-crystal is only slightly affected.

The ternary phase diagram will additionally change with temperature. Two examples in the form of a contour map are represented in Figure 12.5 based on the thermodynamic data from Figure 12.1(a) and (b). The solubility lines of the individual components and the co-crystal extend into curved surfaces and the invariant points become *eutectic grooves* on both sides of the co-crystal phase boundary surface. Relatively symmetric cross sections can be seen at all temperatures for the phase diagram derived from the system with a dystectic. Very asymmetric temperature cross sections are found in the example with a peritectic melting of the co-crystal.

More complex changes in the ternary phase diagram with temperature will be encountered in reality as a result of more pronounced differences in the entropy of the crystalline species and numerous non-idealities such as the temperature dependence of their enthalpy and entropy or finite interaction energies in the liquid phase. As a result of this, combinations of congruent melting and incongruent dissolution or vice versa may occur (Figure 12.6).

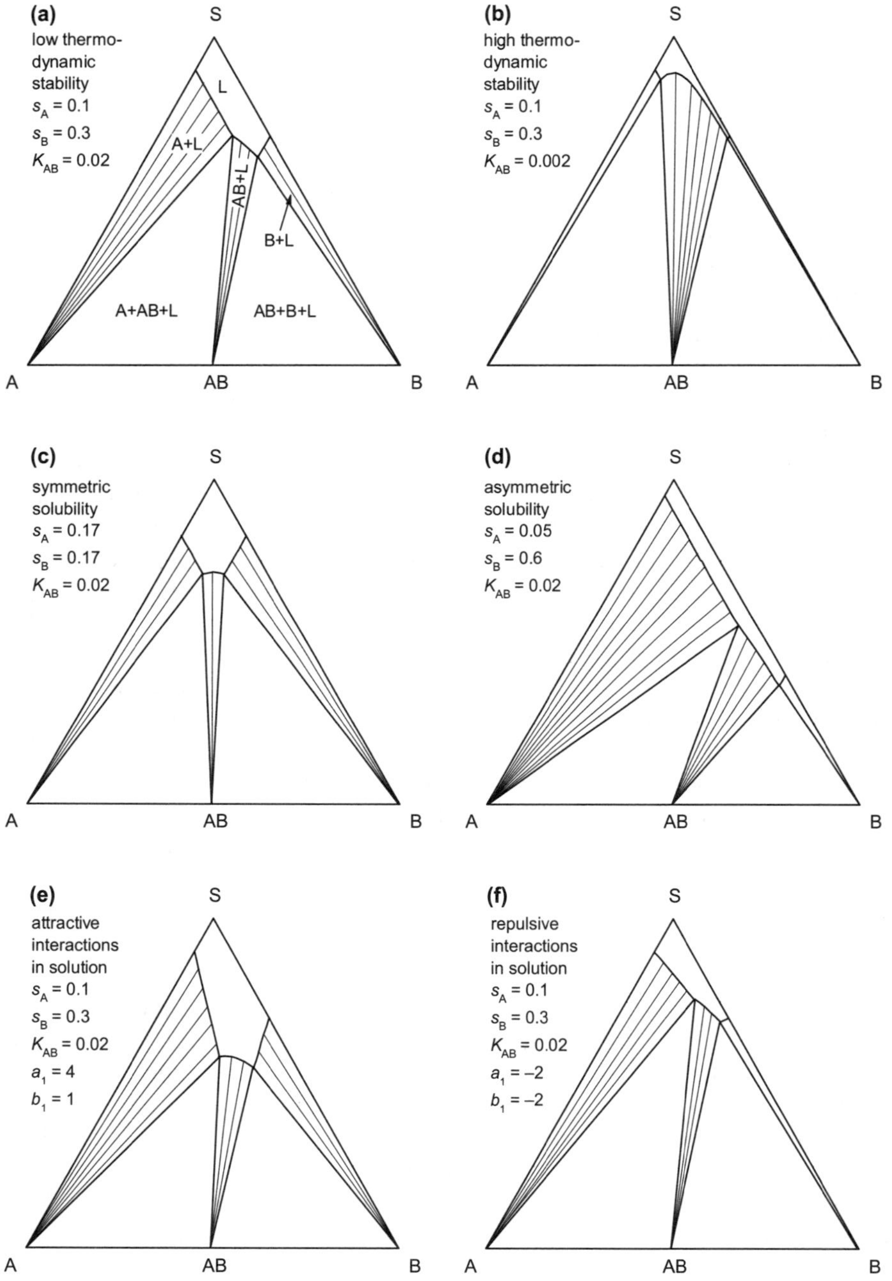

Figure 12.3 Shape and location of the co-crystal domain in the ternary phase diagram depending on: (a) and (b) the relative thermodynamic stability of the co-crystal, (c) and (d) the relative solubility of the two components, and (e) and (f) the types of interaction in solution. The solubilities of the components, the solubility product of the co-crystal, and the virial coefficients for calculating the solubility curves are indicated on the left of each graph (*cf.*, Equations (12.5) to (12.7)).

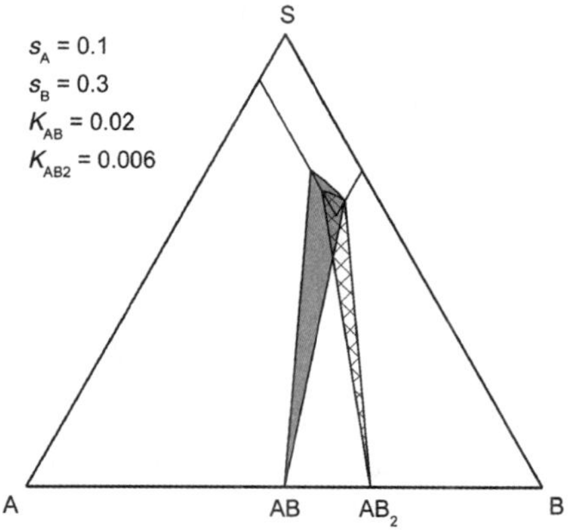

Figure 12.4 Overlay of the phase diagrams of co-crystals with different stoichiometry. The stability domains of the co-crystals overlap near the crossing point of the solubility curves of A and B.

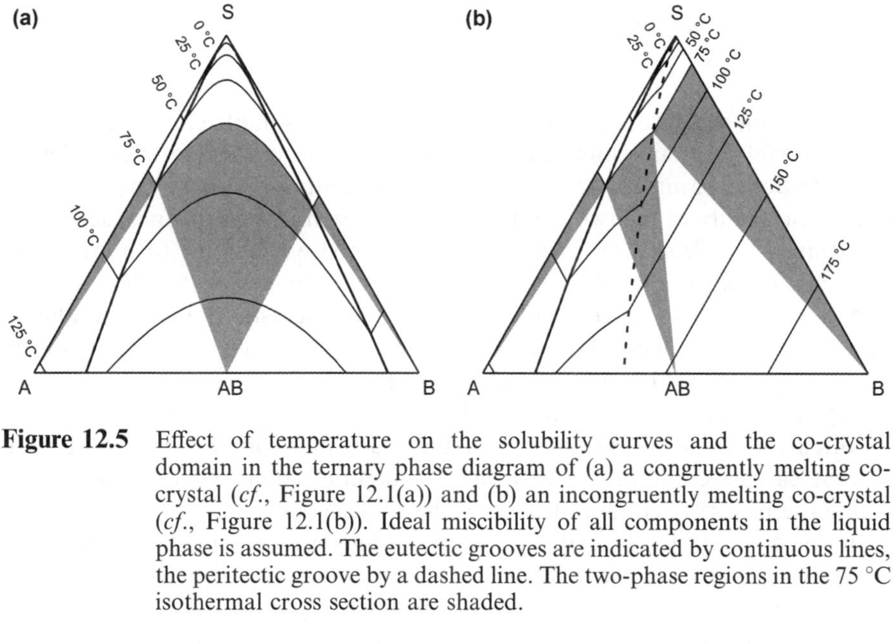

Figure 12.5 Effect of temperature on the solubility curves and the co-crystal domain in the ternary phase diagram of (a) a congruently melting co-crystal (*cf.*, Figure 12.1(a)) and (b) an incongruently melting co-crystal (*cf.*, Figure 12.1(b)). Ideal miscibility of all components in the liquid phase is assumed. The eutectic grooves are indicated by continuous lines, the peritectic groove by a dashed line. The two-phase regions in the 75 °C isothermal cross section are shaded.

12.5 Thermodynamic Selection Criteria for Suitable Preparation Conditions

For co-crystallizations in the presence of a solvent, it is not appropriate under all circumstances to provide the components at the stoichiometry of the co-crystal. As can be seen from Figure 12.3, starting from a 1:1 mixture, a pure solid phase of the AB co-crystal will only be obtained in the two examples Figure 12.3(b) and 12.3(c) under equilibrium conditions. In all other cases,

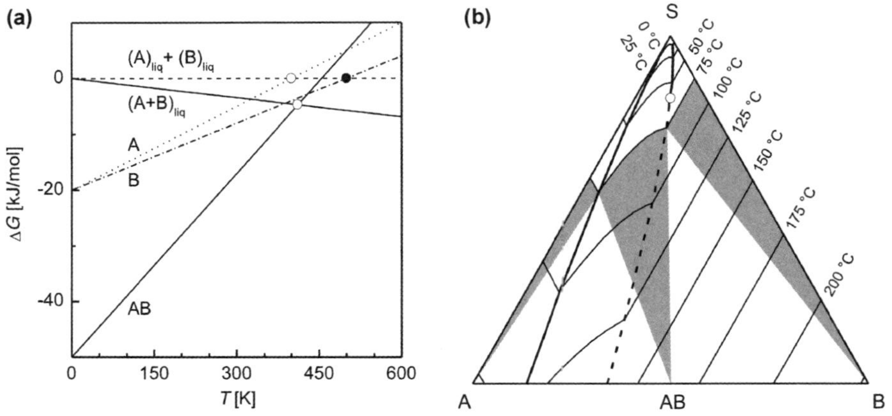

Figure 12.6 (a) Gibbs free energy curves for A, B, and AB with different entropies for A + B and AB. (b) The corresponding temperature-dependent ternary phase diagram with a transition between congruent dissolution and incongruent melting of the co-crystal phase. Ideal miscibility of all components in the liquid phase is assumed. The transition between eutectic and peritectic behavior is indicated by an open circle.

depending on the amount of solvent added, the less soluble component A will either crystallize as an admixture to the co-crystal or even replace it completely. The additional parameter that has to be considered in a ternary mixture is the relative solubility of the two components, because it is the principal factor that determines the degree of asymmetry of the phase diagram. Moreover, the domain of existence of the co-crystal is directly connected to the solubility curves of the individual components. Thus, the co-crystal domain can always be reached by following the solubility curve of one of the components in the ternary system. This applies independently of the stoichiometry of the co-crystal (Figure 12.4). When the solubility curves are not known (which will be often the case), the solubilities of the individual components in the pure solvent can be extrapolated to locate a potential co-crystal domain and to define the preferred conditions for co-crystal discovery.

In this context, it is crucial that the *final* stage of a crystallization process is located in the correct area of the phase diagram, that is, at a composition where only the co-crystal exists as a thermodynamically stable solid phase. The method for accessing the co-crystal domain is not directly affected by this. For example, it can consist of an evaporation, cooling of a hot solution, mixing of two saturated solutions (arrows 1 in Figure 12.7), or suspending one solid component in a saturated solution of the second one (arrow 2 in Figure 12.7). The last method is considered to be particularly reliable and easy to implement because it does not require the explicit determination of solubilities. In addition, in the case of suspension equilibration the system has more time to convert to its thermodynamically stable state than in an evaporation or a cooling crystallization, for example.

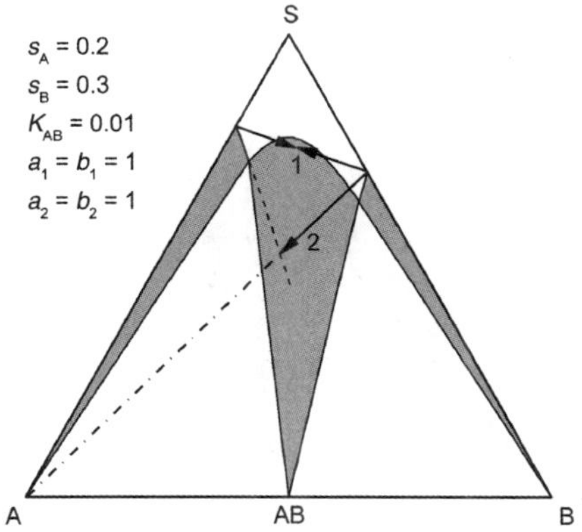

Figure 12.7 Accessing the co-crystal domain based on the solubilities of the individual components by mixing two saturated solutions (arrows 1) or by adding one solid component to a saturated solution of the second component (arrow 2).

12.6 Possible Failures in Co-crystal Search

Co-crystallization experiments may fail for a number of reasons. First of all, the discovery of a co-crystal under equilibrium conditions is excluded when no thermodynamically stable co-crystal exists, that is, when no co-crystal phase is present in the *binary* phase diagram. Under these circumstances, a co-crystal will not appear in the *ternary* system either, no matter which solvent or other crystallization parameters are chosen.

But even when a thermodynamically stable co-crystal exists, it may still be missed in crystallization experiments for various reasons. A factor of outstanding importance is the composition of the ternary mixture as already indicated in Figure 12.3. In the binary system, the preparation of a pure co-crystal phase (*e.g.*, by melting and cooling) requires that the two components A and B are mixed in the ratio at which they are expected to be present in the co-crystal. This correlation is no longer compulsory in a ternary system because different fractions of the components will dissolve in the liquid phase (Figure 12.8). Thus, as already stated, solubilities have to be considered ahead of stoichiometries. This can be done in two ways: either an excess of the more soluble component is added to the system, or a solvent with similar solubility of both components is selected. Obviously, when only minute amounts of solvent are added (*e.g.*, in solvent-drop grinding), the required amounts of the components will again approach the stoichiometry of the co-crystal. The effect of different solubilities becomes less important, but at the same time the stoichiometry needs to be known or assumptions have to be made with regard to it.

The situation may become even more challenging when the phase diagram exhibits a highly non-ideal shape as a result of strong attractive interactions in solution (co-solvation). As can be seen from Figure 12.9, the strategy of adding the solid material of one component to a saturated solution of the second may

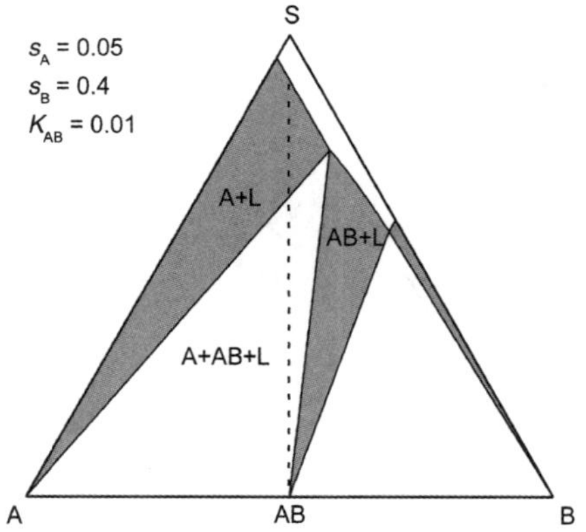

Figure 12.8 Co-crystal formation in a solvent with asymmetric solubility of both components. Crystallizations starting from equimolar mixtures of A and B (dashed line) lead to solid A at low concentration and to mixtures of A and AB at higher concentration. A pure co-crystal solid phase is only obtained when excess B is added to the system or the content of solvent is essentially decreased to zero (solvent-drop grinding).

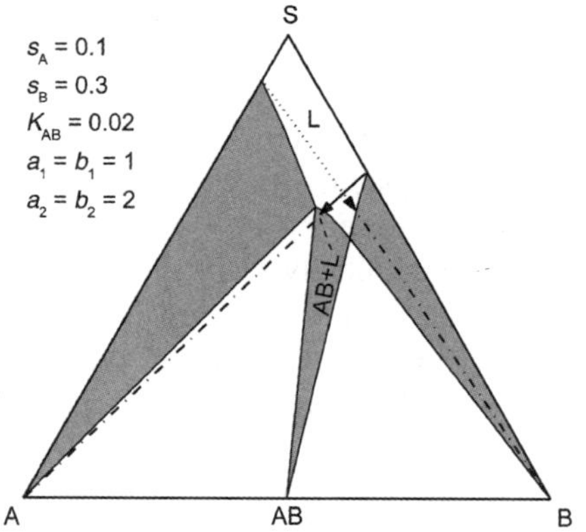

Figure 12.9 Accessing the co-crystal region by extrapolation of the individual solubilities in a highly non-ideal ternary phase diagram. The co-crystal domain is reached by adding A to a saturated solution of B in S (solid arrow), but not by adding B to a saturated solution of A in S (dotted arrow).

fail under these conditions (dotted arrow). At the same time, Figure 12.9 also indicates that the chances of success of the suspension equilibration experiment are higher when the less soluble component is suspended in a saturated solution of the more soluble one (solid arrow) than vice versa.

The formation of a co-crystal may also be compromised by the competition with solvate formation of any of the species under consideration, that is, A, B, or AB. Depending on the concentrations of the components involved, the

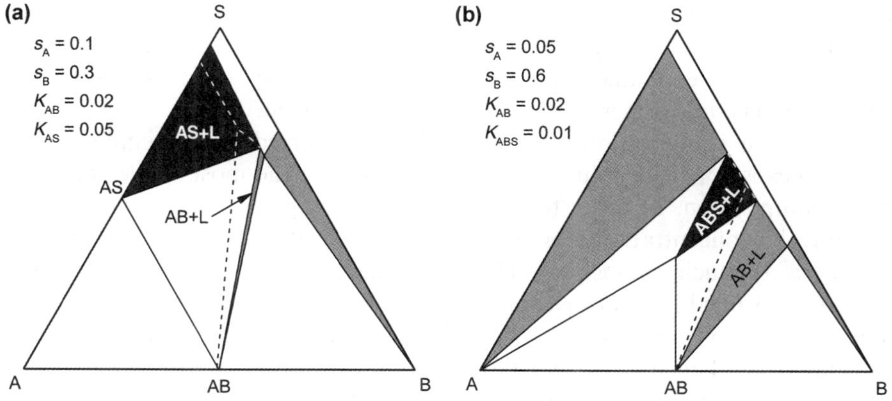

Figure 12.10 Competition of co-crystal formation with solvate formation, (a) of component A and (b) of the co-crystal AB. The dashed lines indicate the situation without solvate formation.

solvate may be the most stable solid state form. It will then supersede other crystalline phases in the corresponding part of the phase diagram. As a result, the co-crystal domain may be narrowed or may even completely disappear. The effect is exemplified for solvate formation with component A (solvate AS) and AB (solvate ABS) in Figure 12.10.

Strategies to lower the risk of solvate formation include an increase in temperature (the desolvated state is typically favored by entropy) and a decrease in the solvent activity. The latter can be achieved by lowering the solvent concentration by using the highest possible concentration of the non-solvate-forming component and employing solvent mixtures.

Finally, the formation of a co-crystal may also fail for kinetic reasons. This factor is considered to be particularly relevant when the co-crystal is accessed though other solid state phases without longer equilibration times. An example of this is the evaporation of a stoichiometric solution in a solvent with asymmetric solubility of both components (Figure 12.3(a), (d), (e) and (f)). The formation of crystalline A is thermodynamically favored at an early stage of the evaporation process and this crystalline phase may not transform to the most stable state at a later stage. A similar situation is encountered in cooling crystallization in systems with peritectic behavior or disproportionation (Figure 12.1(b), (d) and (e)). Crystalline B will separate first from the melt (or solution) and may not convert to the thermodynamically most stable phase upon further cooling.

12.7 Phase Diagrams for Most Efficient Co-crystal Search

In most instances it is not the purpose of a co-crystal search to find the maximum number of co-crystals, but to find at least one suitable co-crystal for further development. However, even if the discovery of a maximum number of co-crystals is intended, it will always be preferable to perform the search with a minimum of experimental effort. Minimizing the number of experiments requires that the reliability of each experiment is optimized. If this is achieved, even one single experiment per co-crystal former may suffice to decide on suitable co-crystal formers with a reasonable confidence level.

Suspension equilibrations will generally be preferred for screening experiments because of their natural tendency to generate the most stable state at a given composition. In order to maximize the driving force for co-crystal formation, the concentrations (or rather activities) of the components should be increased as much as possible. Therefore, the component of interest is preferably suspended in *saturated* solutions of the co-crystal formers to be tested. The reliability of the screening can be further increased when solvent mixtures are employed. This lowers the risk of misinterpretations as a result of solvate formation, can improve the transformation kinetics, and tends to even out solubility differences between different compounds (Figure 12.11). On one hand, this latter effect will lead to more symmetric phase diagrams. On the other hand, it permits the same solvent system to be used for a large set of experiments, which has the potential to simplify the dosing of the suspended component on a microtiter plate.

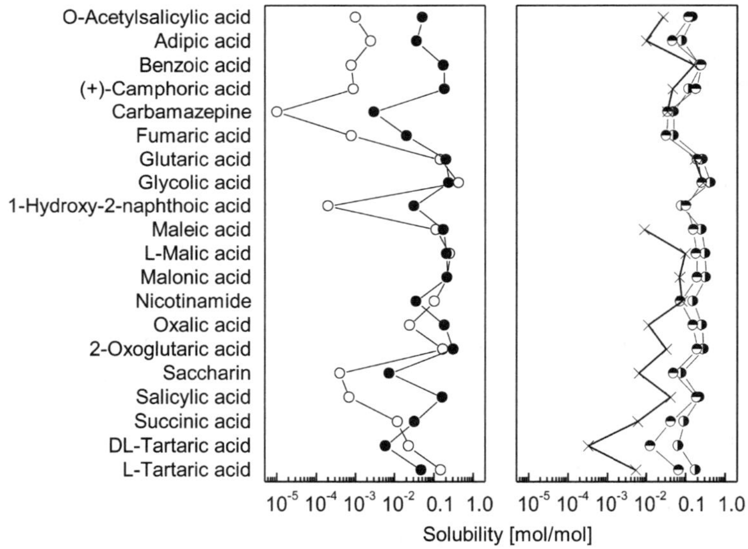

Figure 12.11 Solubilities of carbamazepine and 19 different cocrystal formers in water (○), 2-propanol (●), an equimolar mixture of acetic acid, acetone, *p*-dioxane, DMF, DMSO, ethylene glycol, formic acid, sulfolane, and water (◑), an equimolar mixture of anisole, benzyl alcohol, 2-butanol, *n*-butyl acetate, butyronitrile, decalin, *p*-dioxane, ethylene glycol, methyl isobutyl ketone, NMP, toluene, and water (◒), as well as ideal solubility values calculated from differential scanning calorimetry (DSC) data with the Schröder–van Laar equation (×). Part of the experimental solubility data are taken from Rager and Hilfiker (2010), see Further Reading, ref. 24.

12.8 Phase Diagrams for Optimized Co-crystal Preparation

The primary purpose of a crystallization optimization is to maximize the yield and purity of the crystallized compound. The yield requirements will typically refer to the drug substance (component A) only, which is in most cases significantly more valuable than the co-crystal former. The price per kilogram of the latter may often be comparable with that of a solvent.

From the fact that the solubility of a co-crystal is defined by the solubility product (*cf.*, Equation (12.7)) it follows immediately that an increase in the concentration of one of the components will lower the concentration of the second one (Figure 12.12). Thus, adding an excess of the co-crystal former will typically result in the crystallization of a higher percentage of the drug substance in the co-crystal phase. This is illustrated in Figure 12.13 based on the idealized phase diagram already used in Figure 12.2. At an equimolar ratio of A and B and a molar solvent content of 60%, little more than 40% of component A is recovered in the form of a co-crystal. By adding excess B to the mixture (while keeping the A-to-solvent ratio constant), the yield can be increased to almost 80% (arrow 1). When substituting part of the solvent by component B (arrow 2), even a yield of almost 90% may be reached before pure component B begins to crystallize as well.

In real systems, the effect may not always be as pronounced as this. In particular, a co-solvating effect of the co-crystal former may weaken or even reverse the consequences of the solubility product on yields. Such effects are particularly likely to occur when ionic interactions between components prevail in solution (*e.g.*, as a result of proton transfer). For systems that do not deviate too far from ideality, it can be stated that the concentration of the co-crystal former should be chosen as high as possible in order to increase

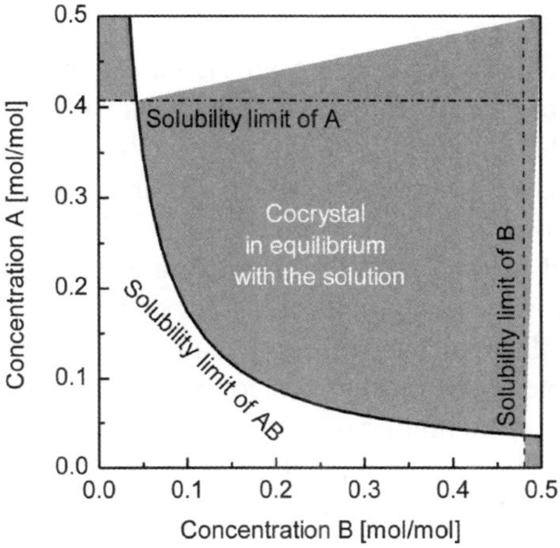

Figure 12.12 Interdependence of the concentrations of drug substance and co-crystal former represented by rectangular coordinates.

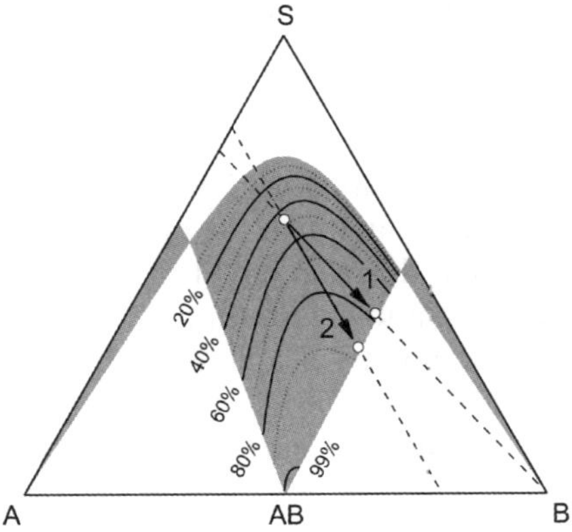

Figure 12.13 Percent recovery of substance A in the co-crystal phase as a function of composition of the ternary mixture. The phase diagram is based on the thermodynamic data in Figure 12.1(a) at 75 °C and assumes ideal miscibility in the liquid phase.

yields. To go one step further, this also provides a selection criterion for the solvent. The co-crystal phase can only be in equilibrium with a large excess of B in solution when the solvent has a much higher solubilizing power for B than for A, or, in other words, a highly asymmetric phase diagram as represented in Figure 12.3(d) will be preferred.

With regard to purity, the relevant aspect in the present context concerns the uniformity of the solid phase, that is, the absence of any excess of A or B. This is best ensured when the co-crystal is the only and thermodynamically most stable solid state form over the whole crystallization process. Therefore, in the case of cooling crystallization, the location of the co-crystal domain as a function of temperature becomes relevant for a systematic optimization of the crystallization process. Ideally the complete phase diagram of the system should be known with some accuracy.

To exemplify this, the two phase diagrams from Figure 12.5 are re-evaluated in Figure 12.14. In the case of an almost symmetric phase diagram (Figure 12.14(a)), the whole area of the co-crystal domain at 25 °C is accessible via cooling of a liquid phase of the same molar composition. On the other hand, in the case of a more asymmetric phase diagram (Figure 12.14(b)), the co-crystal phase at 25 °C can only be reached directly via cooling in dilute systems. In cooling crystallization with lower solvent content, it is thermodynamically preferred that component B crystallizes first. Although the co-crystal will become the thermodynamically stable phase in a certain concentration range at lower temperature, its formation may be hindered kinetically at this stage.

Thus, from the point of view of purity, a relatively symmetric phase diagram similar to Figure 12.3(c) will be preferred for a cooling crystallization process. This reveals that the requirements of yield and purity are to some extent opposed and the optimal crystallization process will often be a compromise between these two goals. Obviously, additional parameters will also have to be

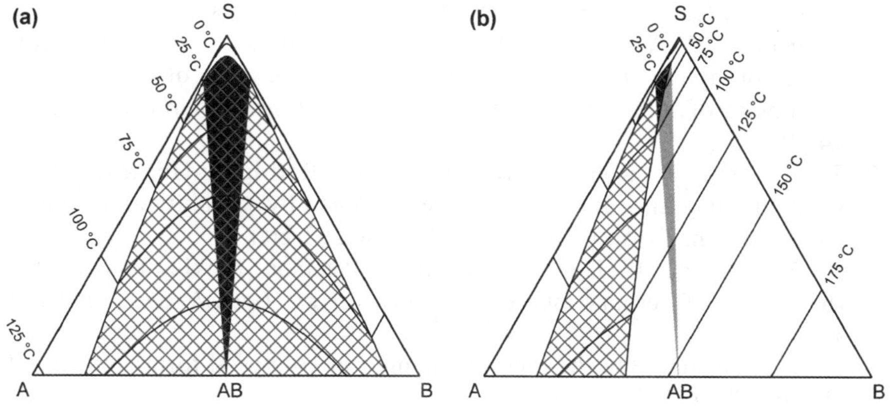

Figure 12.14 Overlap between the co-crystal domain at the temperature of isolation (arbitrarily set at 25 °C) and the liquid-to-solid phase boundary through which the co-crystal domain is initially accessed (cross-hatched area). In the black areas the co-crystal phase can be accessed directly via cooling, whereas in the grey area the co-crystal phase is only reached after passing through the solid state phase of crystalline B.

considered in a real process such as toxicological aspects, the efficient removal of impurities, the crystallization kinetics, the temperature limitations caused by the boiling point of the solvent or the degradation of the solutes, and a sufficient temperature gradient of the solubility.

12.9 Conclusion

Phase diagrams are a powerful and intuitive tool for planning co-crystal screenings and crystallization optimizations.

Further Reading

Phase Diagrams and Solubility in General

1. A. Findlay, A. N. Campbell and N. O. Smith, *Die Phasenregel und ihre Anwendungen*, Verlag Chemie, Weinheim, 9th edn, 1958.
2. J. Nývlt, *Solid-Liquid Phase Equilibria*, Elsevier, Amsterdam, 1977.
3. D. J. W. Grant and T. Higuchi, *Solubility Behavior of Organic Compounds*, John Wiley & Sons, New York, 1990.
4. M. Hillert, *Phase Equilibria, Phase Diagrams and Phase Transformations, Their Thermodynamic Basis*, Cambridge University Press, Cambridge, 2nd edn, 2008.

Co-crystal Phase Diagrams

5. J. W. Poole and T. Higuchi, Complexes formed in aqueous solutions by sarcosine anhydride; interactions with organic acids, phenols, and aromatic alcohols, *J. Am. Pharm. Assoc.*, 1959, **48**, 592–601.

6. K. Ito and K. Sekiguchi, Studies on the molecular compounds of organic medicinals. II. Application of the solubility product principle and consideration by the phase rule to the solubility phenomena of the molecular compound of sulfanilamide and sulfathiazole, *Chem. Pharm. Bull.*, 1966, **14**, 255–262.
7. S. J. Nehm, B. Rodríguez-Spong and N. Rodríguez-Hornedo, Phase solubility diagrams of co-crystals are explained by solubility and solution complexation, *Cryst. Growth Des.*, 2006, **6**, 592–600.
8. R. A. Chiarella, R. J. Davey and M. L. Peterson, Making co-crystals – the utility of ternary phase diagrams, *Cryst. Growth Des.*, 2007, **7**, 1223–1226.
9. R. S. Schartman, On the thermodynamics of co-crystal formation, *Int. J. Pharm.*, 2009, **365**, 77–80.
10. A. Jayasankar, L. S. Reddy, S. J. Bethune and N. Rodríguez-Hornedo, Role of co-crystal and solution chemistry on the formation and stability of co-crystals with different stoichiometry, *Cryst. Growth Des.*, 2009, **9**, 889–897.
11. K. Chadwick, R. Davey, G. Sadiq, W. Cross and R. Pritchard, The utility of ternary phase diagram in the discovery of new co-crystal forms, *CrystEngComm.*, 2009, **11**, 412–414.
12. A. Ainouz, J.-R. Authelin, P. Billot and H. Lieberman, Modeling and prediction of co-crystal phase diagrams, *Int. J. Pharm.*, 2009, **374**, 82–89.
13. T. Rager and R. Hilfiker, Stability domains of multi-component crystals in ternary phase diagrams, *Z. Phys. Chem.*, 2009, **223**, 793–813.
14. K. Guo, G. Sadiq, C. Seaton, R. Davey and Q. Yin, Co-crystallization in the caffeine/maleic acid system: lessons from phase equilibria, *Cryst. Growth Des.*, 2010, **10**, 268–273.
15. Y. Corvis, P. Négrier, M. Lazerges, S. Massip, J.-M. Léger and P. Espeau, Lidocaine/L-menthol binary system: co-crystallization versus solid-state immiscibility, *J. Phys. Chem. B*, 2010, **114**, 5420–5426.
16. A. Jayasankar, L. Roy and N. Rodríguez-Hornedo, Transformation pathways of co-crystal hydrates when coformer modulates water activity, *J. Pharm. Sci.*, 2010, **99**, 3977–3985.

Phase Transformations in Suspension

17. E. Shefter and T. Higuchi, Dissolution behavior of crystalline solvated and nonsolvated forms of some pharmaceuticals, *J. Pharm. Sci.*, 1963, **52**, 781–791.
18. P. T. Cardew and R. J. Davey, The kinetics of solvent-mediated phase transformations, *Proc. R. Soc. London, Ser. A*, 1985, **398**, 415–428.
19. C. H. Gu, V. Young Jr. and D. J. W. Grant, Polymorph screening: influence of solvents on the rate of solvent-mediated polymorphic transformation, *J. Pharm. Sci.*, 2001, **90**, 1878–1890.

Co-crystal Formation in Suspension

20. N. Rodríguez-Hornedo, S. J. Nehm, K. F. Seefeldt, Y. Pagan-Torres and C. J. Falkiewicz, Reaction crystallization of pharmaceutical molecular complexes, *Mol. Pharmaceutics*, 2006, **3**, 362–367.

21. G. G. Z. Zhang, R. F. Henry, T. B. Borchardt and X. Lou, Efficient co-crystal screening using solution-mediated phase transformation, *J. Pharm. Sci.*, 2007, **96**, 990–995.
22. S. L. Childs, N. Rodríguez-Hornedo, L. S. Reddy, A. Jayasankar, C. Maheshwari, L. McCausland, R. Shipplett and B. C. Stahly, Screening strategies based on solubility and solution composition generate pharmaceutically acceptable co-crystals of carbamazepine, *CrystEngComm.*, 2008, **10**, 856–864.
23. J. H. ter Horst and P. W. Cains, Co-crystal polymorphs from a solvent-mediated transformation, *Cryst. Growth Des.*, 2008, **8**, 2537–2542.
24. T. Rager and R. Hilfiker, Co-crystal formation from solvent mixtures, *Cryst. Growth Des.*, 2010, **10**, 3237–3241.

Co-crystal Formation by Cooling Crystallization

25. E. Gagnière, D. Mangin, F. Puel, A. Rivoire, O. Monnier, E. Garcia and J. P. Klein, Formation of co-crystals: kinetic and thermodynamic aspects, *J. Cryst. Growth*, 2009, **311**, 2689–2695.
26. Z. Q. Yu, P. S. Chow and R. B. H. Tan, Operating regions in cooling co-crystallization of caffeine and glutaric acid in acetonitrile, *Cryst. Growth Des.*, 2010, **10**, 2382–2387.

Co-crystal Formation by Solvent-drop Grinding

27. N. Shan, F. Toda and W. Jones, Mechanochemistry and co-crystal formation: effect of solvent on reaction kinetics, *Chem. Commun.*, 2002, 2372–2373.
28. W. Jones, W. D. S. Motherwell and A. V. Trask, Pharmaceutical co-crystals: an emerging approach to physical property enhancement, *MRS Bulletin*, 2006, **31**, 875–879.

CHAPTER 13

Limits of the Co-crystal Concept and Beyond

GERARD COQUEREL

Unité de Cristallogenèse, EA3233 IMR 4114, Université de Rouen, F-76821 Mont Saint Aignan, France

In this chapter, views of co-crystals will be broadened. Indeed, any classification has its limits and, clearly, the thermodynamics of heterogeneous equilibria can help to replace the concept of the co-crystal in a more general perspective.

13.1 Most Frequent Heterogeneous Equilibria Related to Co-crystals

When a solid phase is characterized, the independent components of the related system must be defined together with the physical variable(s) applicable to the system. For instance, let us suppose that A, B and C are three independent components – C stands for a solvent – and any new crystallized entities different from the polymorphs of A, B and C are called compounds.[1] These chemical entities, <1-1-0> for <AB> and in general <*n*-*m*-*p*> for any ternary compound $<A_nB_mC_p>$, can have different attributes such as those listed below:

- congruent melting/non congruent melting (Figure 13.1)
- polymorph/no polymorph detected (Figure 13.2)
- stable/metastable: another compound or mixture of components can be more stable than compound <*n*-*m*-*p*> (Figure 13.3)
- stoichiometric compound/non-stoichiometric compound (Figure 13.4)

RSC Drug Discovery Series No. 16
Pharmaceutical Salts and Co-crystals
Edited by Johan Wouters and Luc Quéré

Published by the Royal Society of Chemistry, www.rsc.org

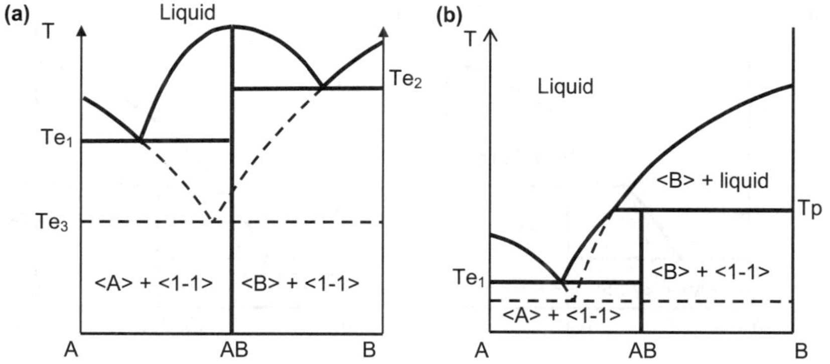

Figure 13.1 (a) <AB> is a stoichiometric compound with congruent fusion, (b) <AB> is a stoichiometric compound with non-congruent fusion – it decomposes reversibly at *T*p via a peritectic invariant. Dashed lines represent metastable equilibria; full lines represent stable equilibria.

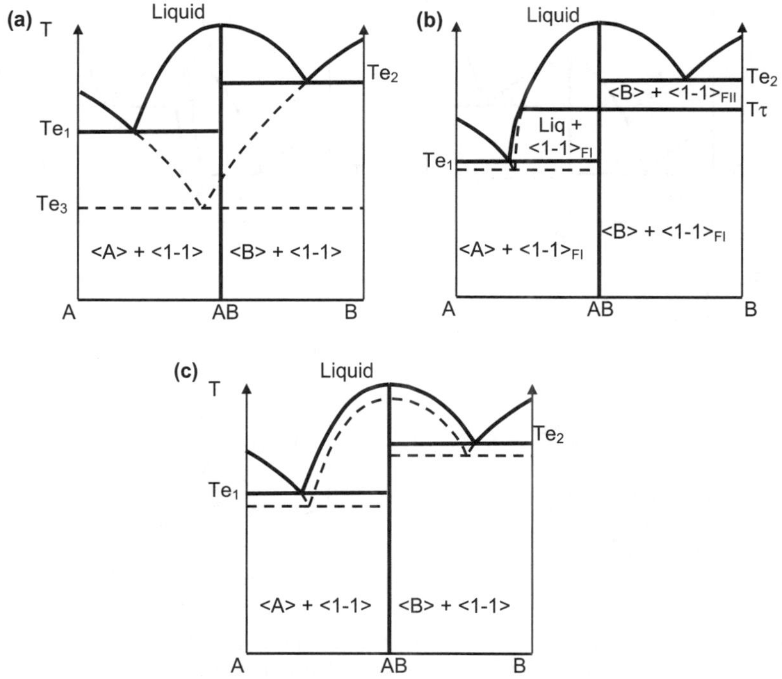

Figure 13.2 (a) Experimentally, no polymorphism has been detected for <AB>. (b) <AB> crystallizes in two varieties. $T\tau$ is the temperature of transition between these two forms: they are said 'enantiotropic'. The metastable eutectic between A and B at Te_3 is not represented for reasons of clarity. (c) <AB> crystallizes in two varieties. There is no temperature of reversible transition between these two forms: the metastable form is said to have a monotropic character under this pressure. The metastable eutectic between A and B at Te_3 is not represented for reasons of clarity.

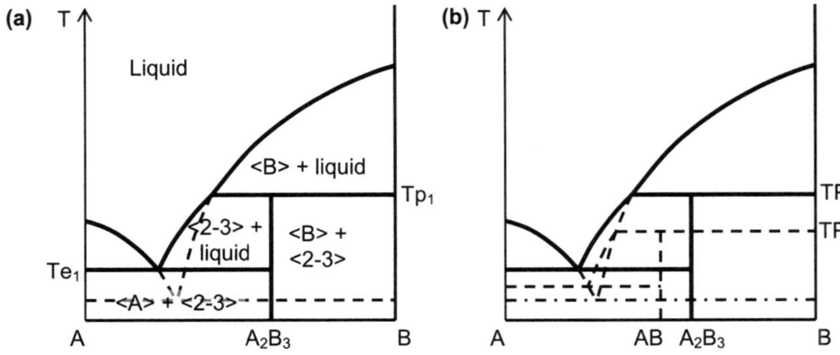

Figure 13.3 (a) <2-3> is a stable stoichiometric compound up to its non congruent melting point at Tp_1. (b) <1-1> has a metastable character when compared to <2-3>; it also shows non-congruent fusion at TP_2.

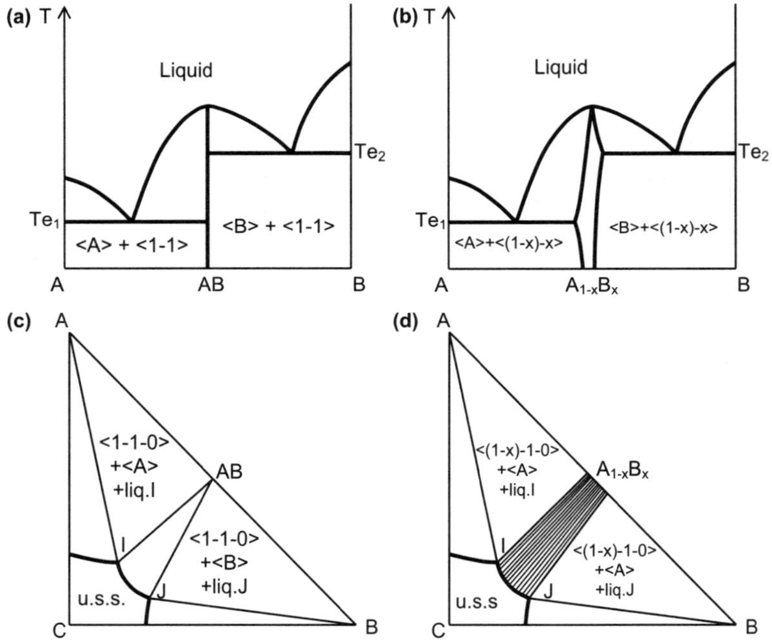

Figure 13.4 (a) <AB> is a stoichiometric compound with congruent fusion; the metastable eutectic between A and B is not represented. Conversely, (b) shows <$A_{1-x}B_x$> a non-stoichiometric compound with congruent fusion (note that the composition of the melting point does not necessarily correspond to a stoichiometric composition); the metastable eutectic between A and B is not represented. (c) <AB> is a stoichiometric compound in the isothermal section of the ternary system A-B-C. (d) <AB> is a non-stoichiometric compound in the isothermal section of the ternary system A-B-C. The thin segments, representing the so-called tie lines, connect the compositions of the two phases in equilibrium: crystals <$A_{1-x}B_x$> and their saturated liquid phase. u.s.s. is an unsaturated solution.

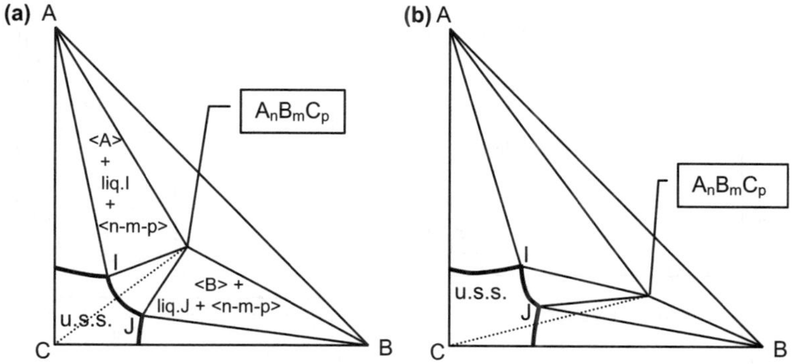

Figure 13.5 (a) shows a ternary compound <*n-m-p*> with congruent solubility. This means that the segment connecting the points that represent the solid phase and the pure solvent C intersect the solubility curve of the compound (curve IJ). (b) shows a ternary compound <*n-m-p*> with non-congruent solubility. This means that the segment connecting the points that represent the solid phase and the pure solvent C do not intersect the solubility curve of the compound (curve IJ). A similar case is described in Ref. 2, but $p = 0$.

- congruent solubility/non-congruent solubility of a ternary compound (Figure 13.5)
- hygroscopic character up to deliquescence/efflorescent (Figure 13.6).

It is worth noting that thermodynamics of heterogeneous equilibria is rather precise about these attributes, regardless on the nature of the bonds that link the different components. There is also no information on the structural features of these crystallized phases. For example, featuring a channel hydrate can be consistent with a non-stoichiometric compound. But this consistency emerges from a cross curricular approach to thermodynamic data (dynamic vapour sorption (DVS) for instance) and structural information (resolution of single crystal structures).[3] In other words, from a thermodynamic point of view, there are components, which define the system, and compounds, characterized by their attributes, as detailed above. Thus, in essence, there is no difference between salts, hydrogeno-salts, abnormal salts (see below), solvates, host–guest associations, co-crystals, adducts, clathrates, hybrids of all sorts and compounds.

The series of examples given below have been selected for their systematic and general aspects.

13.2 Compounds that could be Defined as Hybrid Salt–Co-crystals

Compounds that have been classified as 'abnormal salts' could be labelled now as hybrid salt–co-crystals (HSCCs). These compounds have compositions

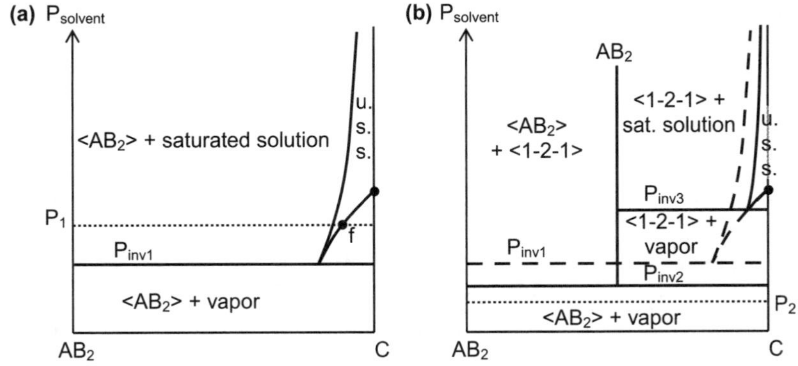

Figure 13.6 Isothermal section of the ternary system A-B-C; the physical variable is pressure (it is assumed that AB_2 is a co-crystal and P_{AB2} is negligible). (a) shows that for $P_{solvent}$ greater than P_{inv1} (*e.g.* $P1$), AB_2 will uptake solvent C so that it will at first partially and then ultimately completely dissolve. If an infinite reservoir of solvent vapour is assumed, the final point of evolution is 'f', an undersaturated solution. If C stands for water, compound $AB_2 = <1\text{-}2\text{-}0>$ is said to be hygroscopic up to deliquescence. (b) Due to the crystallization of the solvate $<1\text{-}2\text{-}1>$, P_{inv1} has become a metastable invariant. P_{inv2} corresponds to the equilibrium between the $<1\text{-}2\text{-}0>$, $<A>$ and C_{vapor}. P_{inv3} corresponds to the three phase invariant between $<1\text{-}2\text{-}1>$, its saturated solution and C_{vapor}. At $P_{solvent} = P_2$, the solvated co-crystal should irreversibly decompose (because $P_2 < P_{inv2}$) and $<1\text{-}2\text{-}1>$ is said to be efflorescent. Because of the stability domain of the solvate, the hygroscopic character of the stable phase has receded to $P_{solvent} > P_{inv3}$. In other words, deliquescence means that the system is missing a solvate.

which deviate from the 'natural' acido-basic ratio that undergraduate chemistry textbooks teach. For instance, there are an unbalanced number of acidic moieties/acidic components with respect to the number of basic functions:

1. $R–NH_3^+$ —$^-$OOC–R'
2. $R–NH_3^+$ —$^-$OOC–R''–COOH
3. $R–NH_3^+$ —$^-$OOC–R''–COOH—HOOC–R''–COOH

(1) represents a conventional salt, (2) represents a conventional hydrogeno salt, and (3) illustrates the 'abnormal' salt or hybrid salt–co-crystal.

Although it is difficult to predict the possibility of isolating a HSCC, it is probably more a rule than an exception that these species, which are not obvious candidates and which have 'exotic' stoichiometries with respect to conventional titrations ratios in solution, should be crystallized. One has to keep in mind that the solid state has constraints which do not need to match those prevailing in solution. The existence of these 'chimeras' clearly shows that there are borderline cases which resist any clear cut classification. So far, few researchers have extended their studies sufficiently to be able to put themselves in a situation to detect these HSCCs.

13.2.1 Example 1: Resolution of Fenfluramine

In order to access to its eutomer, racemic fenfluramine has been resolved by using D-camphoric acid. Mixtures of (+) and (–) fenfluramine (monobase) with variable amounts of this resolving agent (diacid) have been investigated.[4] *S*(+)-Fenfluramine gives a hydrogeno-D-camphorate and a D-camphorate. *R*(–)-Fenfluramine give a trihydrogeno-di-D-camphorate and a D-camphorate which is isostructural with the other diastereomer that has the same acid/base stoichiometry. Thus, there is an important contrast between these compounds in terms of chiral discrimination. The *S*(+)-fenfluraminium hydrogeno-D-camphorate and the HSCC:*R*(–)-fenfluraminium trihydrogeno-di-D-camphorate are stoichiometric phases without a detectable solid solution domain for the chiral base. The crystal lattices offer full recognition of the chiral amine. By contrast, when the two acidic functions of the D-camphoric acid are engaged in ionic bonds, the crystal lattice loses its stereoselectivity, that is, it can accommodate almost indifferently the *S*(+) and *R*(–) fenfluraminium cations. Therefore, this phase is a complete solid solution which can be symbolized by:

$S(+)\text{-fenfluraminium}_{(2-x)}\ R(-)\text{-fenfluraminium}_x\text{-D-camphorate with } 0 \leq x \leq 2.$

13.2.2 Example 2: Trimebutine Maleate

Trimebutine (monobase) forms a hydrogenomaleate and also a trihydrogenodimaleate with maleic acid (di-acid).[5] The two hydrogeno maleic acid and the doubly protonated maleic acid are linked by a classical dimeric –COOH—HOOC– bond, identical to the one found in the crystal structure of the pure maleic acid (Figure 13.7). It has been tempting to test the pentahydrogenotrimaleate or an even higher homologue in this series but no other solid phase has been detected.

13.2.3 Example 3: a Hybrid Salt–Co-crystal

A benzathine-like dibase: *trans-N,N'*-dibenzyldiaminocyclohexane (B) crystallizes with four molecules of 2,3-dichlorophenylacetic acid (AH) as a stable conglomerate without a detectable partial solid solution.[6] The crystal structure, resolved by single crystal X-ray diffraction, revealed an odd stoichiometry composed of: B^{2+}; two A^- and two AH. (Figures 13.8 and 13.9)

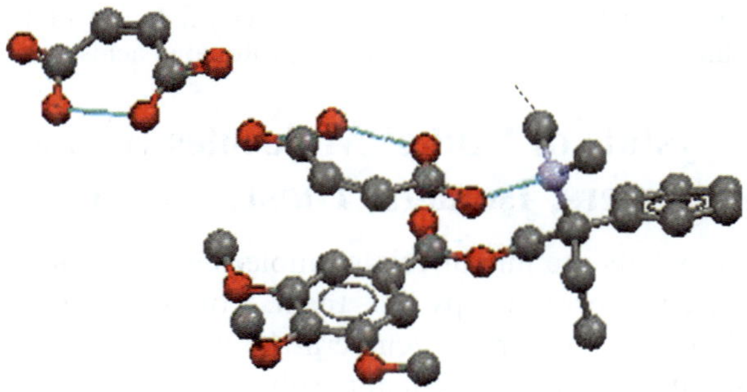

Figure 13.7 Trimebutine trihydrogenodimaleate (H-atoms not represented for clarity).

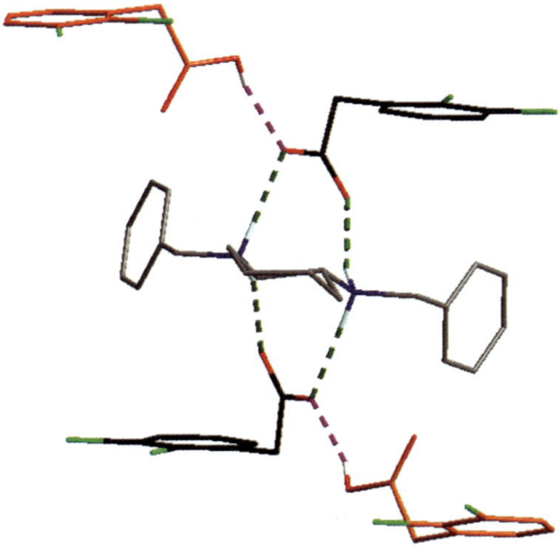

Figure 13.8 Building unit formed by one molecule of B^{2+} (grey), two molecules of A^- (black) and two molecules of AH (orange) adapted from Ref. 6.

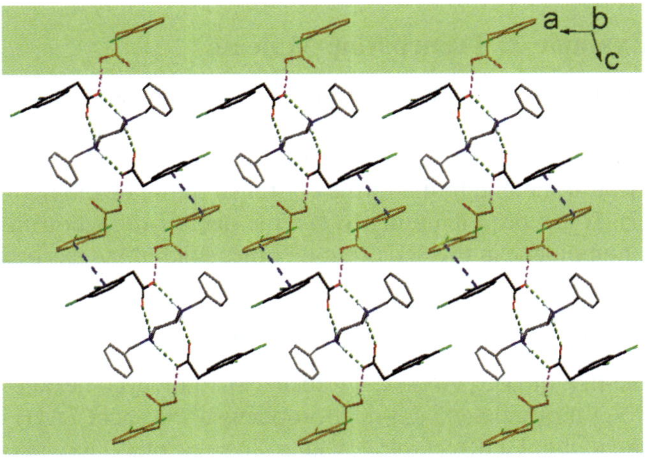

Figure 13.9 Projection along the *b* axis of *trans*-*N*,*N*'-dibenzyldiaminocyclohexane – (2,3-dichlorophenylacetic acid)$_4$. The green strips of d_{003} thickness correspond to non-chiral layers of protonated acid. The white strips of $2xd_{003}$ thickness correspond to the chiral layers of the diammonium chiral organic base and the two deprotonated acids, adapted from ref. 6.

13.3 Co-crystals of 'Alike' Molecules (including Enantiomers, Isomers, Diastereomers . . .)

When two components are mirror related molecules (*i.e.* optical antipodes or enantiomers), special relationships link them. For instance, they have equal non-vectorial intensive properties (temperature and enthalpy of fusion, *etc.*) and are opposite in sign but have equal absolute values of vectorial properties (*e.g.* α°). The heterogeneous equilibria are also symmetrical which allows modification of the Gibbs phase rule into the Gibbs–Scott rule.[7] When the two components can interconvert, that is to say, there is fast racemization in the liquid phase (rapid dynamic equilibrium between the two enantiomers), there is no need for the modified phase rule.[8] Moreover, if there is discrimination in the solid state, recent experiments suggest that the ultimate state of the system is homochirality, that is, a spontaneous break of symmetry.[9–12]

Since Pasteur's pioneer work, several generations of researchers, representing more than 150 years of experience, have shown the very strong trend towards the formation of stoichiometric <1-1> compounds (co-crystals). Even if significant deviations appear in some series of molecules (*e.g.* 5-aryl-5methylhydantoin[13]), it is accepted that 90–95% of equimolar mixtures of small chiral organic molecules, including of course the chiral APIs, crystallize as racemic compounds.[14,15] This statement is still valid for racemizable enantiomers in the liquid state. Due to the virtual unlimited number of asymmetric organic molecules, one can therefore contemplate an 'infinite' number of racemic co-crystals.

Although there is *a priori* no restriction in symmetry applicable to the space group of the racemic compound crystal lattice, a clear majority of these racemic compounds crystallize in one of the 92 centrosymmetric space groups; among them $P2_1/c$, P-1, $C2/c$, *Pbca*, *Pnma* are by far the most popular.[16,17].

Some chiral racemic compounds crystallize in the 73 non-chiral and non-centrosymmetric space groups such as *Pc*, $Pna2_1$ (form III of rac_modafinil[18,19]), $Pca2_1$ *Fdd2* (form IV rac-modafinil[20]), and so on. Very rare examples of racemic compounds crystallizing in one if the 65 chiral space groups have been spotted so far (obviously *Z'* is greater or equal to 2).[21]

These observations are perfectly consistent with Kitaigorodskii rules[22] about preferred symmetry for crystal lattices of organic molecules which are in the following decreasing order of preference: (1) translations (mandatory for any crystallized material), (2) centres of symmetry, (3) glide mirrors and (4) screw axes. There are several reasons for this: the flexibility of the centre of symmetry without impairing the highest density of the crystals, the opposite electrostatic charges on the face to face parts of neighbouring molecules, cancellation of permanent dipoles, and so on.

By contrast the pure enantiomers can only crystallize in space groups deprived of second order symmetry operators such as centre of symmetry (= –1), mirror (m = –2, any glide mirror of type: *a, b, c, n, d*) and inverted axes (–3, –4 and –6). Thus only 65 space groups out of 230 remain with a strong deviation from an even distribution. Indeed the cumulative top 9 occurrences – $P2_12_12_1$ (58%); $P2_1$ (30%); $P1$, $C2$, $P2_12_12$, $P4_1$, $P4_3$, $P4_12_12$, $P4_32_12$ – represent more than 95%.

These restrictions might be the cause of some difficulties in crystallizing pure enantiomers. In the case of direct asymmetric synthesis (without prior synthesis of the racemic mixture), it might be interesting to synthesize the antipode and then mix the two enantiomers in order to increase the chance of crystallization. The racemic compound, that is, the racemic co-crystal, is likely to appear; it can be used later on as template for the crystallization of the desired pure enantiomer.

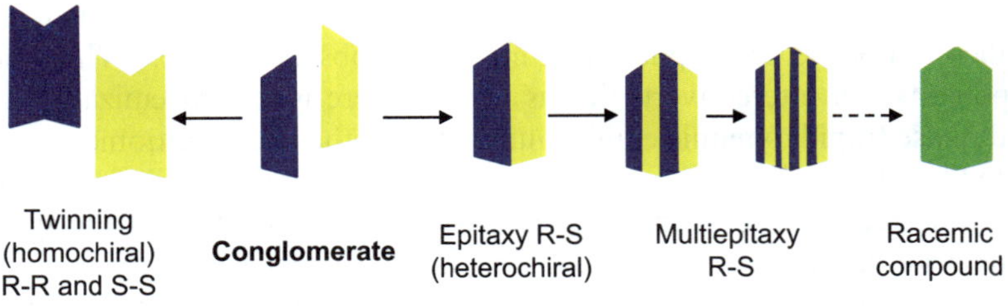

Figure 13.10 Schematic representation of twinning, epitaxy and multi-epitaxy among enantiomers.

It is worth noting that some chiral compounds are able to alternate, at variable scales (from nm to mm), with layers of pure antipodes.[23–26] These multi-epitaxies between enantiomers create bi-dimensional racemic compounds which, for reasons not yet understood, do not expand in the third dimension. Actually there is no conceptual limit between these repetitive epitaxies and the corresponding racemic co-crystal (Figure 13.10).

The so-called chiral switch represents a way back from co-crystal to the pure chiral component (*i.e.* from the racemic compound to the pure enantiomer). The usual motivations for such a substitution are:

- possible improved therapeutic activity (sometime there is a non-linear effect *versus* enantiomeric excess (e.e.) such as 8 – 15 – 90 for the distomer, racemic compound and the eutomer, respectively)
- less toxic effects mainly due to the distomer
- more simple and selective pharmacodynamic and/or pharmacokinetic profiles
- reduced risk of drug interactions
- more simple relationship between plasma concentration, clearance and effect.
- extended intellectual properties

Table 13.1 illustrates some examples of the chiral switch among APIs.[27]

The chiral discrimination in the solid state of modafinil is particularly interesting because the API, the *n*-1 and *n*-2 precursors (Figure 13.11) are respectively representative of the three binary systems represented in Figure 13.11. This case is also representative of the extensive problem of polymorphism and solvate formation for both the racemic compound and the pure enantiomer.[28–30]

When two couples of enantiomers have a high degree of similarity, the *S* enantiomers can fit in the same crystal lattice; that is, a complete solid solution. This mechanism of substitution expands across the binary system (idem for the two *R* enantiomers). By contrast, the couples of antipodes (*S*1-*R*2 and *S*2-*R*1) lead to the formation of the so-called quasi-racemic compounds: <*S*1-*R*2> and <*S*2-*R*1>, which can be considered as co-crystals[33–35] (Figure 13.12). Prior to the emergence of the X-ray diffraction technique using the anomalous

Table 13.1 Racemic APIs and their eutomers, adapted from ref. 27.

Drug as Racemic Mixture or Racemic Compound	*Corresponding Drug as Pure Enantiomer*
Albuterol Rac	Levalbuterol (Xopenex)
Amlodipine (Norvasc) Rac	(*S*)-Amlodipine
Budesonide (Rhinocort)	Dexbudesonide
Cetirizine (Zyrtec)	Levocetirizine (Xyzal/Zuzal)
Citalopram (Celexa)	Escitalopram (Lexapro)
Fenfluramine rac and polymorph	Dexfenfluramine (Isomeride) polymorph
Formoterol (Foradil)	Arformoterol
Ketoprofen (Actron) Rac	Dexketoprofen (Trometamol)
Lansoprazole (Prevacid) Rac	Dexlansoprazole
Methylphenidate (Ritalin)	Dexmethylphenidate (Focalin)
Modafinil (Provigil) Rac and polymorph	Armodafinil (Nuvigil) polymorph
Nirvanol (conglomerate)	*Never developed*
Omeprazole (Prilosec) Rac hydrates	Esomeprazole (Nexium) hydrates
Oxybutynin (Detropan)	Esoxybutynin
Sotalol (Betaspace) Rac	Dexsotalol
Thalidomide (teratogenic)	Teratogenic effect from a single enantiomer
Venlafaxine (Effexor) rac et conglomerate	Desvenlafaxine
Zopiclone (Imovane) Rac hydrate	Eszopiclone (Lunesta)

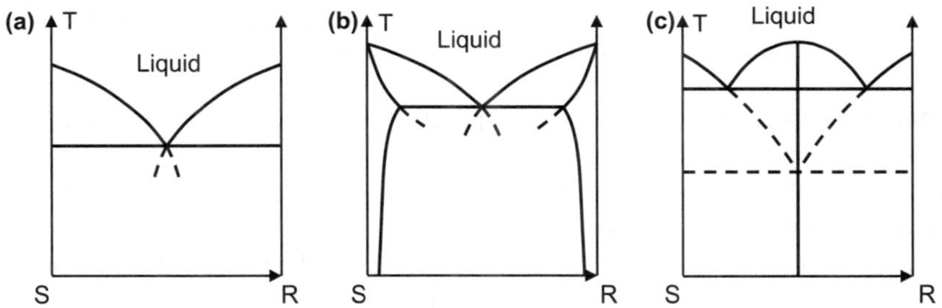

Figure 13.11 (a) R = OH: (–) and (+) modafinic acids do not form a co-crystal, they form a stable conglomerate [29,31](*i.e.* a eutectic mixture without detectable domain of partial solid solution). (b) R = OMe: (–) and (+) antipodes do not form a co-crystal, they form a stable conglomerate with a partial miscibility in the solid state[32] (*i.e.* a eutectic mixture with a partial solid solution). (c) R = NH_2: racemic compound modafinil (*i.e.* co-crystal), polymorphism of the enantiomer and the racemic co-crystal is omitted.

scattering effect, the systematic aspect of these co-crystals formations was used to assign the absolute configuration of similar chiral organic molecules. Nevertheless, the unaddressed question has always been what is the degree of similarity necessary for observation of this quasi-racemic compound?

Alike molecules also means that all sorts of isomerism (*syn-anti*, atropisomerism, *etc.*) can also lead to co-crystals. 6-oxo-6,7,8,9,10,11-hexahydrocyclohepta[*c*]chromen-3-yl sulfamate (or sulfamic acid 6,7,8,9,10, 11-hexahydro-6-oxobenzo[*b*]cyclohepta[*d*]pyran-3-yl ester (Figure 13.13) is a

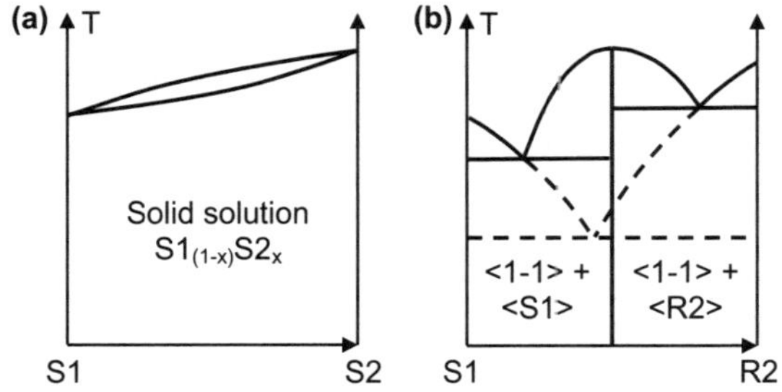

Figure 13.12 Binary systems of 'very' similar couples of enantiomers. (a) *S*1-*S*2 and *R*1-*R*2 (not represented) with an identical absolute configuration lead to complete solid solutions. (b) *S*1-*R*2 and *R*2-*S*1 (not represented) lead to quasi racemic compounds.

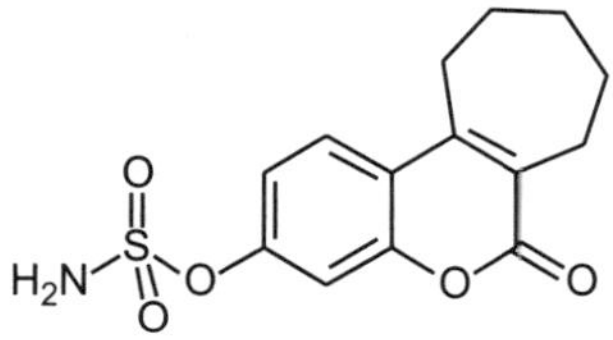

Figure 13.13 Scheme of 6-oxo-6,7,8,9,10,11-hexahydrocyclohepta[*c*]chromen-3-yl sulfamate (or sulfamic acid 6,7,8,9,10,11-hexahydro-6-oxobenzo[*b*]-cyclohepta[*d*]pyran-3-yl ester (molecule A)).

steroid sulfatase inhibitor under development for different pathologies including hormone-dependent breast cancer.[36,37]

This molecule exhibits polymorphism but, more interestingly, it is possible to crystallize a co-crystal between the API and one of its side products, easily obtained after hydrolysis of the sulfamic moiety.[38] The ORTEP view presents the configuration of the two molecules (Figure 13.14).

The structure is formed by stacking double layers of molecules A and molecules B along <001> (Figure 13.15). These layers of molecules A and B interact via hydrogen bonds and π–π interactions.

Figure 13.16 illustrates how systematic studies of heterogeneous equilibria are able to spot co-crystals and the diversity of these intermediate compounds. One stoichiometric compound (*C*1, $X_{mol.} = 0.11$) and three non-stoichiometric compounds (*C*2, *C*3 and *C*4) have been identified.[39]

*C*2, a non-stoichiometric compound, stable at low temperature, decomposes reversibly through a peritectoid invariant at 320 K in the solid solution of the high temperature form of $C_{15}H_{29}OOH$ and a non-stoichiometric intermediate compound (*C*3) stable in a narrow domain of temperature (*ca.* 319.4–321.6 K). Another non-stoichiometric compound (*C*4) whose composition expands around 50% emerges from a eutectoid invariant at *ca.* 317.5 K and decomposes reversibly through a peritectic invariant at *ca.* 323.3 K. Other systematic studies of these long chain carboxylic acids compounds reveal similar formation of non-stoichiometric compounds.

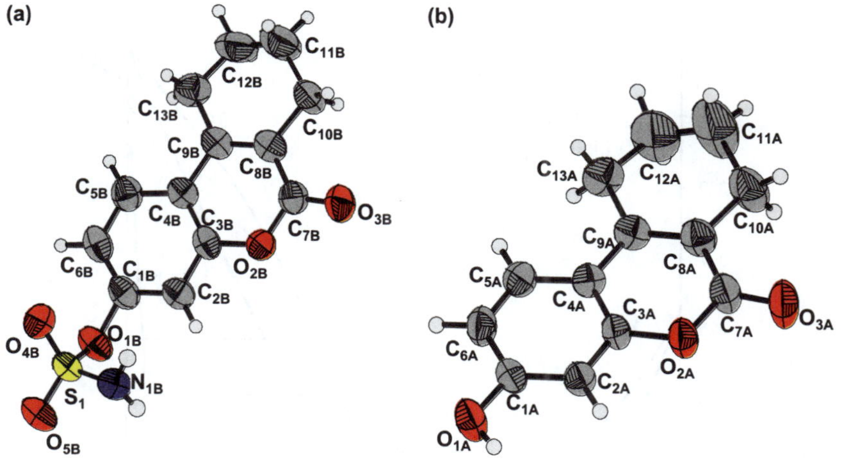

Figure 13.14 ORTEP views of: (a) molecule A and (b) its side product (molecule B) which forms a co-crystal.

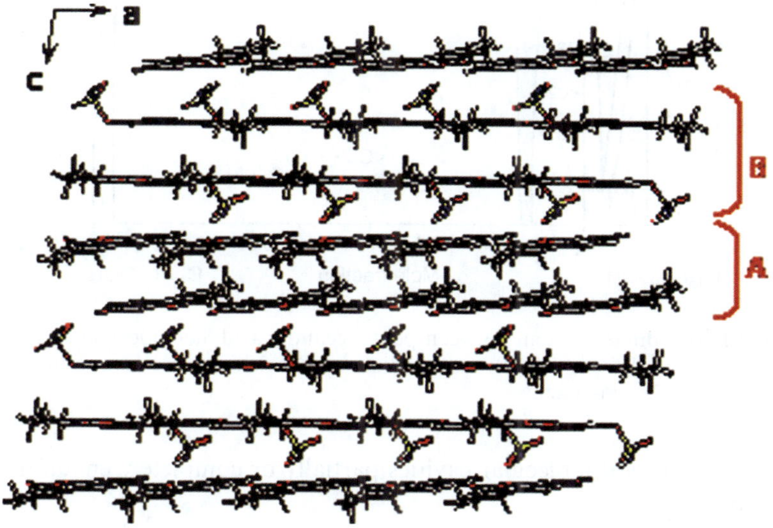

Figure 13.15 Stacking of alternate double layers of molecules A and molecules B along 001.

13.4 Host–guest Compounds: Co-crystals at least Partially Driven by Inclusion Phenomena at the Molecular Level

When crystallized, this class of compounds can be considered to result from a double supramolecular recognition:

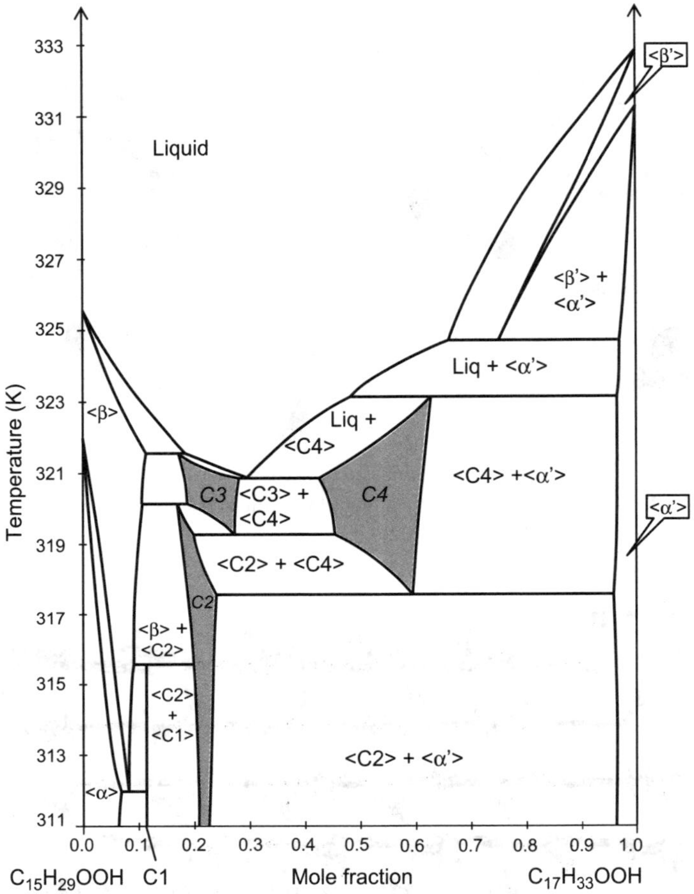

Figure 13.16 Binary system between pentadecanoic acid–heptadecanoic acid (adapted from ref. 39; intermediate non-stoichiometric compounds in grey, *i.e.* co-crystals)

1. One or several molecular cavities partially or completely engulf a 'guest' molecule.
2. These host–guest associations are assembled in a three-dimensional (3D) periodic array.

Calix-*n*-arenes, crown *n*-ether, cyclopeptides and cyclodextrines are non-exhaustive examples of these molecular cavities. Moreover, such topological relationships between different partners in a crystal can also be between (i) a first periodic supramolecular assembly of one component, which creates a low density packing with empty channels, interlayer spaces or interconnected tubes and (ii) a second component filling those empty spaces. Urea and thiourea, along with a lot of linear guest molecules, make these kinds of co-crystals with hexagonal symmetry[40] (Figure 13.17). A MOF (metal–organic framework) can

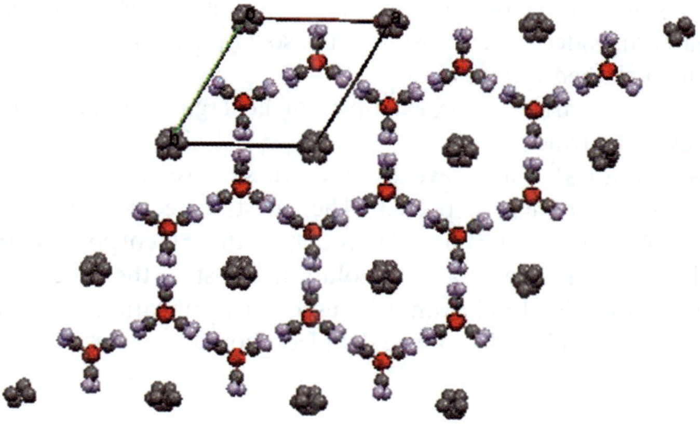

Figure 13.17 1, 10-decanedicarboxylic acid urea inclusion compound showing hexagonal symmetry, adapted from ref. 40.

Table 13.2 Examples of host–guest associations between APIs and cyclodextrines, adapted from refs. 46–48.

INN	*Nature of the CD*	*Indication*
Cetirizine	Beta CD	Allergy
Piroxicam	Beta CD	Non-steroidal anti-inflammatory
Ethylestradiol (and drospirenone)	Beta CD	Progestogen and estrogen
Nicotine	Beta CD	De-addiction from tobacco
Perindopril	Hydroxypropyl Beta CD	Anti-hypertension
Pramipexole	Beta CD	Parkinsons
Cefotiam	Alpha CD	Antibiotic (cephalosporine 3rd generation)

also illustrate this concept.[41,42] Therefore, one can clearly see that there is a continuum between channel solvates and this class of compounds (co-crystals). In these channels, the guest molecules can be irregularly or periodically located. In the latter case, the periodicity of the guest molecule can match or not match that of the host molecule.[43]

In pharmaceutical field the most studied cavities by far are the cyclodextrines (CDs).[44,45] These cavities are obtained by fermentation of starch. The most popular CDs are the α-6-membered ring, β-7-membered ring and γ-8-membered ring. They can be used as native species or derivatized.

Usually, associations between API and CDs are expected to improve one or several of the following problems: (i) enhanced solubility and dissolution rate, (ii) bioavailability, (iii) stability, including photostability, prevention of hydrolysis, prevention of self reaction, (iv) masking bitterness, (v) enhancing stability, (vi) prevention of crystallization (*i.e.* long lasting amorphous therapeutic material; indeed such host–guest associations can be difficult to crystallize), (vii) sustained release.

Table 13.2 summarizes some examples of host–guest associations between APIs and cyclodextrines.

Systematic studies[49] have revealed that these host–guest associations are usually less selective than expected. The guest molecules of several 'alike' molecules such as API and some side products can be incorporated at the same time in the crystal lattice as a solid solution. Most of the time this encapsulation at the molecular level cannot be used for purification purposes but for resolving other difficulties like those detailed above.

13.5 Conclusions

Co-crystals are nothing other than intermediate compounds. Thus, the term might only be justified by the nature of the bonds connecting the partners that constitute the crystal. Nevertheless, the concept suffers from intrinsic uncertainties due to the numerous, difficult to categorize, situations in which they can occur, such as the continuum between solvate and co-crystals, hybrid salt–co-crystals, multi-epitaxy and so on. It is even possible to find the same partners with simultaneously salt-like and co-crystal-like relationships in a given crystal structure.[50]

Therefore, rather than creating questionable terms[†] and maintaining endless semantic debates with poor added value.[51] the scientific community should concentrate more on the three long lasting problems related to intermediate compounds listed below:

1. Prediction of intermediate compounds: indeed, apart from statistical trends (*e.g.* racemic compounds as detailed in this chapter), there is no way to predict the existence of an intermediate compound with acceptable reliability.
2. Prediction of the attributes of these intermediate compounds: to date the connection between structures and properties is very weak despite the evident interest.
3. How to design a process to obtain those intermediate phases with a good yield and minimum side products? Moreover these processes should fulfil the specifications for the final products such as crystallinity, crystal size distribution, morphology, and so on.

These questions are really meaningful from an industrial perspective and for advancing knowledge in the organic solid state.

[†] 'Co' from the Latin 'cum' means 'with'. Crystallization of any eutectic mixture is thus etymologically a co-crystallization but of course has nothing to do with co-crystals since it corresponds to the simultaneous crystallization of several phases.

Acknowledgements

We thank Dr Marie-Noëlle Petit for help with the illustrations and IPSEN (Dr Christian Diolez; Les Ulis France) for its support.

References

1. J. E. Ricci, *The Phase Rule and Heterogeneous Equilibrium*, Dover Publications Inc., New York, 1966.
2. R. A. Chiarella, R. J. Davey and M. L. Peterson, *Cryst. Growth Des.*, 2007, **7**(7), 1223–1226.
3. T. Steiner and G. Koellner, *J. Am. Chem. Soc.*, 1994, **116**(12), 5122–5128.
4. G. Coquerel, N. Mofaddel, M.-N. Petit and R. Bouaziz, *Bull. Soc. Chim. Fr.*, 1991, **128**, 418–422.
5. C. Gicquel-Mayer, G. Coquerel, M.-N. Petit and R. Bouaziz, *Anal. Sci.*, 1993, **9**, 25–27.
6. J. Mahieux, S. Gonella, M. Sanselme and G. Coquerel, Crystal structure of a hybrid salt–cocrystal and its resolution by preferential crystallization: (±) (*trans*-N, N'-dibenzyldiaminocyclohexane)(2, 3-dichlorophenylacetic acid)4. *CrystEngComm*, Submitted.
7. R. L. Scott, *J. Chem. Soc., Faraday Trans*, 2 1977, **3**, 356–360.
8. G. Coquerel, *Enantiomer*, 2000, **5**, 481–498.
9. S. Gonella, G. Levilain and G. Coquerel, *J. Therm. Anal. Calor.*, 2011, **103**(1), 125–129.
10. G. Levilain and G. Coquerel, *CrystEngComm*, 2010, **12**(7), 1983–92.
11. W. L. Noorduin, E. Vlieg, R. M. Kellogg and B. Kaptein, *Angew. Chem., Int. Ed.*, 2009, **48**(51), 9600–9606.
12. C. Viedma, *J. Cryst. Growth*, 2004, **261**, 118–121.
13. G. Coquerel, M.-N. Petit and F. Robert, *Acta Crystallogr.*, 1993, **C49**, 824–825.
14. G. Coquerel, *Preferential crystallization*, in *Novel Optical Resolution Technologies*, ed. K. Sakai, N. Hirayama and R. Tamura, Springer, Berlin, 2007, vol. 269, pp. 1–51.
15. J. Jacques, A. Collet and S. H. Wilen, *Enantiomers, Racemates and Resolutions*, Krieger Publishing Company, Malabar (Florida), 1994.
16. T. Steiner, *Acta Crystallogr.*, 2000, **B56**, 673–576.
17. P. M. Zorky, *J. Mol. Struct.*, 1996, **374**, 9–28.
18. M. Pauchet, C. Gervais, L. Courvoisier and G. Coquerel, *Cryst. Growth Des.*, 2004, **4**(6), 1143–1151.
19. M. Pauchet and G. Coquerel, *Cryst. Growth Des.*, 2007, **7**(9), 1612–1614.
20. J. Mahieux, M. Sanselme and G. Coquerel, Structure of rac-Modafinil form IV obtained in Gels. Comparisons between form I, II and IV molecular conformations, mechanical and thermal stabilities. *J. Mol. Struct.*, paper to be submitted.
21. N. G. Panina, M. Pauchet, F. Dufour, P. M. Zorky, J. Maddaluno and G. Coquerel, BIWIC 2003, 10th International Workshop on Industrial

Crystallization, Sept. 3 and 4 2003, Université de Rouen, Ed. G. Coquerel, Mainz Verlag, Aachen, pp. 196–200.
22. A. I. Kitaigorodski, *Molecular Crystals and Molecules*, Academic Press, New York, 1973.
23. P. A. Levkin, V. Y. Torbeev, D. A. Lenev and R. G. Kostyanovsky, Homo- and heterochirality in crystals, in *Topics in Stereochemistry*, ed. S. E. Denmark and J. S. Siegel, John Wiley & Sons, Inc., 2006, vol. 25, pp. 81–134.
24. C. Gervais, S. Beilles, P. Cardinaël, S. Petit and G. Coquerel, *J. Phys. Chem. B*, 2002, **106**(3), 646–652.
25. V. Y. Torbeev, K. A. Lyssenko, O. N. Kharybin, M. Y. Antipin and R. G. Kostyanovsky, *J. Phys. Chem. B*, 2003, **107**(48), 13523–13531.
26. J. T. H. van Eupen, W. W. J. Elffrink, R. Keltjens, P. Bennema, R. de Gelder, J. M. M. Smits, E. R. H. van Eck, A. P. M. Kentgens, M. A. Deij, H. Meekes and E. Vlieg, *Cryst. Growth Des.*, 2008, **8**, 71–79.
27. A. J. Hutt and J. Valentova, *Acta Fac. Pharm. Univ. Comenianae*, 2003, **50**, 7–23.
28. M. Broquaire, L. Courvoisier, F. Mallet, G. Coquerel and A. Frydman, Modafinil polymorphic forms. *WO* 02004014846 19/02/2004, 2004.
29. O. Neckebrock, L. Courvoisier, S. Graf, G. Serrure, G. Coquerel, S. Rose, C. Besselièvre, F. Mallet and A. J. V. Langevelde, Method for the production of crystalline forms crystalline forms of optical enantiomers of modafinil. *US Pat.*, 2006135621, 22-06-2006, 2006.
30. A. Ceausu, A. Lieberman and J. Aronhine, Crystalline forms of modafinil. *US Pat.*, 2005034652, 17-02-2005, 2005.
31. T. Prisinzano, J. Podobinski, K. Tidgewell, M. Luo and D. Swenson, *Tetrahedron Asymmetry*, 2004, **17**(7), 1053–1058.
32. L. Renou, T. Morelli, S. Coste, M.-N. Petit, B. Berton, J.-J. Malandain and G. Coquerel, *Cryst. Growth Des.*, 2007, **7**(9), 1599–1607.
33. M. Delépine, *Bull. Soc. Chim. Fr.*, 1921, **29**, 656.
34. A. Fredga, *Tetrahedron*, 1960, **8**, 126.
35. J. Timmermans, *Recl. Trav. Chim. Pays-Bas*, 1929, **48**, 890–894.
36. B. V. L. Potter and M. J. Reed, Dérivés du type sulfamate non stéroïdiens à noyau polycyclique, leur préparation et leur utilisation comme inhibiteurs de l'oestrone sulfatase. *Eur Pat.*, 0880514, 12-02-1998, 1998.
37. S. J. Stanway, A. Purohit, L. W. L. Woo, S. Sufi, D. Vigushin, R. Ward, R. H. Wilson, F. Z. Stanczyk, N. Dobbs, E. Kulinskaya, M. Elliott, B. V. L. Potter, M. J. Reed and R. C. Coombes, *Clin. Cancer Res.*, 2006, **12**(5), 1585–1592.
38. Thanks are due to Ipsen Company, Les Ulis France. The crystal structure of this co-crystal was solved by single crystal X-ray diffraction (space group $P21/a$, with cell parameters: $a = 13.041$ Å, $b = 12.911$ Å, $c = 14.297$ Å, $b = 98.52°$), unpublished results.
39. G. Gbabode, P. Négrier, D. Mondieig, E. Moreno, T. Calvet and M. A. Cuevas-Diarte, *Chem. Phys. Lipids*, 2008, **154**, 68–77.
40. L. Yeo, K. D. M. Harris and F. Guillaume, *J. Solid State Chem.*, 1997, **128**, 273–281.

41. P. Horcajada, C. Serre, M. Vallet-Regi, M. Sebban, F. Taulelle and G. Férey, *Angew. Chem. Int. Ed.*, 2006, **45**, 5974–5978.
42. A. C. McKinlay, R. E. Morris, P. Horcajada, G. Férey, R. Gref, P. Couvreur and C. Serre, *Angew. Chem. Int. Ed.*, 2010, **49**, 6269–6266.
43. G. Chapuis, M. Farkas-Jahnke, J. M. Pérez-Mato, M. L. Senechal, W. Steurer, C. Janot, D. Pandey and A. Yamamoto, *Acta Crystallogr.*, 1997, **153**, 95–100.
44. K. Harata, *Chem. Rev.*, 1998, **98**(5), 1803–1828.
45. J. Szejtli, *Cyclodextrin Technology*, Kluwer Academic Dordrecht, 1988.
46. R. Challa, A. Ahuja, J. Ali and R. K. Khar, *AAPS PharmSciTech*, 2005, **6**(2), E329–E357.
47. T. Loftsson and M. Brewster, *J. Pharm. Sci.*, 1996, **85**(10), 1017–1025.
48. L. A. Miller, R. L. Carrier and I. Ahmed, *J. Pharm. Sci.*, 2007, **96**(7), 1691–1707.
49. A. Grandeury, E. Condamine, L. Hilfert, G. Gouhier, S. Petit and G. Coquerel, *J. Phys. Chem. B*, 2007, **111**(25), 7017–7026.
50. B. Olenik, T. Smolka, R. Boese and R. Sustmann, *Cryst. Growth Des.*, 2003, **3**(2), 183–188.
51. A. D. Bond, *CrystEngComm*, 2007, **9**, 833–834.

CHAPTER 14

Co-crystals: Commercial Opportunities and Patent Considerations†

MARCEL HOFFMAN[a] AND JEFFREY A. LINDEMAN[b]

[a] Crystallics, Zekeringstraat 29, 1014 BV, Amsterdam, The Netherlands; [b] J.A. Lindeman & Co. PLLC, 3190 Fairview Park Drive, Suite 480, Falls Church, VA 22042, USA

The first co-crystal patent ever granted may be US Patent No. 3,028,420, "Compound of Betaine and Chloral and Method for Preparing Same", April 3, 1962. This patent reports the "unexpected discovery that chloral and its hydrate combine exothermally with betaine, a physiologically inactive, naturally occurring base to form a new compound with the empirical formula, $C_7N_{14}O_4NCl_3$, which may be obtained in crystalline form from solvents such as, for example, water or ethanol". The patent goes on to report that infrared examination of the new compound indicated strong hydrogen bonding between the hydroxyl groups of chloral hydrate and the COO^- group of betaine which is charge stabilized by the cationic amine $-N^+(CH_3)_3$ group of

†DISCLAMER: The views expressed in this chapter are those of the authors and not those of their employers, clients or customers. Because this chapter surveys the patent and potential commercial landscape for co-crystals, the discussion by its own nature is general. Specific co-crystals and the circumstances surrounding them may, and are likely to, require further and other considerations. The chapter also presents a legal discussion of patents and co-crystal patents in general terms. This discussion is for educational purposes only. It is not legal advice and, knowing that the legal landscape is an ever-changing one, should not be substituted for obtaining current legal counsel in regard to a specific co-crystal or fact situation.

RSC Drug Discovery Series No. 16
Pharmaceutical Salts and Co-crystals
Edited by Johan Wouters and Luc Quéré

Published by the Royal Society of Chemistry, www.rsc.org

betaine. According to the patent, "[t]his new compound is probably 6-trichloromethyl-4,7-dihydra-1,3,5,7-tetracycloöct-2-ato-trimethylammoniomethane of the formula:

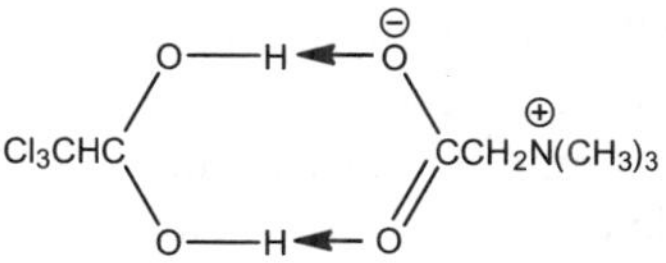

which formula represents a structural type hitherto unknown in organic chemistry". Today, such compounds are known as co-crystals and have become an area of intense pharmaceutical research.[1]

14.1 Introduction

Co-crystals, the crystalline combination of an active pharmaceutical ingredient (API) and at least one co-former, have emerged in pharmaceutical research and formulation as a possible way not only to engineer new properties or characteristics but also to achieve new patent protection and new commercial value for the API. The chapter considers (1) how engineering pharmaceutical properties using co-crystals can create new commercial value for an API, (2) the definition of "co-crystal", (3) the patentability of co-crystals, and (4) the information needed for effective co-crystal patents, and by considering these topics engages the unique commercial and patent opportunities raised by co-crystals. In short, the chapter discusses how the chemistry of co-crystals may translate into new patents and commercial value. Obtaining new commercial value does not necessarily require obtaining a patent, but for pharmaceuticals a new patent can mean greater and longer exclusivity or even new exclusivity where none previously existed, which is a most meaningful way to obtain (and to protect) commercial value.

There are three basic criteria for the patentability of any invention. First, the invention must be useful, that is, possess a practical utility, which often ties to the invention's commercial value. Second, the invention must be novel and, third, the invention must be non-obvious or, as it is termed in Europe, possess an inventive step. The exact nature of these basic criteria depends upon the country or regional law where the patent application is filed. Internationally, each of the criteria has more commonality than difference. For the purpose of this chapter, we will primarily draw from the law of the United States (Title 35 of the United States Code, 35 U.S.C.) and the European Patent Convention (EPC).[2]

14.2 Co-crystals – Engineering Pharmaceutical Properties/Creating Commercial Value

Although therapeutic efficacy is the primary concern for an active pharmaceutical ingredient (API), the solid state form (*i.e.*, the crystalline or amorphous form) of an API can be critical to its pharmacological properties, such as bioavailability, and to its development as a viable drug candidate. Recently, crystalline forms of APIs have been used to alter the physicochemical properties of a particular API. Each crystalline form can have different solid state (physical and chemical) properties. The differences in physical properties exhibited by a novel solid form of an API, for example a new polymorph, affect pharmaceutical parameters such as storage stability, compressibility and density (important in formulation and product manufacturing), and solubility and dissolution rates (important factors in determining bioavailability). Because these practical physical properties are influenced by the solid state properties of a crystalline form of the API, they can have a significant impact on the selection of a compound as a drug candidate, the ultimate pharmaceutical dosage form, the optimization of manufacturing processes, and absorption in the body. Moreover, finding the most adequate solid state form for further drug development can reduce the time and the cost of that development.

While polymorphs may open up new properties for an API, forming a co-crystal from the API and a co-former can make it possible to achieve other desired properties beyond those of the particular API itself. Co-crystals offer the opportunities and benefits that come from a new chemical composition as well as from its crystalline form. A co-crystal allows the properties of an API to be engineered on the basis of the composition of the co-crystal. Different co-formers can, and probably do, yield co-crystal composition with differing properties. The choice of the co-former then leads to a different composition and different properties for the particular co-crystal. The crystalline nature or form of the co-crystal can also provide its own unique properties to the co-crystal. Thus, the composition and crystalline nature of a co-crystal provide two paths to "engineering" the properties of the co-crystal and the API within it.

A co-crystal is likely to possess unique properties in comparison with the API and co-former(s) which make up its constituent parts. These properties are generally speaking not predictable but rather are determined experimentally when characterizing and studying a particular co-crystal. These new properties can often establish the patentability of the co-crystal and open new avenues of commercial opportunity for an API. A co-crystal may possess unique properties, such as those discussed above, in relation to crystalline forms but which are also unique due to its particular chemical composition. For example, a co-crystal may possess more favorable pharmaceutical and pharmacological properties or be easier to process than known forms of the API. A co-crystal may have different dissolution and solubility properties than the API itself and can be used to deliver APIs therapeutically. *In vivo*, the co-crystal may have greater bioavailability that the API alone. New drug formulations comprising a co-crystal of a given API may have superior properties over its existing drug formulations. A co-crystal may simply be able to be formulated where the API alone cannot. They may also have better storage stability. In other words, the co-crystal may possess other properties and advantages not necessarily related to therapeutic considerations but which nonetheless enable the successful commercialization of the API.

Examples of manufacturing properties and advantages include better filtering, improved flow, less caking, better drying, lower hygroscopicity, greater

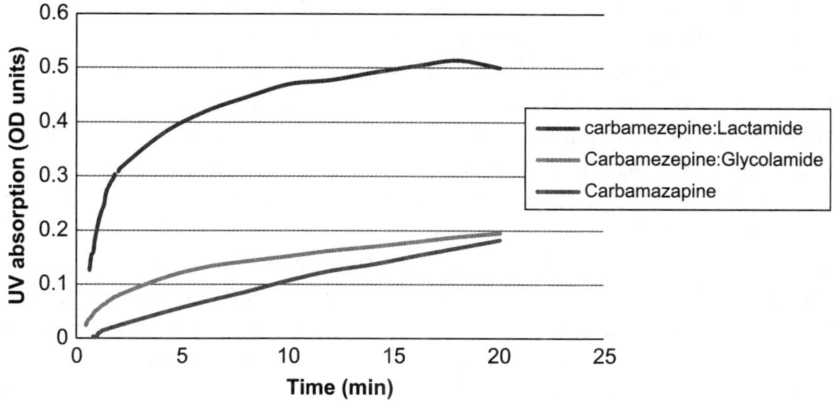

Figure 14.1 Dissolution rate of carbamazepine and two co-crystals of carbamazepine. Carbamazepine:lactamide has a rapid dissolution rate and reaches high solubility levels quickly. Carbamazepine:glycolamide in contrast has a lower dissolution rate and its profile is similar to that of carbamazepine itself.[5]

processing stability, and so on. These manufacturing properties point to another advantage of co-crystals – their use as intermediates in manufacturing. Even where the ultimate product may be the API alone, a co-crystalline intermediate or a co-crystal of the API as the penultimate intermediate provides unique advantages and opportunities and the ability to patent new compositions and methods.

Examples of co-crystals having improved solubility and dissolution rates are found in the work of Remenar *et al.* published in 2003.[3] In this publication Remenar demonstrated that some of the 1,4 dicarboxylic acid co-crystals of *cis*-itraconazole have solubilities and dissolution rates equal to those of the amorphous form of *cis*-itraconazole in the commercial formulation which is much higher than those of the crystalline free base. In another example Trask *et al.* describes an oxalic acid co-crystal of theophylline with improved physical stability.[4] Anhydrous theophylline exposed for 3 days to 75% relative humidity (RH) will convert to its monohydrated form, the anhydrous oxalic acid co-crystal remains stable even at 98% RH.

Figure 14.1 is an example of improved dissolution properties achieved via a co-crystal and shows the differences in properties between two co-crystals of the same API, carbamazepine.

14.3 Definition of a Co-crystal

Co-crystals are crystalline compositions of matter based on the combination of an active pharmaceutical ingredient (API) and at least one co-former which interact with one another through a binding interaction. Many definitions of a "co-crystal" have been offered. Typically, a definition of "co-crystal" seeks to distinguish a co-crystal composition from a crystalline salt or a crystalline solvate. Differing definitions may be acceptable in some contexts. Definitions are important in patents, however, because the definition of a term, especially a term in a patent claim, is a definition of property rights.

The words of a patent claim define the scope of the intellectual property right represented by the patent. Under US patent law, "[t]he specification shall conclude with one or more claims particularly pointing out and distinctly claiming the subject matter which the applicant regards as his invention".[6] Similarly, under the EPC, "[t]he claims shall define the matter for which protection is sought. They shall be clear and concise and be supported by the description".[7] The claims, then, are like a deed to a piece of property defining what intellectual property is within the patent and what intellectual property is outside the patent.

Quite often, the technical description, or patent specification, defines the terms used in a patent claim setting forth the patented invention. In fact, the definition of a term in the technical description typically controls the meaning of the term as it is used in the patent and in its claims.[8] If then the patent claims a "co-crystal" of API A and co-former B, then the definition of co-crystal provided in the patent specification will control what is a "co-crystal" according to the patent and what is not.

By one definition, for example, a co-crystal is a crystalline entity in which more than one molecular substance is incorporated into the unit cell. According to this definition, by convention, this normally excludes salts and solvates. Salts are distinguished by proton transfer, giving electrostatic linkage between oppositely charged ions. Solvates are associations of substrates with solvents from which they are crystallized. Bonding mechanisms can be similar to those in co-crystals. This definition stipulates that both (or all) molecular components are solid at room temperature and pressure.[9]

Another definition from a patent application states that the term "co-crystals" is meant to define crystalline phase wherein at least two components of the crystal interact by hydrogen bonding and possibly by other non-covalent interactions rather than by ion pairing. The primary difference is the physical state of the pure isolated compound. If one component is liquid at room temperature, the crystals are referred to as solvates; if both components are solids at room temperature, the products are referred to as co-crystals.[10] Claims 1 and 2 of that patent application recite:

1. A pharmaceutical co-crystal comprising an active pharmaceutical ingredient; and a co-crystal agent having the structure R^1–C(=O)XH, wherein X is O, N(C_{1-6}alkyl) or NH and R^1 is a C_{3-8} alkyl group containing at least one *trans*-oriented double bond and being substituted by 0, 1, 2, 3 or 4 groups independently selected from halo, phenyl and hydroxyl.
2. A pharmaceutical co-crystal according to claim 1, wherein the co-crystal agent is selected from sorbic acid, *trans*-2-hexenoic acid, *trans*-3-hexenoic acid, *trans*-4-hexenoic acid, *trans*-2-butenoic acid, *trans*-2-pentenoic acid, *trans*-3-pentenoic acid, *trans*-2,4-pentadienoic acid.[11]

It does not necessarily matter that every inventor or patent employs the same definition of "co-crystal" although a common accepted definition may take hold in the pharmaceutical art. A patentee may choose to define a term uniquely in the patent specification. If so, the patentee's definition controls the meaning of the term in the claim. The patentee acting as his or her own lexicographer overcomes the presumption favoring the term's ordinary meaning or a dictionary definition where the patentee has clearly set forth an explicit definition of the term different from its ordinary meaning.[12]

The point is this. Use of the term "co-crystal" in the introduction, or preamble, of a patent claim defines the type of composition being claimed. First, defining the claimed composition as a "co-crystal" refers to the solid state and excludes amorphous materials. Second, it is a co-crystal and therefore distinct from a non-crystalline mixture or a salt, a crystalline solvate, or other prior art combinations of the API and co-former (should they exist). While a general preamble may not affect the scope of a claim, the preamble of a claim when used to define a substance to the exclusion of others may in fact limit the scope of the claim.[13] Thus, the purpose and effect of an inventor using the term "co-crystal" to define the relationship between the API and co-former in the composition should be considered not only for what it includes, but also for what it excludes. This can be a useful defitional boundary to focus on what is the patentable subject matter and why it is patentable.

14.4 Patentability of Co-crystals

Co-crystals, as a class of compounds, represent compositions of matter which are within the statutory classes of patentable subject matter.[14] Most countries also allow claims for methods of making a co-crystal and often methods of using them, as well as pharmaceutical compositions containing them. The basis for this patentability is that a new co-crystal is a new composition of matter between an API and a co-former(s) with a unique binding interaction between them. The crystalline nature of the co-crystal composition may, and in most countries does, add to co-crystal patentability.

For an invention, such as a composition of matter, to be patentable it must first be new or novel.[15] Under the EPC, an invention shall be considered to be new if it does not form part of the state of the art. The state of the art comprises everything made available to the public by means of a written or oral description, by use, or in any other way, before the date of filing of the European patent application.[16] In the United States, novelty is defined by categories within the patent statute with reference to, for example, the date of invention or the application filing date, whether or not the prior disclosure was by the inventor or another, and the type of disclosure, for example a publication, a public sale or a patent application. In the United States, novelty-defeating prior art is said to "anticipate" a claimed invention. An invention is anticipated only if *each and every element* in the claim is found, expressly or inherently, in the prior art.[17,18] Thus, when assessing the novelty of an invention, the differences between the invention and the prior art are important.

The unique combination of an API and a co-former in a co-crystal, with its unique chemical bonding between the API and co-former, typically does not occur in nature and generally means that a new co-crystal, as a chemical composition, would generally be expected to be novel.[19] The stoichiometry of the API and the co-former within the co-crystal distinguishes it from a simple physical mixture of the API and co-former. This then generally satisfies the novelty requirement for patentability. Although that is not to say that any co-crystal is *per se* patentable.

In addition to being new or novel, to be patentable an invention must possess an inventive step or be non-obvious.[20] This criterion of patentability looks at what the prior art would have taught or suggested to one of ordinary skill. The focus here is again on the differences between the prior art and the patent claims, but not just on what is specifically described in the prior art but also on what one of ordinary skill would have learned or understood from it. The EPC states that an invention shall be considered as involving an inventive step if, having regard to the state of the art, it is not obvious to a person skilled in the art.[21] In the United States, a patent may not be obtained, even though the invention is not identically disclosed or described, if the differences between the subject matter sought to be patented and the prior art are such that ... the subject matter would have been obvious to a person having ordinary skill in the art.[22]

Because a co-crystal brings together an API and a co-former via a bonding interaction in a specific stoichiometry, one can think of co-crystal formulation as an endeavor of chemical synthesis, of reaction chemistry. Unlike polymorphs, co-crystals are not the products of recrystallization but more the result of "supra-molecular synthesis". Attempting to engineer the desired properties, a chemist may choose to combine a given API with a particular co-former. This is not to say that a co-crystal will form or that a predictable result will occur, only that there is some level of design in the experiment. When there is design in the experiment, the patent law simply asks the question of whether the resulting invention would have been obvious, or if it possesses an inventive step.

To determine inventive step, the European Patent Office (EPO) employs the "problem-and-solution" approach. In this approach, the EPO takes the view that every invention is the solution to a technical problem. The general analysis of the "problem-and-solution" approach (1) determines the closest prior art; (2) then starting from the closest prior art, establishes the objective technical problem to be solved; and (3) assesses whether, in light of the prior art, it would have been obvious to solve the objective technical problem in the way claimed in the patent application.[23]

The American test for obviousness requires factual analysis of: (1) the scope and content of the prior art; (2) the differences between the prior art and the claimed invention; (3) the level of ordinary skill in the art; and (4) any objective evidence of non-obviousness.[24] In its recent decision concerning *KSR International*, the United States Supreme Court articulated a broad inquiry for determining obviousness. The proper question is whether one of ordinary skill, facing the wide range of needs created by developments in the field of endeavor would have seen the benefit of making the changes in the invention.[25,26]

The focus in both the European inventive step approach and the American obviousness analysis is ultimately the differences between the claimed invention and the prior art. Often the prior art is the API itself, a salt of the API or a crystalline form of the API. The bonding combination of the API and co-former(s) in the co-crystal can provide a difference in chemical composition that is a starting point for establishing inventive step or non-obviousness. The unique properties and characteristics of the co-crystal can serve further to establish the patentability of the co-crystal with regard to these criteria. Often this information is provided in the patent specification which discloses and describes the inventive co-crystal and its properties.

14.5 Information for Effective Co-crystal Patents

Effective patenting of co-crystals focuses primarily on two considerations: (1) the characterization and proof of actual co-crystal and (2) the properties of the co-crystal. The first is perhaps more important than the second, although as discussed above, the co-crystal must possess practical utility and usefulness. That practical utility is often directly related to the properties of the co-crystal and, sometimes, even the existence of the co-crystal, as a solid-state form of the API, solves the problem of handling, processing, and/or formulating the API.

Co-crystals are, as mentioned above, found by experimentation. Although carefully planned, the outcome of that experiment, whether or not a co-crystal will form, is not predictable. Nor is the exact composition of the co-crystal or its properties predictable. Often many experiments must be completed in the search for a co-crystal. To accelerate the experimental timetable, high throughput screening approach may be used. To find possible co-crystals, high throughput screening makes use of a library of co-formers and a variety of solvent combinations. Co-crystal formation is not only dependent on compatibility of the API and co-former but also on relative solubility of the API and co-former and the solubility of the co-crystal after formation. This large degree of freedom in co-crystal formation makes high throughput screening a good method for identifying co-crystals or potential co-crystals, although it must be noted that often, as a result from a high throughput-screen, it is hints of co-crystal formation that are found. These screening hits often may need to be further investigated, meaning more experiments, before a good "recipe" can be provided to make a co-crystal.

Co-crystal patents usually contain experimental examples that describe the preparation of the co-crystal and the characterization of the co-crystal. Characterization of the co-crystal describes the co-crystal itself and its various properties which include its solid state characteristics and stoichiometry. Typically, the solid state characteristics of a crystalline solid are shown by one or more of the following analytical techniques: X-ray powder diffraction (XRPD), single crystal X-ray diffraction (SCXD), Raman spectroscopy, infrared (IR) spectroscopy, solid state nuclear magnetic resonance spectroscopy (SSNMR), and differential scanning calorimetry (DSC). The stoichiometry of a co-crystal may be established through solution techniques such as comparison of peak integrations in a solution ^{1}H NMR spectrum, data regarding pharmaceutically relevant properties such as solubility data, dissolution rate data, pharmacokinetic data, and so on.

To show that a co-crystal has formed, it is typical to show comparative data that indicate the existence of the co-crystal. For example, a comparison of the XRPD pattern or the ^{13}C SSNMR spectra of the API, the co-former, and the co-crystal can show the differences or changes that have occurred upon co-crystal formation. These differences can then be used to claim the co-crystal as a unique composition of matter based upon its characteristic peaks. Figure 14.2[27] is an example of such a comparison using a stack plot of X-ray powder diffraction (XRPD) patterns. The differences in peaks can be seen. For a patent, a unique set of peaks differentiating the co-crystal from the API and co-former could be used to form the basis of a patent claim for the co-crystal.

Single crystal X-ray diffraction is one technique that does not require comparative data to establish the existence of a co-crystal. Quality single crystal X-ray diffraction (SCXRD) data can show the exact arrangement of the API and co-former in the co-crystal. Figure 14.3 is an ORTEP drawing of a co-crystal based on SCXRD data.[28]

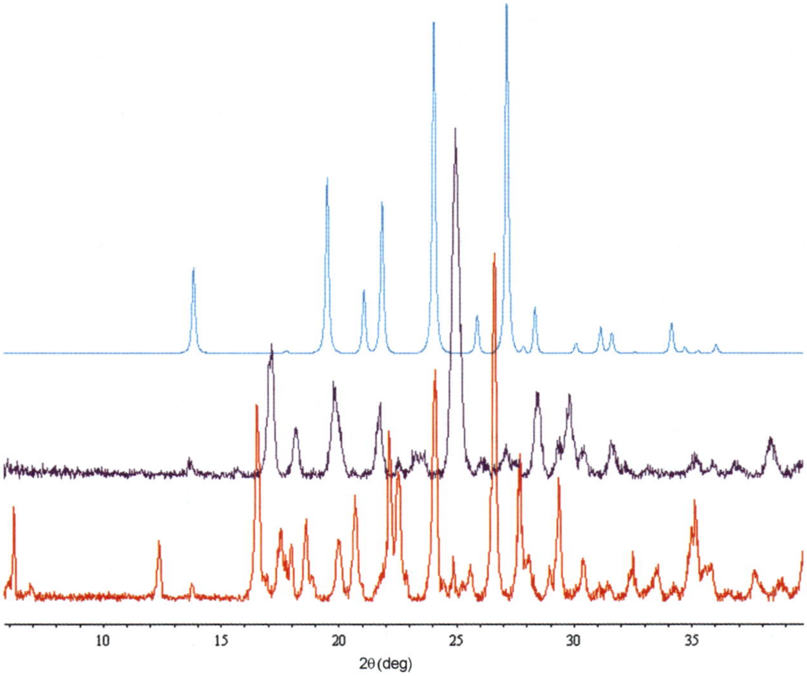

Figure 14.2 XRPD patterns of an API, co-former and the co-crystal can show the differences or changes that have occurred upon co-crystal formation. The XRPD pattern of the API is shown on top in blue. The XRPD pattern of the co-former, glutaric acid, is shown in the middle in purple and the XRPD pattern of the co-crystal is on the bottom in red.

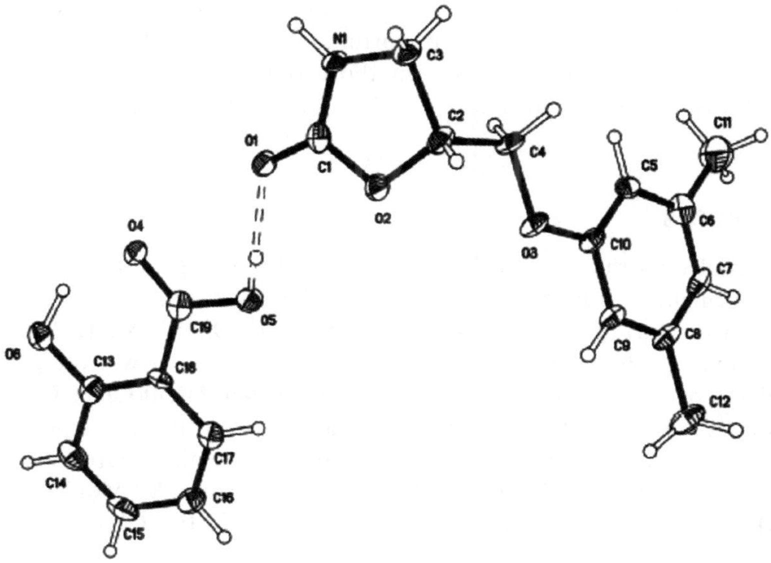

Figure 14.3 ORTEP drawing of a co-crystal based on single crystal X-ray diffraction data.

Co-crystal patents, then, describe and disclose a co-crystal as a new composition of matter. Similar to the traditional composition of matter patents, they establish the existence and identity of the co-crystal. Because of a co-crystal's unique bonding interaction between the API and co-former(s), a co-crystal patent takes as prior art its constituent parts[29] and establishes those by comparison. The properties and advantages of the co-crystal and the arguments for its patentability are also made in comparison with the API and its properties. It is from these comparisons that effective co-crystal patents are constructed.

14.6 Conclusions

This chapter has merely surveyed some ways in which co-crystals may translate into new patents and commercial value and has discussed some considerations for accomplishing this. It is clear that the success of co-crystals in commercialization and as new patents is tied to their unique chemical composition, crystalline structure, and properties. Engineering pharmaceutical properties using co-crystals can create new commercial value for an API and extend its patent life. This value comes from the composition of the co-crystal as well as from its crystalline form. Patenting co-crystals captures that value in the form of intellectual property. Yet, co-crystal patents have their own unique aspects to be addressed as can be seen from the discussion of the definition of "co-crystal", the patentability of co-crystals, and the information needed for effective co-crystal patents, each of which just touched the surface. What is clear is that co-crystals do present opportunities, commercially and in the form of patents, and the way to take up those opportunities is to engage the chemistry of the co-crystals and take advantage of that chemistry.

References

1. J. Lindeman thanks Paul Burgess, M.S., J.D. of Segrub Pharmaceuticals for bringing US Patent No. 3,028,420 to his attention.
2. Relevant portions of the US Patent Trademark Office's Manual of Patent Examining Procedure (MPEP) available at http://www.uspto.gov/web/offices/pac/mpep/index.htm and of the European Patent Office Guidelines for Examination (EPO Guidelines) available at http://www.epo.org/law-practice/legal-texts/html/guiex/e/foreword.htm are also referenced.
3. J. F. Remenar, S. L. Morissette, M. L. Peterson, B. Moulton, J. M. MacPhee, H. R. Guzmán and O. Almarsson, *J. Am. Chem. Soc.*, 2003, **125**(28), 8456–8457.
4. A. V. Trask, W. D. Motherwell and W. Jones, *Int. J. Pharm.*, 2006, **320**(1–2), 114–123.
5. Avantium Pharma, in house research project.
6. 35 U.S.C. § 112, ¶ 2.
7. EPC Article 84.
8. *See* MPEP § 2111.01 and EPO Guidelines Part C, § 4.14.
9. P. Visheweshwar, J. A. McMahon, J. A. Bis and M. J. Zaworotko, *J. Pharm. Sci.*, 2006, **95**(3), 499–516.
10. A. Bak and D. Ostovic, *US Pat.*, 2008/0051453A1, Sorbic Acid Analog Co-Crystals, published 28 Feb, 2008.
11. *Id.*, claims 1 and 2.
12. MPEP § 2111.01.
13. MPEP § 2111.02.
14. 35 U.S.C. § 101; EPC, Part II, Chapt. 1, Art. 52.
15. 35 U.S.C. § 102; EPC, Part II, Chapt. 1, Art. 54.
16. EPC, Part II, Chapt. 1, Art. 54.
17. EPO Guidelines, Part C, Chapt. IV, Art. 9.2; MPEP § 2112.
18. Under "inherent anticipation", a reference may anticipate without disclosing a feature of the claimed invention if that missing characteristic is necessarily present, or inherent, in the single anticipating reference. *Continental Can Co. v. Monsanto Co., 948 F.2d 1264, 1268 (Fed. Cir. 1991).* Inherent anticipation does not require that a person of ordinary skill in the art at the time would have recognized the inherent disclosure. *In re Cruciferous Sprout Litig.*, 301 F.3d 1343, 1351 (Fed. Cir. 2002).
19. This assumes that the co-crystal, once discovered, has not itself been publicly disclosed prior to filing a patent application. Such a public disclosure defeats the novelty requirement from the patent law perspective.
20. EPC Art. 56, 35 U.S.C. § 103.

21. EPC, Part II, Chapt. 1, Art. 56.
22. 35 U.S.C. § 103(a).
23. EPO Guidellines Part c. Chapt IV, Art. 11.5.
24. *Graham v. John Deere Co.*, 383 U.S. 1, 148 USPQ2d 459 (1966).
25. *KSR Int'l Co. v. Teleflex, Inc.* 550 U.S. 398, (2007).
26. *See also* MPEP §§ 2141–2143 for full discussion of the USPTO approach to obviousness.
27. McNamara *et al.*, *Pharm. Res.*, 2006, **23**(8), 1888–1897.
28. Nuformix Limited (Cambridge, UK) *WO* 2011/077252 A2, Metaxalone Co-crystals, published 30 June, 2011.
29. This assumes that the API and co-former are known compounds prior to filing of the co-crystal patent application.

CHAPTER 15

Concluding Remarks using Piracetam as a Learning Model

JOHAN WOUTERS,[a] ANAELLE TILBORG[a] AND LUC QUÉRÉ*[b]

[a] Department of Chemistry, University of Namur (FUNDP), 61 Rue de Bruxelles, B-5000 Namur, Belgium; [b] UCB Pharma s.a., Chemin du Foriest, B-1420 Braine-L'Alleud, Belgium

The previous chapters of this book should have convinced the reader of the great potential of co-crystallising active pharmaceutical ingredients (APIs) with pharmaceutically acceptable co-crystal formers. The resulting pharmaceutical co-crystals potentially improve the physical properties (solubility, dissolution rate, melting point, *etc.*) of the final products.[1,2]

As suggested several times in the book, one should not get carried away too much with semantic debates nor ascribe too much scientific significance to the term co-crystal. Nevertheless the discussion in itself is valuable in highlighting the complex reality of the solid state chemistry of multi-component systems (Figure 15.1). Coquerel, in Chapter 13, very clearly demonstrated the wide scope associated with co-crystals and the "salt to co-crystal continuum" is only one aspect of this complexity.

However a common feature presented throughout the book is that co-crystals are multi-component systems in which the entities are held by reversible bonds.

RSC Drug Discovery Series No. 16
Pharmaceutical Salts and Co-crystals
Edited by Johan Wouters and Luc Quéré

Published by the Royal Society of Chemistry, www.rsc.org

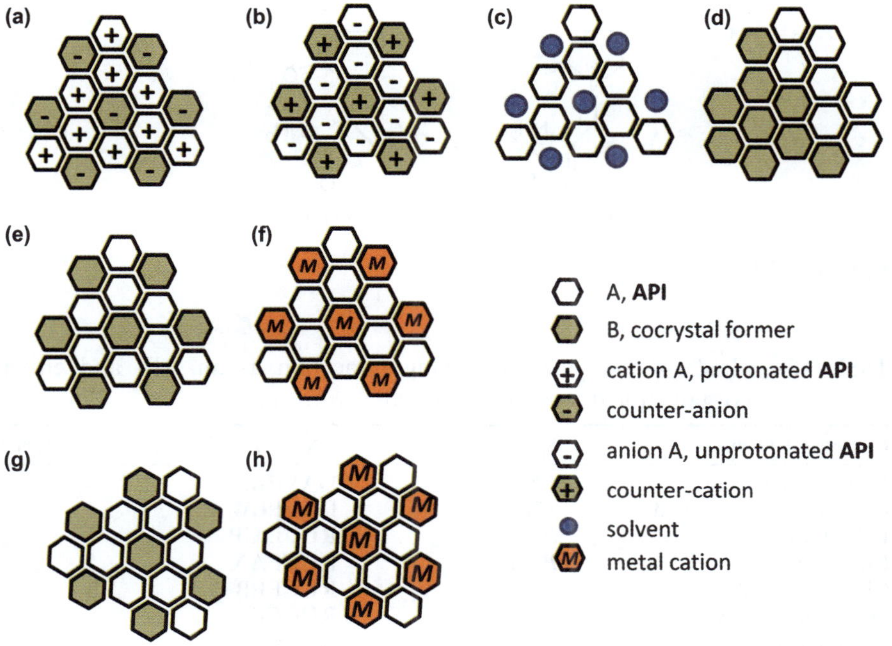

Figure 15.1 A few examples of multi-component systems: pharmaceutical salts (a) and (b), solvates (c), and co-crystals (e) and (f). These systems are different from a physical solid mixture (d). Polymorphism can affect these solid states: polymorphism of the API or the co-crystal former, polymorphism of the co-crystals (g) and (h).

H-bonds[†] are the most frequently discussed stabilising interactions. Halogen bonding is another paradigm that complements the hydrogen bond and a striking parallelism exists between the properties of these two types of interactions.

Metal coordination is an alternative, almost unexplored in the pharmaceutical world, even though it is another way to stabilise an API within a co-crystal. The resulting complexes are either metallo-drugs, in which the metal ion is also a biologically active component, or pharmaceutical coordination polymers (also coined "ionic co-crystals" by Braga and co-workers), in which the metal ion plays the role of a carrier for the API molecule, similar to the organic co-former in a more classical pharmaceutical co-crystal definition.[3] Generally, the central metal (cation) is bonded to a surrounding array of molecules and/or anions that serve as ligands through coordinate (dative, dipolar) covalent bonds.

[†] IUPAC definition (2011). A hydrogen bond is an attractive interaction between a hydrogen atom from a molecule or a molecular fragment X–H in which X is more electronegative than H, and an atom or a group of atoms in the same or a different molecule, in which there is evidence of bond formation. Etter (*Acc. Chem. Res.*, 1990, **23**, 120) proposed several hydrogen bonding rules, one of which states: "the best hydrogen-bond donor and the best hydrogen-bond acceptor will preferentially form hydrogen bonds to one another".

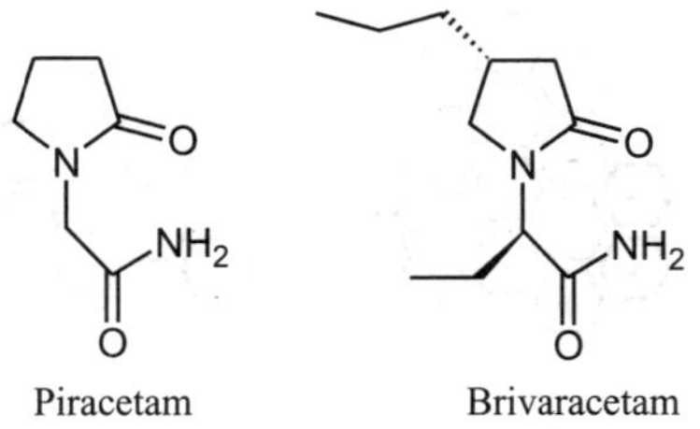

Scheme 15.1

Table 15.1 List of co-crystals involving piracetam as an API and either organic (acids) or inorganic salts co-formers.

Co-crystal Former	*Space Group*	*CSD ref code*[a]	*Reference*
Gentisic acid	$C2/c$	DAVPAS	4
p-Hydroxybenzoic acid	$P2_1/n$	DAVPEW	4
L-Tartaric acid	$P2_12_12_1$	RUCDUP	5
Citric acid (1:1)	$P2_1/a$	RUCFAX	5
Citric acid (3:2)	$P2_1/a$	RUCFEB	5
DL-Mandelic acid	$C2/c$	RUCFIF	5
L-Mandelic acid	$P2_12_12_1$	XOZSOV	5
Cobalt (Co (II))	P-1	YEDJIA	6a
Cobalt (Co (II))	P-1	HAFSAI	7a
Copper (Cu (II))	$P2_1/n$	HEPWII	6b
Copper (Cu (II))	$P2_1/c$	JOFJAP	8
Copper (Cu (II))	$P2_1/n$	WAJPAY	7b
Zinc (Zn (II))	$P2_1/n$	WAJSOP	7c
Zinc (Zn (II))	P-1	WEGBEP	6c
Manganese (Mn (II))	P-1	LEDZUP	6d
Nickel (Ni (II))	P-1	HEPWUU	6e
Nickel (Ni (II))	P-1	OHECEK	9

[a]CSD is the Cambridge Structural database.

Piracetam, (2-oxo-1-pyrrolidineacetamide), is a good example to illustrate the possibility of formation of co-crystals using both H-bonds or coordination to a metal cation (Table 15.1; Figure 15.2). Table 15.1 lists co-crystals involving piracetam as an API and either organic acids or inorganic salts co-formers.

The selected structures of co-crystals involving piracetam presented in Figure 15.2 illustrate distinct patterns of interactions (supramolecular synthons) within the crystal packing. In the presence of organic acid co-formers (Figure 15.2, left), H-bonds patterns correspond to both homo (amide–amide) or hetero (amide–acid) supramolecular synthons (Figure 15.3). In the presence of inorganic salts (Figure 15.2, right), coordination to the cation involves the oxygen atom of either the amide C=O (*e.g.* HEPWII[6b] and YEDJIA[6a]) or the endocyclic C=O (*e.g.* HAFSAI[7a]).

One reason for the successful use of co-crystals is thus the discovery of reliable supramolecular synthons for co-crystal design. Identification of these

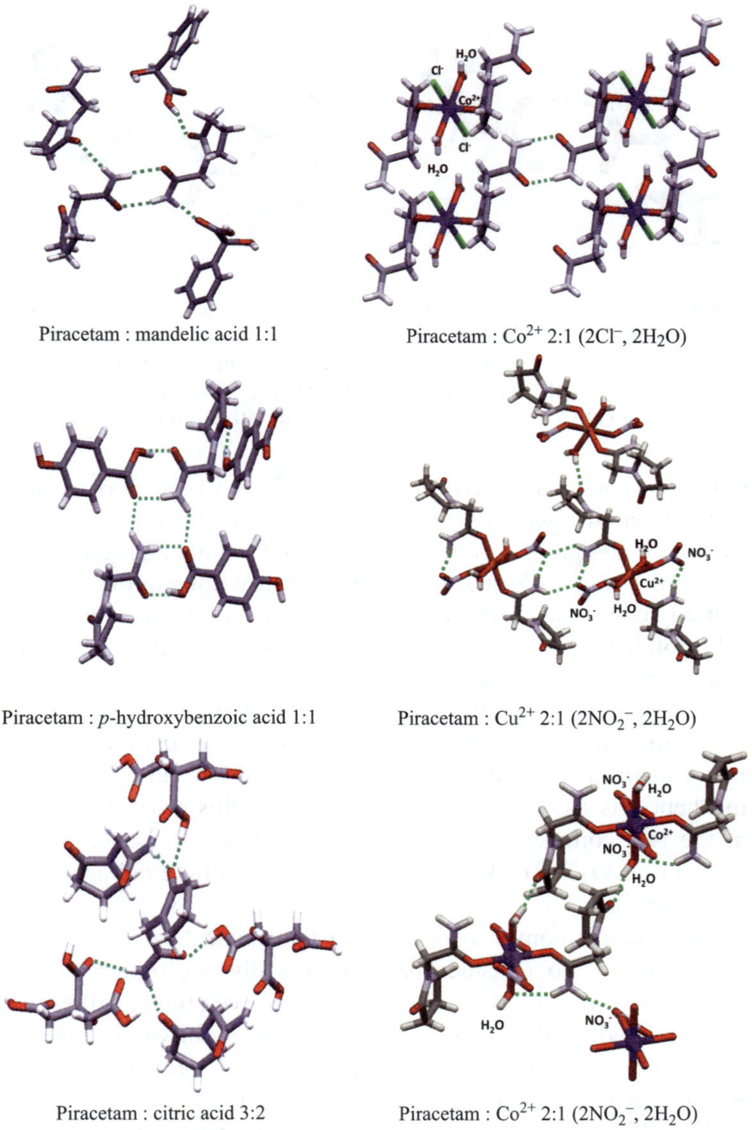

Figure 15.2 Selected examples of structures of co-crystals involving piracetam. Top: co-crystals with mandelic acid (left, XOZSOV[5]) and $Co^{2+}/Cl^-/H_2O$ (right, HAFSAI[7a]) are built around a dimeric amide–amide (homosynthon) H-bonding pattern. Middle: co-crystals with *p*-hydroxybenzoic acid (left, DAVPEW[4]) and $Cu^{2+}/NO_3^-/H_2O$ (right, HEPWII[6b]) lead to a tetrameric arrangement involving two molecules of API and two co-crystal former moieties (carboxylic acid (left) and NO_3^- ion (right)) held together by H-bonds. Bottom: co-crystals with citric acid (left, RUCFEB[5]) and $Co^{2+}/NO_3^-/H_2O$ (right, YEDJIA[6a]) form a 3D network of H-bonds involving all the donor (NH_2 of the amide) and acceptor (amide C=O and endocyclic C=O) sites of the API. Note the different conformations of piracetam in those co-crystal structures. (Pictures generated with Mercury, CCDC).[10]

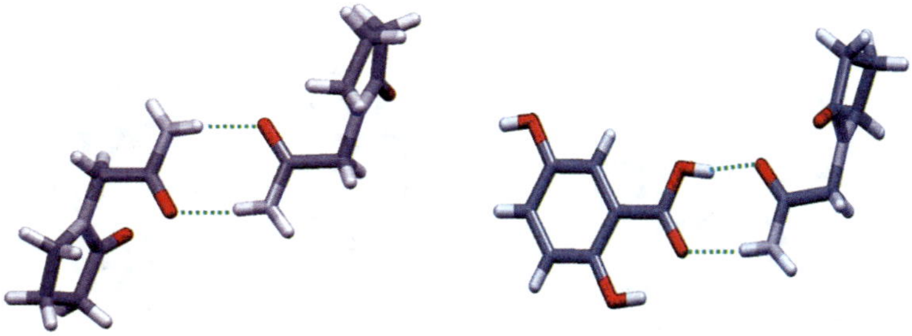

Figure 15.3 Illustration of the concept of homo (amide–amide) (left) and hetero (amide–acid) (right) supramolecular synthons applied to the structural characterisation of co-crystal structures involving piracetam (left, co-crystal with mandelic acid (XOZSOV[5]); right, co-crystal with gentisic acid (DAVPAS[4]). (Pictures generated with Mercury, CCDC).[10]

supramolecular synthons for co-crystal synthesis has been pursued with great intensity by different research groups.[11,12] Examples have been provided in this book. While the design of co-crystals can rest on rational design, a strategy in which the choice of a co-former is neither arbitrary nor random, no guarantee exists that any co-crystal designed on the basis of supramolecular synthons will actually be obtained. This issue of non-obviousness is particularly sensitive in the context of patentability and has been addressed in Chapter 14 of this book.

In the case of the coordination complexes involving piracetam, it is important to underline that the nature of other ligands (*i.e.* counter-anions of the metal cation) can also influence the structure of the resulting co-crystal solid form and hence its associated properties. This is illustrated by the difference between the two complexes with Co^{2+} for which the presence of either Cl^- (HAFSAI[7a]) or NO_3^- (HEPWII[6b]) deeply influences the final geometry of the molecular assembly.

Of course, when it comes to developing pharmaceutical salts or co-crystal candidates, one has to consider pharmaceutically acceptable cations. New chemical entities with a safer carrier have been successfully synthesised with a structurally related analogue of piracetam, that is, brivaracetam, a second generation antiepileptic drug (AED) (Figure 15.4). Indeed, it was shown that brivaracetam (currently in phase III trials), under the right conditions, readily reacts with $MgCl_2$, for example, leading to well shaped prismatic crystals with drastically improved melting point.[3a]

This last example of crystal engineering on piracetam and its analogues shows that we can expand from the H-bond paradigm for the co-crystal and use the specific features provided by the coordination chemistry to hunt for new materials with new assets. This may be where the two distinct fields of research in crystal engineering underlined by Desiraju in his introductory commentary (Chapter 1) can be nicely combined, that is, the metalloorganic framework (MOF) and the co-crystal one.

Getting new insights into mechanisms of molecular recognition and crystal packing constitutes a clear shared interest between the academic and

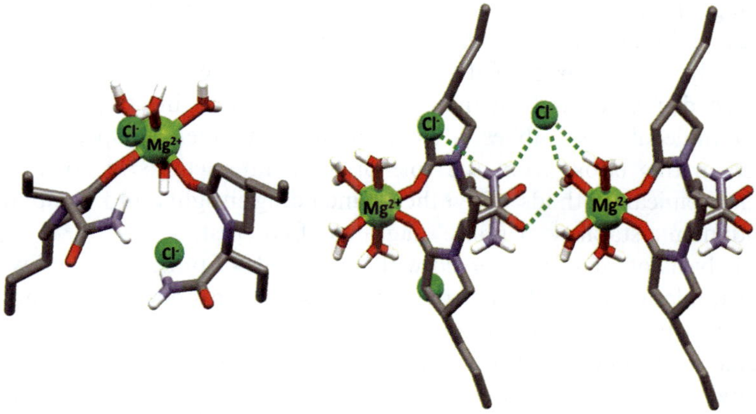

Figure 15.4 Crystal structure of the 2:1:4 co-crystal brivaracetam–$MgCl_2$–H_2O. This analogue of piracetam forms a coordination complex in which the oxygen atom (endocyclic C=O) of the API coordinates Mg^{2+} metal cation in a 2:1 ratio (left). The coordination sphere of the cation is completed by four water molecules (octahedral geometry). The two Cl^- counterions participate in the stability of the co-crystal and are engaged in H-bonds (right, dotted lines). (Pictures generated with Mercury, CCDC).[10]

pharmaceutical worlds. In this respect, *a posteriori*, the crystal structures of co-crystals of piracetam appear to make sense in terms of hydrogen bonds and electrostatic interactions. Indeed, in those co-crystal structures, the donor (NH_2 of the amide) and acceptor (amide C=O and endocyclic C=O) sites of the API are engaged in H bonds or coordination links leading to versatile end structures. As stated in Chapter 1, *a priori* prediction of the interaction patterns of the API–co-former combination is indeed a challenging endeavour. A large number of factors influence crystallisation outcome. Crystal structure prediction is progressing in tackling such complex systems as co-crystals and salts. These simulations demand a high level of precision in the lattice energy calculations and require substantial computational resources. With the ever increasing power of modern computers and the development of methods, such simulations should become accessible in the future, as suggested by several authors in this book who have reviewed recent advances in the field.

A combination of logic driven synthon-based design and high-throughput crystallisation is most probably key, as Desiraju made clear. In this context, new strategies combining screening crystallisation tools and the use of well designed libraries of co-formers are expected to yield more co-crystals with advantageous properties.

In the last section of the book, the reader will find a representative list of useful and pharmaceutically acceptable co-crystal formers.

Certainly, a primary concern of the pharmaceutical industry is to find new opportunities and better ways of answering the various medical needs. This book has addressed essential aspects of this quest for novel forms such as considering co-crystals in the pharmaceutical development continuum, improving pharmacokinetics, solving process and scale up issues, as well as analytical techniques to support such investigations.

The study of pharmaceutical co-crystals is also a re-emerging and exciting field of academic research in the context of crystal engineering and rational design of molecular assemblies. Most, if not all, fundamental aspects have been reviewed in this book, from the role of weak interactions in co-crystals to mechanochemical methods of synthesis, including insights into solubility and thermodynamic stability or phase diagrams of co-crystals.

On the basis of what is observed with co-crystals of piracetam, and with the many examples that have been provided throughout this book, it is tempting to generalise in parallel to ideas about polymorphism[‡] and to state that "every API could lead to potential co-crystals and the number of co-crystals obtained with a given compound is proportional to the time and money spent in research on that compound". Indeed, in essence, the nature of molecular interactions that allow an API to bind to its biological target is the same as that which stabilises the API in the solid state (either as a single or a multi-component system). Most (if not all) of the APIs thus contain in their structure the functional group(s) that could lead to co-crystallisation. This exciting perspective fully justifies both academic and industrial interest in these systems.

References

1. (a) J. F. Remenar, S. L. Morissette, M. L. Peterson, B. Moulton, J. M. MacPhee, H. R. Guzman and O. Almarsson, *J. Am. Chem. Soc.*, 2003, **125**, 8456; (b) D. P. McNamara, S. L. Childs, J. Giordano, A. Iarriccio, J. Cassidy, M. S. Shet, R. Mannion, E. O'Donnel and A. Park, *Pharm. Res.*, 2006, **23**, 1888; (c) M. B. Hickey, M. L. Peterson, L. A. Scoppettuolo, S. L. Morrisette, A. Vetter, H. Guzman, J. F. Remenar, Z. Zhang, M. D. Tawa, S. Haley, M. J. Zaworotko and O. Almarsson, *Eur. J. Pharm. Biopharm.*, 2007, **67**, 112; (d) A. Bak, A. Gore, E. Yanez, M. Stanton, S. Tufekcic, R. Syed, A. Akrami, M. Rose, S. Surapaneni, T. Bostick, A. King, S. Neervannan, D. Ostovic and A. Koparkar, *J. Pharm. Sci.*, 2008, **97**, 3942; (e) A. V. Trask, W. D. S. Motherwell and W. Jones, *Int. J. Pharm.*, 2006, **320**, 114; (f) S. L. Childs, L. J. Chyall, J. T. Dunlap, V. N. Smolenskaya, B. C. Stahly and G. P. Stahly, *J. Am. Chem. Soc.*, 2004, **126**, 13335.
2. (a) P. Vishweshwar, J. A. McMahon, M. Oliveira, M. L. Peterson and M. J. Zaworotko, *J. Am. Chem. Soc.*, 2005, **127**, 16802; (b) M. R. Cai, *Mol. Pharmaceutics*, 2007, 4, 310; (c) S. L. Childs, *Int. Pat. Number*, WO2007/067727 A2, 2007; (d) N. Shan and M. J. Zaworotko, *Drug Discovery Today*, 2008, **13**, 440; (e) W. W. Porter III, S. C. Elie and A. J. Matzger, *Cryst.*

[‡] "Every compound has different polymorphic forms, and in general, the number of forms known for a given compound is proportional to the time and money spent in research on that compound." W. C. McCrone, in *Physics and Chemistry of the Organic Solid State*, ed. D. Fox, M. M. Labes and A. Weissberger, Interscience Publishers, London, 1965, vol. 2, pp. 725–767.

Growth Des., 2008, **8**, 14; (f) S. Aitipamula, P. S. Chow and R. B. H. Tan, *CrystEngComm*, 2009, **11**, 1823.

3. (a) L. Quere, Co-crystals of pyrrolidinones, WO 2007-EP5009 20070606; (b) D. Braga, F. Grepioni, L. Maini, S. Prosperi, R. Gobetto and M. R. Chierotti, *Chem. Commun.*, 2010, **46**, 7715; D. Braga, F. Grepioni, L. Maini, R. Brescello and L. Cotarca, *CrystEngComm*, 2008, **10**, 469.
4. P. Vishweshwar, J. A. McMahon, M. L. Peterson, M. B. Hickey, T. R. Shattock and M. J. Zaworotko, *Chem. Commun.*, 2005, **36**, 4601.
5. M. Viertelhaus, R. Hilfiker and F. Blatter, *Cryst. Growth Des.*, 2009, **9**(5), 2220.
6. (a) V. Sabirov, M. A. Porai Koshits and Y. T Struchkov, *Koord. Khim.*, 1993, **19**, 143; (b) V. Sabirov, M. A. Porai Koshits and Y. T. Struchkov, *Koord. Khim.*, 1993, **19**, 642; (c) V. Sabirov, M. A. Porai Koshits and Y. T. Struchkov, *Koord. Khim.*, 1993, **19**, 81; (d) V. Sabirov, M. A. Porai Koshits and Y. T. Struchkov, *Koord. Khim.*, 1993, **19**, 38; (e) V. Sabirov, M. A. Porai Koshits and Y. T. Struchkov, *Koord. Khim.*, 1993, **19**, 637.
7. (a) V. Sabirov, M. A. Porai Koshits, Y. T. Struchkov and A. Yunushkhodzhaev, *Koord. Khim*, 1992, **18**, 614; (b) V. Sabirov, M. A. Porai Koshits, Y. T. Struchkov, A. F. Dusmatov and A. N. Yunushkhodzhaev, *Koord. Khim.*, 1992, **18**, 292; (c) V. Sabirov, M. A. Porai Koshits, Y. T. Struchkov, K. A. Potekhin and A. N. Yunushkhodzhae, *Koord. Khim.*, 1992, **18**, 307.
8. X. Delacruz, A. Martinez-Balbas, J. Tormo and N. Verdaguer, *Acta Crystallogr.*, 1992, **C48**, 167.
9. D. Braga, F. Grepioni, V. André and M. T. Duarte, *CrystEngComm*, 2009, **11**, 2618.
10. C. F. Macrae, I. J. Bruno, J. A. Chisholm, P. R. Edgington, P. McCabe, E. Pidcock, L. Rodriguez-Monge, R. Taylor, J. van de Streek and P. A. Wood, Mercury CSD 2.0 – new features for the visualization and investigation of crystal structures, *J. Appl. Crystallogr.*, 2008, **41**, 466–470.
11. (a) B. K. Saha, A. Nangia and M. Jaskolski, *CrystEngComm.*, 2005, **7**, 355; (b) C. B. Aakeröy, A. M. Beatty and B. A. Helfrich, *Angew. Chem., Int. Ed.*, 2001, **40**, 3240; (c) G. R. Desiraju, *Angew. Chem., Int. Ed. Engl.*, 1995, **34**, 2311; (d) B. K. Saha, A. Nangia and M. Jaskolski, *CrystEngComm*, 2005, **7**, 355.
12. (a) C. B. Aakeroy, A. M. Beatty and B. A. Helfrich, *J. Am. Chem. Soc.*, 2002, **124**, 14425; (b) C. B. Aakeroy, J. Desper and B. A. Helfrich, *CrystEngComm*, 2004, **6**, 19; (c) T. Friscic and W. Jones, *Faraday Discuss.*, 2007, **136**, 167.

CHAPTER 16

Monographs of most Frequent Co-Crystal Formers

JOHAN WOUTERS,*[a] SANDRINE ROME[b] AND LUC QUÉRÉ[b]

[a] Department of Chemistry, University of Namur (FUNDP), 61 Rue de Bruxelles, B-5000 Namur, Belgium; [b] UCB Pharma s.a., Chemin du Foriest, B-1420 Braine L'Alleud, Belgium

16.1 Introduction

Whilst salts of active pharmaceutical ingredients (APIs) are restricted to (acceptable) counter-ions, the potentially available pharmaceutical space around co-crystals is much broader. Indeed the list of co-formers is long and the API (even if un-ionized) can theoretically be co-crystallized either with acidic, basic or neutral co-formers in multiple combinations thanks to a more adaptive stoichiometry. As one can easily foresee the type of advantageous properties for a solid form that may be accessed in such enlarged space, it is of paramount importance for the pharmaceutical stakeholder to design an efficient strategy for exploring it.

Although choosing logical combinations to start with, both in terms of molecular recognition and crystal packing, it is almost impossible to know in advance whether a salt or co-crystal will effectively form. As we have seen in this book, methods for prediction of the structure and its properties are still under development and it is common practice to resort to extensive screening to discover co-crystals.

Both the screening and the supramolecular synthon-based retrosynthetical approaches benefit from a better knowledge of the physicochemical properties

RSC Drug Discovery Series No. 16
Pharmaceutical Salts and Co-crystals
Edited by Johan Wouters and Luc Quéré

Published by the Royal Society of Chemistry, www.rsc.org

Table 16.1 List (alphabetical order) of selected co-crystal formers. Details of compounds in **bold** are given in the next section.

o-Acetylsalicylic acid	Glycine	N-methyl-D-glucamine
Adipic acid	Glycolamide	Orcinol
4-Aminobenzoic acid	Glycolic acid	**Oxalic acid**
4-Aminobenzamide	Hippuric acid	2-Oxoglutaric acid
Anthranilic acid	4-Hydroxybenzamide	Palmoic acid
Arabinose	**4-Hydroxybenzoic acid**	Pimelic acid
L-Arginine	1-Hydroxy-2-naphthoic acid	Piperazine
L-Ascorbic acid	Imidazole	L-Proline
L-Aspartic acid	Isonicotinamide	L-Pyroglutamic acid
Benzamide	Ketoglutaric acid	Resorcinol
Benzenesulfonic acid	L-Lactamide	**Saccharin**
Benzoic acid	Lactic acid	**Salicylic acid**
Boric acid	Lactose	Sebacic acid
Calcium Chloride	Laurylsulfonic acid	**Sorbic acid**
(+) Camphoric acid	L-Lysine	Sorbitol
Cholic acid	Magnesium chloride	**Stearic acid**
Citric acid	**Maleic acid**	Suberic acid
Cyclamic acid	**L-Malic acid**	**Succinic acid**
Erythritol	**Malonic acid**	Sucrose
Fructose	Maltose	**Tartaric acid**
Fumaric acid	**Mandelic acid**	L-Threonine
Gentisic acid	Mannitol	**Thromethamine**
Glucose	Mannose	*trans*-Cinnamic acid
D-Glucuronic acid	Methyl-4-hydroxybenzoic acid	Trimesic acid
D-Gluconic acid	Neotame	Tyrosine ethyl ester
L-Glutamic acid	**Nicotinamide**	L-Tyrosine
Glutaric acid	Nicotinic acid	**Urea**

of the co-crystal former (or co-former) and it is the aim of this chapter to provide a listing of co-former candidates (Table 16.1). These co-crystal formers generally belong to the list of GRAS (Generally Recognized As Safe) compounds and most of them have been reported to form multi-component entities. For those that are frequently encountered, their physicochemical properties are summarized to help in the subsequent identification of potential co-crystals. Besides more classical properties like p*K*a, melting point and solubility, powder X-ray diffractograms have been simulated in the most representative and informative cases. These characteristics should be affected by the formation of co-crystals and are thus a guide for the further identification of original formulations. Potential polymorphism of the co-crystal former is also flagged and selected examples of co-crystals are provided.

The reader will rapidly recognize the arbitrary aspect of this selection and the authors certainly encourage medicinal and analytical chemists to propose an alternative co-former library. With the assumption that the originality of the final pharmaceutical co-crystals relies equally on the nature of the API itself and on the supramolecular interactions that will be retained in the co-crystal, the sole purpose of such a list is to generate ideas and, it is hoped, a genuine series of trials.

16.2 Monographs on Co-crystal Formers

Physicochemical values were collected from various literature and web sources.[1,2]

CrystalLographic data were retrieved from the Cambridge Structure Database (CSD).[3] References to structures (*REFCODE*) are provided in italic capital letters. Powder diffractograms have been simulated on the basis of the coordinates deposited in the CSD using the Mercury program.[4]

Values of *c*Log*P* and *c*Log*S* were calculated in Chemexper.[2] The calculated value of Log*P*, the Logarithm of the partition coefficient between n-octanol and water, is a well established indicator of a compound's hydrophilicity. High hydrophilicity (and therefore low Log*P*) values cause poor absorption or permeation, whereas high *c*Log*P* values are generally associated with poor aqueous solubility. Calculated Log*S* values provide an estimate of aqueous solubility (mol/liter).

Selected substances can be considered as safe as most of them are used profusely in food and/or pharmaceutical processing. As a measure of acute toxicity, LD50 values are given (most of the time for oral administration in rats).

Acknowledgements

The authors thank Prof C. Wermuth, Dr L. Aerts and Ms S. LeMeur for valuable discussion and comments and B. Norberg for assistance during formatting of CSD data.

4-*AMINOBENZOIC ACID*		
SYNONYMS	*p-Aminobenzoic acid; p-ABA; vitamin BX, anticanitic vitamin.*	
CAS NO.	150-13-0	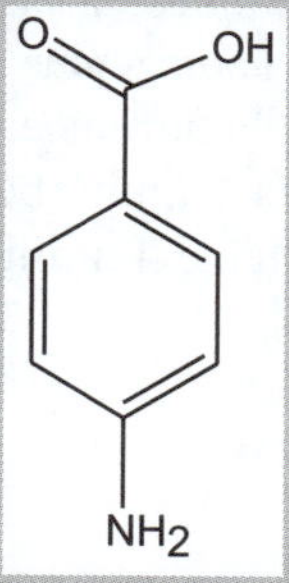
FORMULA	$C_7H_7O_2N$; $(NH_2)C_6H_4COOH$	
MOL WT.	137.14 g/mol	
PHYSICAL STATE	White-grey crystalline powder	
MELTING POINT	187–189 °C	
SOLUBILITY	Slightly soluble in water (0.6 g/100 mL (25 °C)). *c*Log*P*: 0.8; *c*Log*S*: –1.7	
pK_a	2.38; 4.98	
HB DONORS	2	
HB ACCEPTORS	3	
AROMATIC RING	1	
POLYMORPHISM	Two polymorphic forms are reported: monoclinic α ($Z'=2$, P_{21}/n) *AMBNAC06 and* β ($Z'=1$ P_{21}/n) *AMBNAC04.*	
PXRD – 2θ(°)	Polymorphs α *AMBNAC06 and β AMBNAC04* 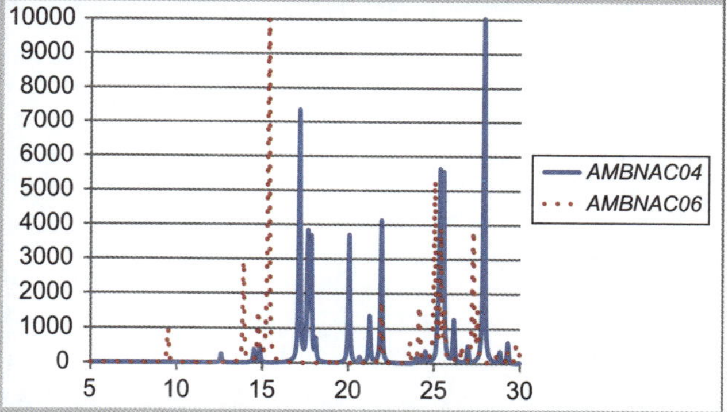	
TOXICITY	Largely non-toxic, oral rat LD50: 6000 mg/kg	
REMARKS		
*p*ABA is used as an UV-blocking ingredient in sun tan cosmetics. It is used in the medical field for preparing local anaesthetics and ointments.		

4-*AMINOBENZOIC ACID* *(Continued)*

The potassium salt is used as a drug against fibrotic skin disorders, such as Peyronie's disease, under the trade name Potaba.

Examples of co-crystals:
Carbamazepine 4-aminobenzoic acid 2:1.[5]
Carbamazepine 4-aminobenzoic acid-hydrate 2:1:1.[6]
Urea 4-aminobenzoic acid 1:2.[7]
Caffeine 4-aminobenzoic acid 1:1.[8]

ANTHRANILIC ACID (2-AMINOBENZOIC ACID)	
SYNONYMS	*o-Aminobenzoic acid; o-ABA; anthranilic acid; vitamin L.*
CAS NO.	118-92-3 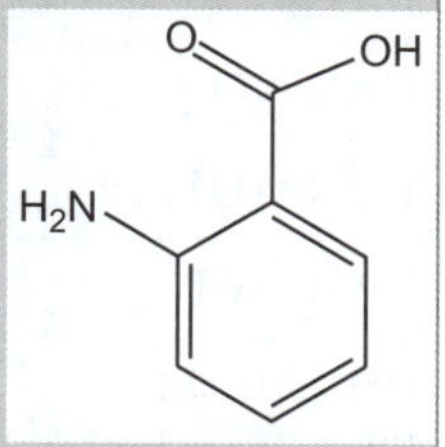
FORMULA	$C_7H_7O_2N$; $(NH_2)C_6H_4COOH$
MOL WT.	137.14 g/mol
PHYSICAL STATE	White to pale yellow crystalline powder
MELTING POINT	146–148 °C
SOLUBILITY	Soluble in water (5.7 g/L (25 °C)). *c*Log*P*: 0.8; *c*Log*S*: –1.7
pK_a	2.11; 4.95
HB DONORS	2
HB ACCEPTORS	3
AROMATIC RING	1
POLYMORPHISM	Three reported polymorphs
PXRD – 2θ(°)	Orthorhombic I ($Z'=2$, $P2_1cn$) *AMBACO07* Orthorhombic II (*Pbca*) *AMBACO03* Monoclinic III ($P2_1/a$) *AMBACO06* 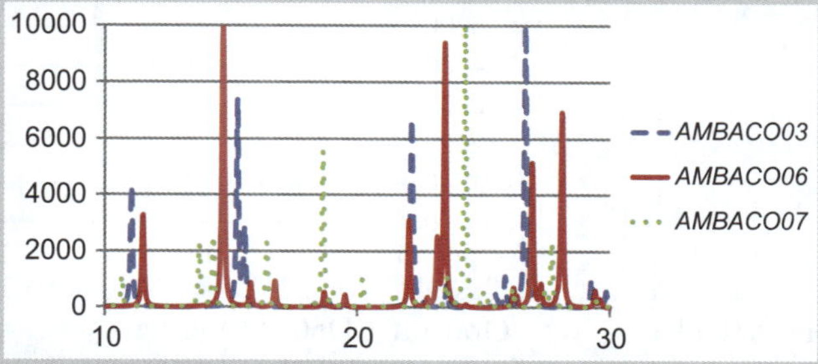
TOXICITY	Oral rat LD50: 5410 mg/kg
REMARKS	

Anthranilic acid is used as an intermediate for production of dyes, pigments, and saccharin. Used in preparing perfumes, pharmaceuticals (*e.g.* loop diuretics furosemide) and UV-absorber.

Examples of co-crystals:
Caffeine 2-aminobenzoic acid 1:1.[9]

L-ASCORBIC ACID	
SYNONYMS	*(+)-L-Ascorbic acid; Vitamin C; (5R)-[(1S)-1,2-dihydroxyethyl]-3,4-dihydroxyfuran-2(5H)-one; E300.*
CAS NO.	50-81-7 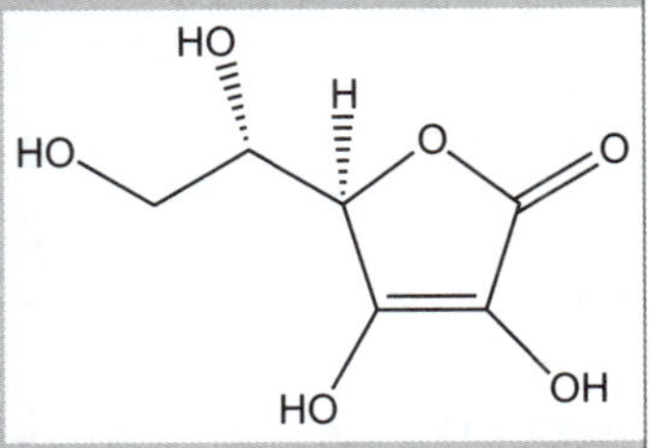
FORMULA	$C_6H_8O_6$
MOL WT.	176.13 g/mol
PHYSICAL STATE	White or light yellow solid
MELTING POINT	190°C
SOLUBILITY	Soluble in water (33 g/100 mL (25 °C)). *c*Log*P*: –2.2; *c*Log*S*: –0.35
pK_a	4.17; 11.57
HB DONORS	4
HB ACCEPTORS	6
PXRD – 2θ(°)	Monoclinic *P*21 (*LASCAC14*) 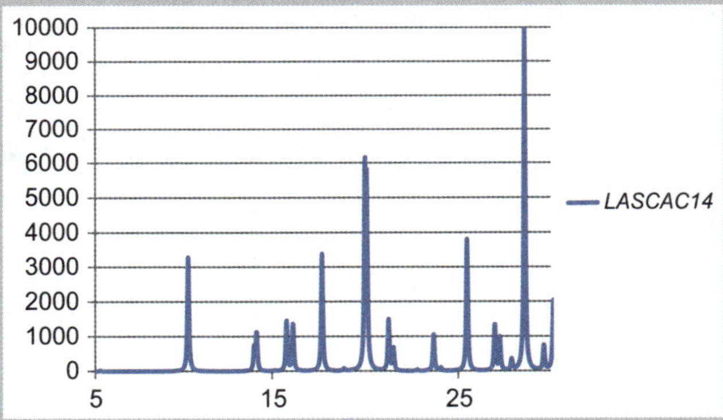
TOXICITY	Oral rat LD50: 11.9 g/kg
REMARKS	

Natural substance (vitamin C) with antioxidant properties, used in foods for this property. Non-hygroscopic, odourless, sensitive to light and oxygen, rapidly oxidized in alkaline solution.

Examples of co-crystals:
Nicotinic acid L ascorbic acid 1:1.[10]
Sarcosine L ascorbic acid 1:1.[11]
Betaine L ascorbic acid 1:1.[12]

CHOLIC ACID	
SYNONYMS	*3α,7α,12α-Trihydroxy-5β-cholanoic acid*
CAS NO.	81-25-4
FORMULA	$C_{24}H_{40}O_5$
MOL WT.	408.57 g/mol
PHYSICAL STATE	White crystalline substance
	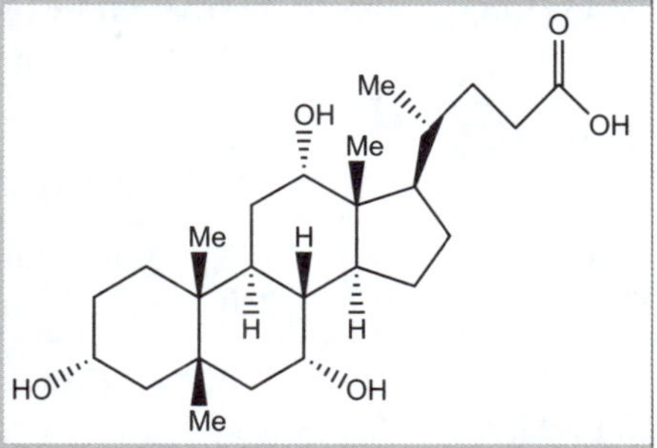
MELTING POINT	198 °C
SOLUBILITY	Sparingly soluble in water (175 mg/L at 20 °C).
pK_a	5.5
HB DONORS	4
HB ACCEPTORS	5
POLYMORPHISM	Anhydrous, monohydrate (*BUGJES*), hemihydrate (*DIPFAJ01*), solvates (methanol, ethanol) and many clathrates.
PXRD – 2θ(°)	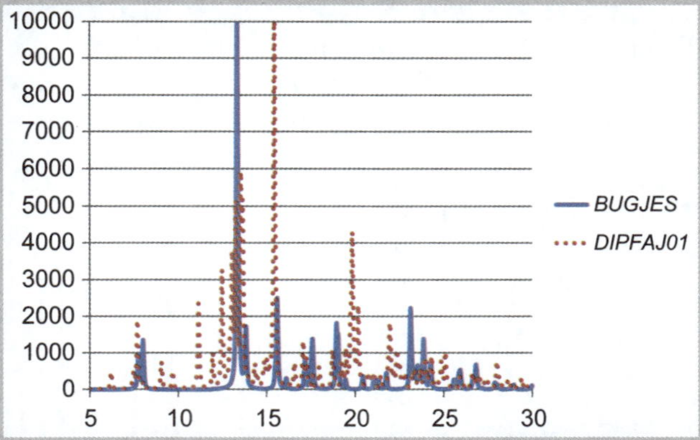
TOXICITY	Oral mouse LD50: 4950.00 mg/kg
REMARKS	

Cholic acid is one of two major bile acids produced by the liver where it is synthesized from cholesterol. Used as alimentary emulsifier (E1000).

Examples of co-crystals:
Thiobenzophenone cholic acid 1:2.[13]
***n*-Alkylammonia** cholic acid.[14]
Melamine cholic acid 2:2.[15]

CITRIC ACID		
SYNONYMS	*2-Hydroxypropane-1,2,3-tricarboxylic acid.*	
CAS NO.	77-92-9 (anhydrous), 5949-29-1 (monohydrate)	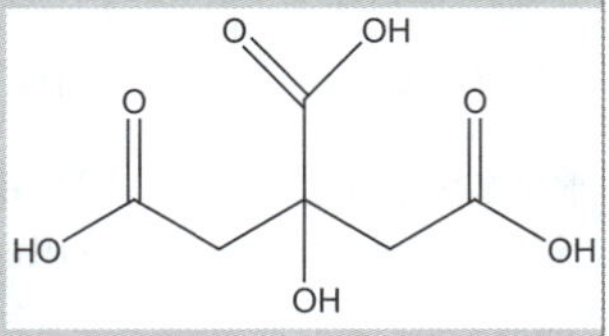
FORMULA	$C_6H_8O_7$; $HOC(COOH)(CH_2COOH)_2$	
MOL WT.	192.12 g/mol (anhydrous) 210.14 g/mol (monohydrate)	
PHYSICAL STATE	White crystals or powder	
MELTING POINT	153 °C (anhydrous)	
SOLUBILITY	Solubility in water (73 g/100 mL (20 °C)). *c*Log*P*: –2.15; *c*Log*S*: 0.07	
pK_a	3.09; 4.75; 6.41	
HB DONORS	4	
HB ACCEPTORS	7	
POLYMORPHISM	Two polymorphs of anhydrous citric acid (*CITRAC10, CITRAC11*) and a crystal structure of citric acid monohydrate have been reported (*CITARC*).	
PXRD – 2θ(°)	Anhydrous: monoclinic (P21/a) *CITRAC10* and (P21/c) *CITRAC11* Monohydrate: *CITARC* 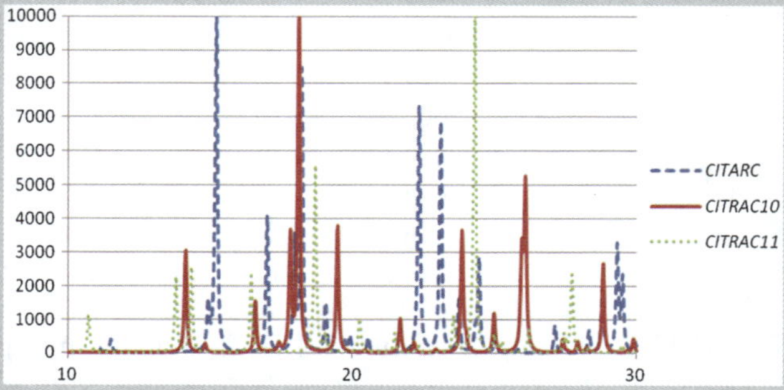	
TOXICITY	Oral rat LD50: 3000 mg/kg	
REMARKS		
Nutraceutical, ingredient used in baking and brewing, natural preservative/ conservative (antioxidant properties), used to add an acidic or sour taste to food and soft drinks.		

***CITRIC ACID** (Continued)*
Citric acid naturally exists in large amounts in a variety of fruits and vegetables (notably citrus fruits). Additive in formulation for effervescent reaction (production of CO_2) with sodium bicarbonate (hydrogenocarbonate).
Examples of co-crystals: **Piracetam** citric acid 2:1.[16] **Piracetam** citric acid 3:1.[17] **Isonicotinamide** citric acid 2:1.[18] **Nicotinamide** citric acid 2:1 dihydrate.[19] **Caffeine** citric acid 2:1.[20] **Theophylline** citric acid 2:1 monohydrate.[21]

FUMARIC ACID		
SYNONYMS	*(E)-2-Butenedioic acid; trans-butenedioic acid; allomaleic acid.*	
CAS NO.	110-17-8	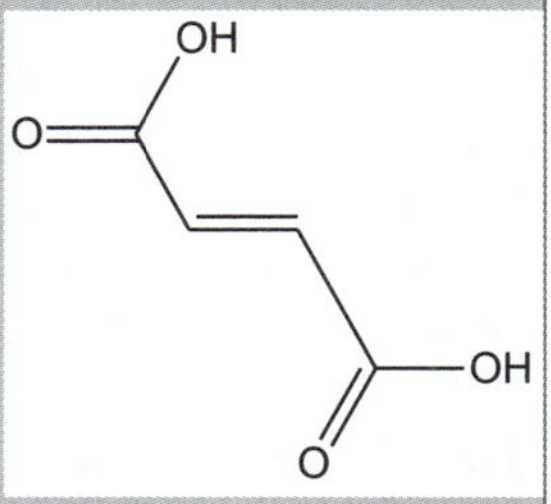
FORMULA	$C_4H_4O_4$; HOOCCH=CHCOOH	
MOL WT.	116.07 g/mol	
PHYSICAL STATE	White crystals	
MELTING POINT	287–302 °C (sublimes)	
SOLUBILITY	Slightly soluble in water (0.63 g/100 mL at 20°C). *c*Log*P*: –0.76; *c*Log*S*: –0.25	
pK_a	3.03; 4.44	
HB DONORS	2	
HB ACCEPTORS	4	
POLYMORPHISM	Two known polymorphic forms (anorthic form obtained from sublimation)	
PXRD – 2θ(°)	Monoclinic α ($P2_1/c$) *FUMACC* and anorthic β (P_{-1}) *FUMACC01* 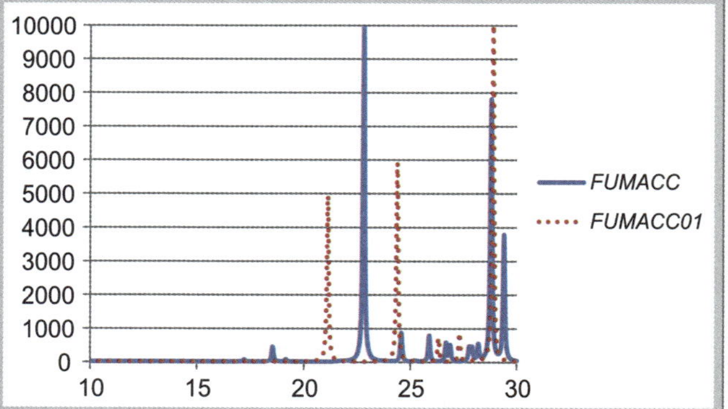	
TOXICITY	Oral rat LD50: 9300 mg/kg	
REMARKS		

Used as non-toxic food acidulent (beverages, baking powders). Fumaric acid esters are used to treat psoriasis.

Examples of co-crystals:
Carbamazepine fumaric acid 2:1.[22]
Fluoxetine HCl (Prozac®) fumaric acid (2:1).[23]
Isonicotinamide fumaric acid 2:1.[24]

L-GLUTAMIC ACID	
SYNONYMS	*2-Aminopentanedioic, acid 2-Aminoglutaric acid, L-Glu.*
CAS NO.	56-86-0 (L-isomer) *6893-26-1 (D-isomer)*
	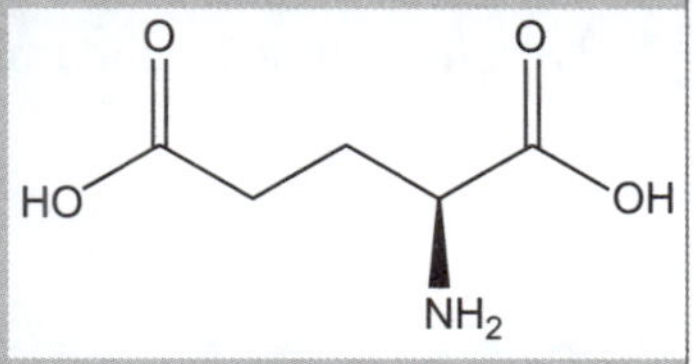
FORMULA	$C_5H_9NO_4$; HOOC-$(CH_2)_2CH(NH_2)COOH$
MOL WT.	147.13
PHYSICAL STATE	White crystalline powder
MELTING POINT	205 °C (decomposes)
SOLUBILITY	Solubility in water: 7.2 g/L (20 °C). *c*Log*P*: –1.488 *c*Log*S*: –0.421
pK_a	2.16; 4.15; 9.58
HB DONORS	3
HB ACCEPTORS	5
POLYMORPHISM	Two known polymorphic orthorhombic forms (α and β)
PXRD – 2θ(°)	Form *α* (*LGLUAC03*) and form *β* (*LGLUAC11*) 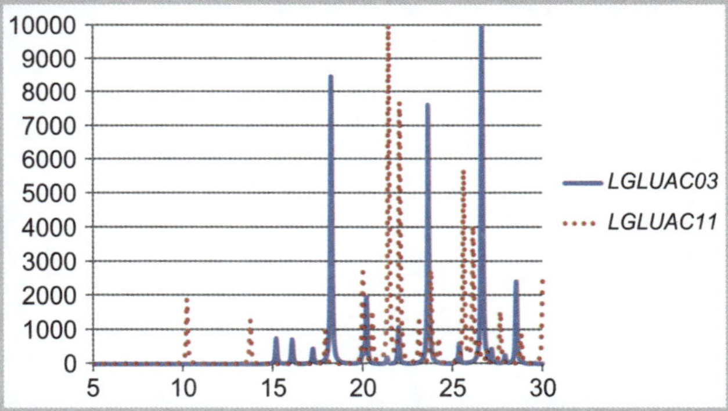
TOXICITY	Oral rat LD50 > 30000 mg/kg
REMARKS	

Non-essential natural amino acid. In neuroscience, glutamate is an important neurotransmitter that plays a key role in long-term potentiation and is important for learning and memory. L-glutamic acid is widely used in the food industry for the production of monosodium glutamate (Umami taste).

L-glutamic acid is a monotropic polymorphic compound. Transformation between the α- and β-polymorphs is solution-mediated. The α-polymorph of L-glutamic acid has a granular consistency whilst the β-form has a needle-like flaky structure.

L-GLUTAMIC ACID (Continued)
Related structures: DL glutamic acid *YUYMOU;* DL glutamic acid monohydrate *CADVUY01;* L glutamic acid. HCl *LGLUTA.*
Examples of co-crystals: **L-Pyroglutamic acid** glutamic acid H_2O 1:1:1.[25]

GLUTARIC ACID		
SYNONYMS	*1,5-Pentanedioic acid; 1,3-propanedicarboxylic acid.*	
CAS NO.	110-94-1	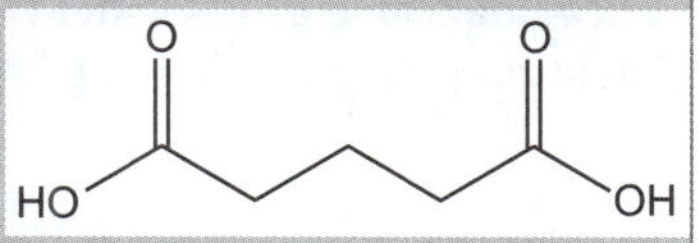
FORMULA	$C_5H_8O_4$; $HOOC(CH_2)_3COOH$	
MOL WT.	132.12 g/mol	
PHYSICAL STATE	White crystalline powder	
MELTING POINT	95–98 °C	
SOLUBILITY	Solubility in water (430 g/L at 20 °C). *c*Log*P*: 0.063; *c*Log*S*: –0.74	
pK_a	4.31; 5.41	
HB DONORS	2	
HB ACCEPTORS	4	
POLYMORPHISM	Two crystalline forms with different thermal stabilities separated by a transition temperature of 63 °C. At room temperature, the β-form is stable, and the α-form is metastable.	
PXRD – 2θ(°)	Monoclinic α (*I*2/*a*) *GLURAC* and monoclinic *β* (*C*2/*c*) *GLURAC04* 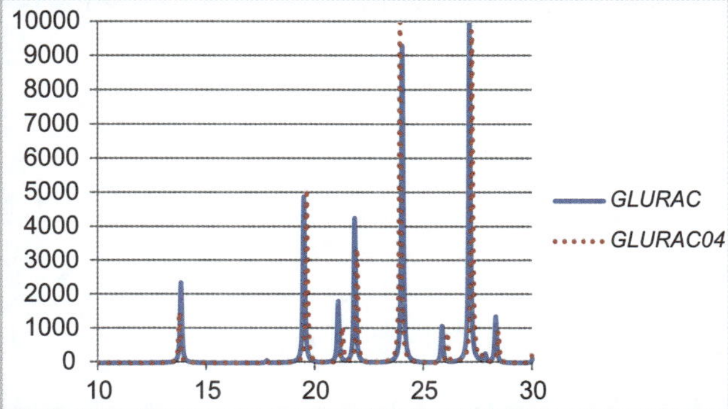	
TOXICITY	Oral mouse LD50: 6000 mg/kg	
REMARKS		
Glutaric acid occurs in plant and animal tissues and is found in the blood and urine. Ketoglutaric acid, found as an intermediate in the Krebs cycle, is used in dietary supplements to improve protein synthesis.		

GLUTARIC ACID *(Continued)*
Examples of co-crystals: **Caffeine** glutaric acid 1:1 (Forms I, II).[26] **Theophylline** glutaric acid 1:1.[27] **Carbamazepine** glutaric acid 1:1.[28] **Nicotinamide** glutaric acid 1:1.[29]

4-*HYDROXYBENZOIC ACID*		
SYNONYMS	*p-Hydroxybenzoic acid; p-HBA.*	
CAS NO.	99-96-7	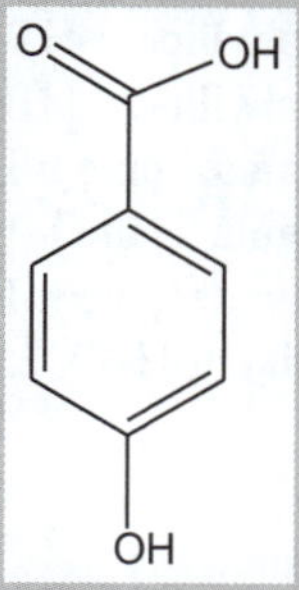
FORMULA	$C_7H_6O_3$; HOC_6H_4COOH	
MOL WT.	138.12 g/mol	
PHYSICAL STATE	White crystalline powder	
MELTING POINT	215–217 °C	
SOLUBILITY	Slightly soluble in water (5000 mg/L at 25 °C). *c*Log*P*: 1.2; *c*Log*S*: –1.3	
pK_a	4.54	
HB DONORS	2	
HB ACCEPTORS	3	
AROMATIC RING	1	
POLYMORPHISM	Two polymorphs are reported (Benson *et al.*, 2000) Indications of polymorphic enantiotropy were found primarily through solubility analysis and FTIR-ATR (Nordstöm and Rasmuson, 2006).	
PXRD – 2θ(°)	Anhydrous (polymorph I, monoclinic *P*21/*a*) *JOZZIH* Monohydrate *PHBZAC02* 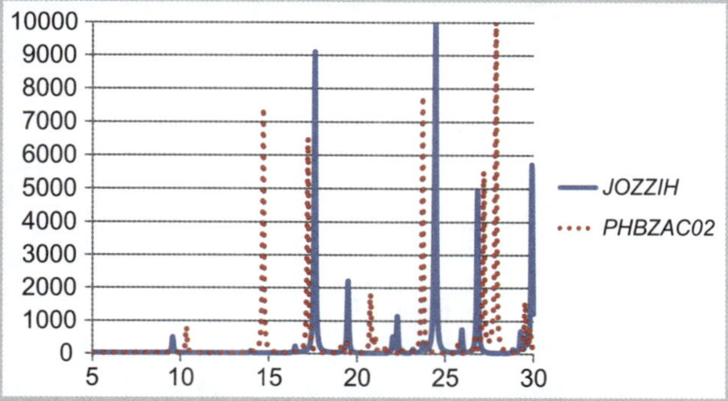	
TOXICITY	Oral rat LD50 > 10 g/kg	
REMARKS		
Used for the preparation of biocides, antiseptics and bacteriostatic agents. Basis for the preparation of its esters, known as parabens, which are used as preservatives in cosmetics. Popular antioxidant in part because of its low toxicity.		

4-*HYDROXYBENZOIC ACID* *(Continued)*
Examples of co-crystals: **Piracetam** *p*-HBA 1:1.[30] **Theophylline** *p*-HBA monohydrate 1:1:1.[31] **Theophylline** *p*-HBA 1:1.[32] **Carbamazepine** *p*-HBA 1:1.[33] **Carbamazepine** *p*-HBA 2:1.[34] **Carbamazepine** *p*-HBA monohydrate 2:1:1.[35] **Caffeine** *p*-HBA 2:1.[36]

MALEIC ACID	
SYNONYMS	*(Z)-2-Butenedioic acid; (Z)-1,2-ethenedicarboxylic acid; malenic acid, maleinic acid, toxilic acid.*
CAS NO.	110-16-7 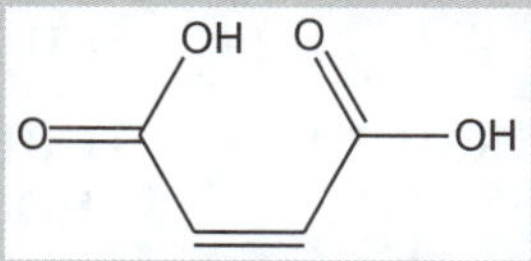
FORMULA	$C_4H_4O_4$; HOOCCH=CHCOOH
MOL WT.	116.07 g/mol
PHYSICAL STATE	White crystals
MELTING POINT	131–139 °C
SOLUBILITY	Soluble in water (pH < 7). *c*Log*P*: –0.76; *c*Log*S*: –0.25
pK_a	1.93; 6.58
HB DONORS	2
HB ACCEPTORS	4
POLYMORPHISM	Form α (*P*21/*c*) *MALIAC12* and form β (*Pc*) *MALIAC13*.
PXRD – 2θ(°)	Monoclinic forms α *MALIAC12* and β (*Pc*) *MALIAC13* 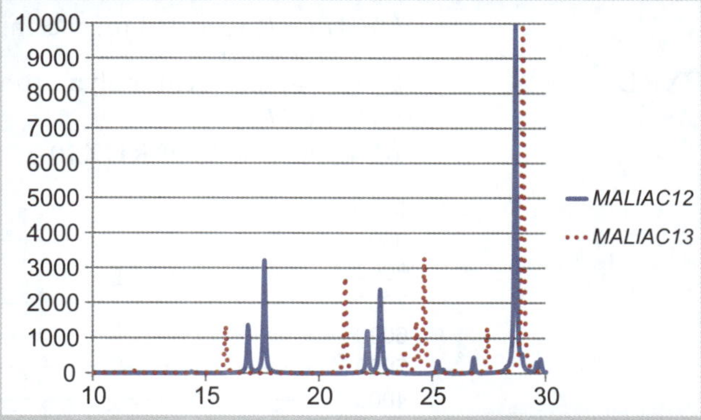
TOXICITY	Oral rat LD50: 708 mg/kg
REMARKS	

Maleic acid is used in the pharmaceutical industry as a pH modifier and a buffering agent. It is also used to prevent rancidity of oils and fats.

Examples of co-crystals:
Caffeine maleic acid 1:1.[37]
Caffeine maleic acid 2:1.[38]
Carbamazepine maleic acid 2:1.[39]
Theophylline maleic acid 1:1.[40]

(D, L, or DL) MALIC ACID	
SYNONYMS	*Hydroxybutanedioic acid; E296.*
CAS NO.	6915-15-7 DL-(+/–)-malic acid 97-67-6 L-(–)-malic acid 631-61-3 D-(+)-malic acid 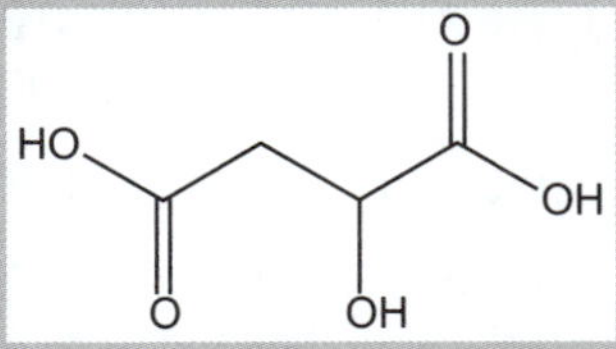
FORMULA	$C_4H_6O_5$; $HOOCCH_2CH(OH)COOH$
MOL WT.	134.09
PHYSICAL STATE	White powder
MELTING POINT	100 °C
SOLUBILITY	Solubility in water 558 g/L (20 °C). *c*Log*P*: –1.4; *c*Log*S*: –0.075
pK_a	3.40; 5.5
HB DONORS	3
HB ACCEPTORS	5
POLYMORPHISM	Two polymorphic forms (α *DLMALC* and β *DLMALC11*) reported for racemic DL-(+/–)-malic acid
PXRD – 2θ(°)	DL-(+/–)-malic acid: polymorphs α *DLMALC* and β *DLMALC11* L-(–)-malic acid COFRUK10 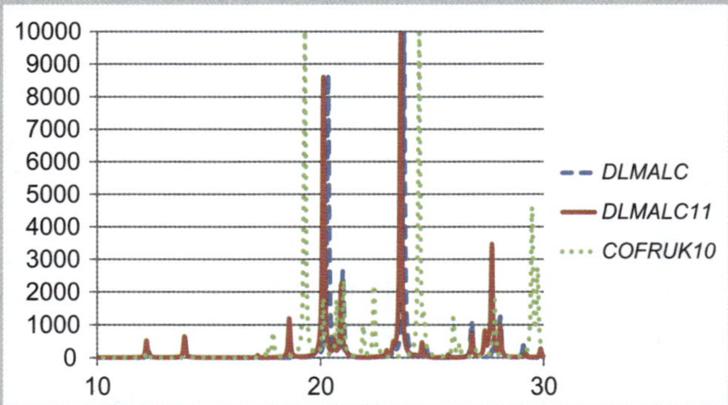
TOXICITY	Oral rat LD50: 4730 mg/kg
REMARKS	
Active ingredient in many sour or tart foods, malic acid is found mostly in unripe fruits. Used with or in place of the less sour citric acid in sour sweets. These sweets are sometimes labelled with a warning stating that excessive consumption can cause irritation of the mouth.	

(D, L, or DL) MALIC ACID *(Continued)*
Examples of co-crystals: **Theophylline** DL-malic acid 1:1.[41] **Theophylline** D-malic acid 1:1.[42] **Theophylline** L-malic acid 1:1.[43] **L-Tartaric** L-malic acid 1:1.[44] **6-Methyl-2-pyridone** L-malic acid 2:1.[45] **Meloxicum** malic acid 1:1.[46]

MALONIC ACID	
SYNONYMS	*1,3-Propanedioic acid.*
CAS NO.	141-82-2
FORMULA	$C_3H_4O_4$; $(HOOC)_2CH_2$
MOL WT.	104.06 g/mol
PHYSICAL STATE	White crystals
MELTING POINT	135 °C
SOLUBILITY	Soluble in water (73 g/L at 20 °C). *c*Log*P*: –0.86; *c*Log*S*: –0.20
pK_a	2.83; 5.69
HB DONORS	2
HB ACCEPTORS	4
POLYMORPHISM	Triclinic β (P_{-1}) (room temperature), orthorhombic α (*Pbcn*), monoclinic γ (P_{21}/c).
PXRD – 2θ(°)	Triclinic β (P_{-1}) *MALNAC02* and orthorhombic α (*Pbcn*) *MALNAC03*

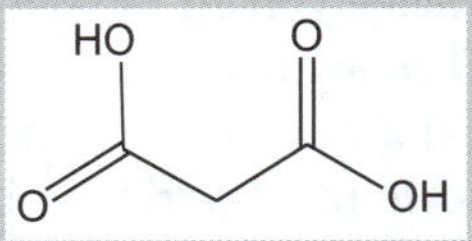

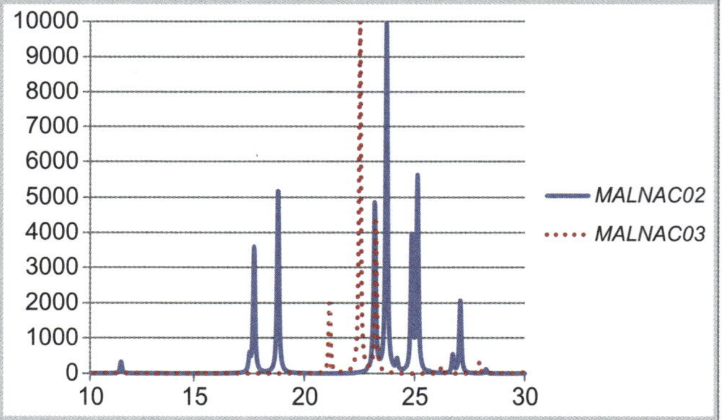

TOXICITY	Oral rat LD50: 1 310 mg/kg High concentrations destroy mucous membranes.
REMARKS	

The name originates from the Greek word μαλον (malon) meaning 'apple'. The calcium salt of malonic acid occurs in high concentrations in beetroot.

Examples of co-crystals:
Carbamazepine malonic acid 2:1.[47]
Caffeine malonic acid 2:1.[48]
Theobromine malonic acid 2:1.[49]

(R, S, R/S) MANDELIC ACID		
SYNONYMS	*2-Hydroxy-2-phenylacetic acid, alpha-hydroxy-benzeneacetic acid, amygdalic acid, uromaline.*	
CAS NO.	611-72-2 *R*(–) mandelic acid, 17199-29 *S*(+) mandelic acid 611-72-3 *R*/*S* mandelic acid	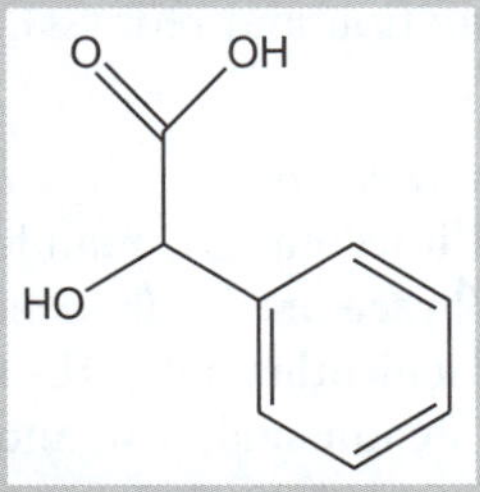
FORMULA	$C_8H_8O_3$; $C_6H_5CH(OH)CO_2H$	
MOL WT.	152.14 g/mol	
PHYSICAL STATE	White crystalline powder	
MELTING POINT	119 °C (*R*/*S* racemic mixture); 132–135 °C (optically pure)	
SOLUBILITY	Solubility in water (15.87 g/100 mL (25 °C) for *R*/*S* mixture). *c*Log*P*: 0.47; *c*Log*S*: –1.05	
pK_a	3.41	
HB DONORS	2	
HB ACCEPTORS	3	
AROMATIC RING	1	
POLYMORPHISM	A stable (orthorhombic *Pbca*) form and a metastable (monoclinic *P*21/*c*) form of racemic *R*/*S* mandelic acid have been reported.	
PXRD – 2θ(°)	*(R/S)* mandelic acid polymorphs: orthorhombic *DLMAND03* and monoclinic *DLMAND02*. *(S)* mandelic acid: monoclinic *FEGHAA*. 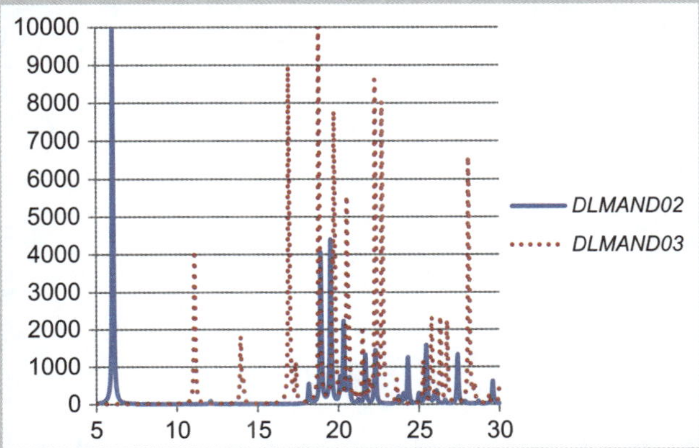	
TOXICITY	Oral rat LD50: 4100 mg/kg	

(R, S, R/S) MANDELIC ACID *(Continued)*	
REMARKS	
Use in the medical community (antibacterial, skin care from acne, inflammation and redness). The racemic (*R*/*S*) mixture is known as *paramandelic acid.* *Examples of co-crystals:* **Piracetam** DL-mandelic acid 2:1.[50] **Piracetam** L-mandelic acid 2:1.[51] **Isonicotinamide** DL-mandelic acid 1:1.[52] **Nicotinamide** L-mandelic acid 1:1.[53]	

NICOTINAMIDE		
SYNONYMS	*3-Pyridinecarboxamide; niacinamide; Vitamin PP.*	
CAS NO.	98-92-0	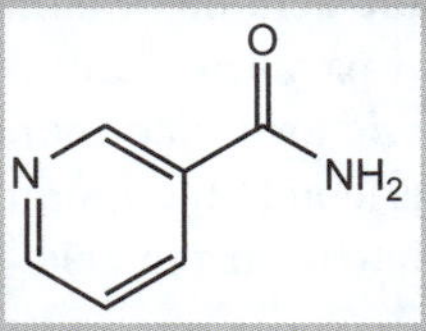
FORMULA	$C_6H_6N_2O$	
MOL WT.	122.12 g/mol	
PHYSICAL STATE	White powder	
MELTING POINT	128–131 °C	
SOLUBILITY	Solubility in water: 500 mg/mL at 25 °C. *c*Log*P*: –0.11; *c*Log*S*: –0.91	
pK_a	3.35	
HB DONORS	1	
HB ACCEPTORS	2	
AROMATIC RING	1	
POLYMORPHISM	No polymorphism reported	
PXRD – 2θ(°)	Monoclinic (*P*21/*c*) *NICOAM01*	

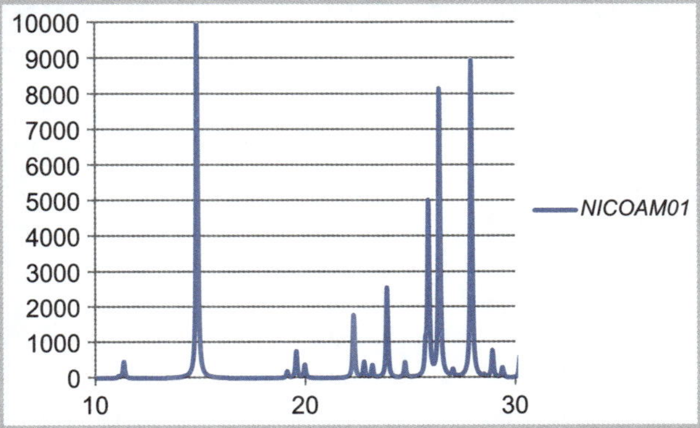

TOXICITY	Oral rat LD50: 3530–3540 mg/kg
REMARKS	

Water-soluble vitamin (part of the vitamin B group). It has demonstrated anti-inflammatory actions. Animal studies show that nicotinamide has anti-anxiety (anxiolytic) properties.

Examples of co-crystals:
Salicylic acid nicotinamide 1:1.[54]
S-Ibuprofen nicotinamide 1:1.[55]
Carbamazepine nicotinamide 1:1.[56]
Celecoxib nicotinamide 1:1.[57]

NICOTINAMIDE (Continued)

Caffeic acid nicotinamide 1:1.[58]
Ferulic acid nicotinamide 1:1.[59]
Ethyl paraben nicotinamide 1:1.[60]
Suberic acid nicotinamide 1:1.[61]
Suberic acid nicotinamide 1:2.[62]
3-Dinitrobenzoic acid, 3 dimethylbenzoic acid nicotinamide 1:1:1.[63]

OXALIC ACID		
SYNONYMS	*Ethanedioic acid.*	
CAS NO.	144-62-7 (anhydrous), 6153-56-6 (dihydrate)	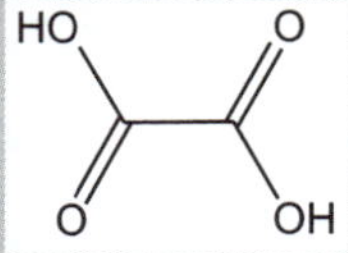
FORMULA	$C_2O_4H_2$; HOOCCOOH	
MOL WT.	90.03 g/mol (anhydrous)	
PHYSICAL STATE	White crystals	
MELTING POINT	182–191 °C (anhydrous) –101 °C (dihydrate).	
SOLUBILITY	Soluble in water (90 g/L at 20 °C) *c*Log*P*: –1.33; *c*Log*S*: 0.065	
pK_a	1.25; 4.14	
HB DONORS	2	
HB ACCEPTORS	4	
POLYMORPHISM	Two anhydrous polymorphs (a low temperature (–260 °C –> 25 °C) stable α form and a high temperature stable β form). The dihydrate form also crystallizes under two polymorphic forms.	
PXRD – 2θ(°)	Anhydrous: monoclinic β *OXALAC04*, orthorhombicα *OXALAC06*. Dihydrate: monoclinic α *OXACDH01*, monoclinic β *OXACDH29*	

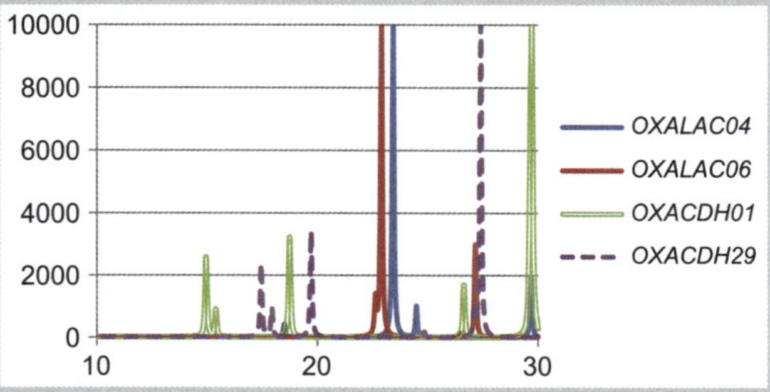

TOXICITY	Rat (oral) LD50: 7500 mg/kg Oxalic acid is rather toxic: (Hazard Symbols: XN, Risk Phrases: 21/22).	
REMARKS		

Oxalic acid and oxalates are present in many plants and occur naturally in animals. Calcium oxalate is the most common component of kidney stones.

OXALIC ACID *(Continued)*
Examples of co-crystals: **Carbamazepine** oxalic acid 2:1.[64] **Caffeine** oxalic acid 2:1.[65]

SACCHARIN		
SYNONYMS	*1,1-Dioxo-1,2-benzothiazol-3-one; o-sulphobenzamide.*	
CAS NO.	81-07-2	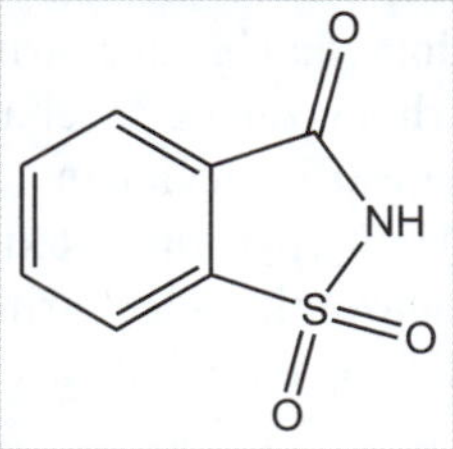
FORMULA	$C_7H_5NO_3S$	
MOL WT.	183.18 g/mol	
PHYSICAL STATE	White crystalline solid	
MELTING POINT	229 °C	
SOLUBILITY	In its acid form, saccharin is not water soluble (0.35 g per 100 mL). Sodium and calcium salts are highly water soluble (67 g per 100 mL at 25 °C). *c*Log*P*: 0.777; *c*Log*S*: –1.535	
pK_a	~2	
HB DONORS	1	
HB ACCEPTORS	3	
AROMATIC RING	1	
PXRD – 2θ(°)	*SCCHRN02* 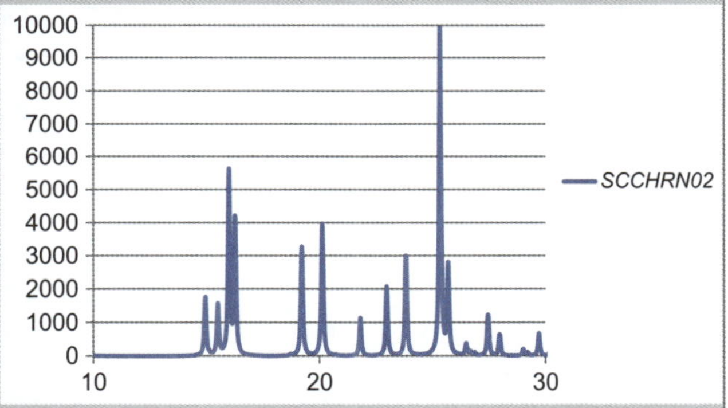	
TOXICITY	Oral mouse LD50: 14200 mg/kg	
REMARKS		
Artificial sweetener (much sweeter than sucrose) but with an unpleasant bitter or metallic aftertaste, especially at high concentrations. It is used to sweeten products such as drinks, candies, biscuits, medicines, and toothpaste. Saccharin is unstable when heated but it does not react chemically. As such, it stores well.		

SACCHARIN *(Continued)*

Examples of co-crystals:
Theophylline saccharin 1:1.[66]
Indomethacin saccharin 1:1.[67]
Carbamazepine saccharin 1:1.[68]
Megestrol saccharin 1:1.[69]
4-Methylpyridin *N*-oxide saccharin 1:1.[70]
Ethenzamide saccharin 1:1.[71]
Adefovir dipivoxil saccharin 1:1.[72]

SALICYLIC ACID (2-HYDROXYBENZOIC ACID)	
SYNONYMS	*o-Hydroxybenzoic acid; o-HBA.*
CAS NO.	69-72-7
FORMULA	$C_7H_6O_3$; HOC_6H_4COOH
MOL WT.	138.12 g/mol
PHYSICAL STATE	White crystalline powder
	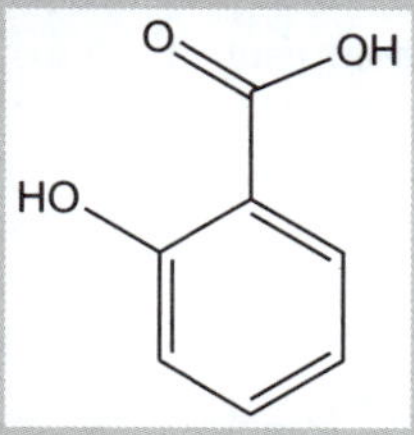
MELTING POINT	159 °C
SOLUBILITY	Slightly soluble in water (0.2 g/100 mL (20 °C)). *c*Log*P*: 1.2; *c*Log*S*: –1.3
pK_a	2.97
HB DONORS	2
HB ACCEPTORS	3
AROMATIC RING	1
PXRD – 2θ(°)	Anhydrous (monoclinic *P*21/*c*) *SALIAC16* 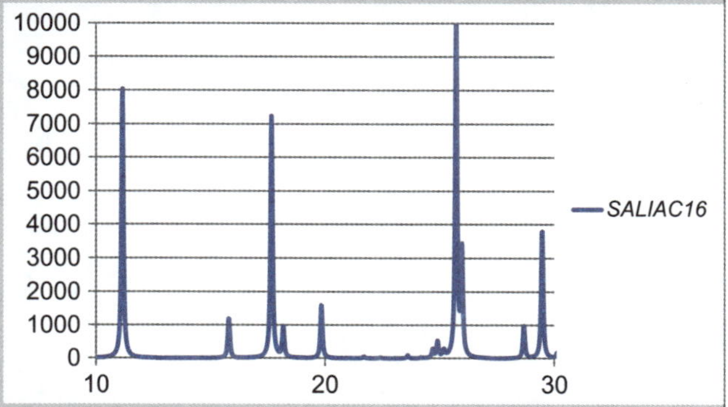
TOXICITY	Oral rat LD50: 891 mg/kg
REMARKS	

Methyl salicylate is used in food flavouring and preservatives. In addition to its analgesic and antipyretic properties, salicylic acid possesses keratinolytic properties and fungicidal properties.

Examples of co-crystals:
Carbamazepine salicylic acid 1:1.[73]
Caffeine salicylic acid 1:1.[74]
Theophylline salicylic acid 1:1.[75]
Isonicotinamide salicylic acid 1:1.[76]

SALICYLIC ACID (2-HYDROXYBENZOIC ACID) *(Continued)*
Sulfadimidine salicylic acid 1:1.[77] **3,5 Dimethyl-1*H*-pyrazole** salicylic acid 1:2.[78] **Phenazine** salicylic acid 1:2.[79]

SORBIC ACID	
SYNONYMS	*(2E,4E)-Hexa-2,4-dienoic acid; E200.*
CAS NO.	110-44-1
FORMULA	$C_6H_8O_2$; $CH_3CH{=}CH{-}CH{=}CH{-}COOH$
MOL WT.	112.13 g/mol
PHYSICAL STATE	White crystals or powder
MELTING POINT	135 °C
SOLUBILITY	Slightly soluble in water (0.25 g/100 mL (30 °C)). *c*Log*P*: 1.165; *c*Log*S*: –1.083
pK_a	4.76
HB DONORS	1
HB ACCEPTORS	2
PXRD – 2θ(°)	Monoclinic (*C*2/*c*) *LEZHUT* 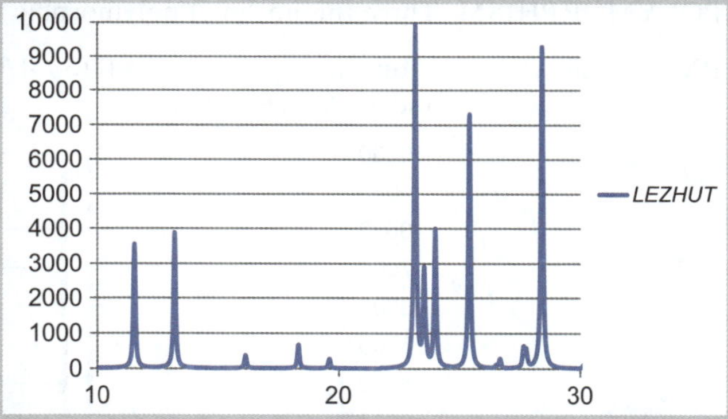
TOXICITY	Oral rat LD50: 7360 mg/kg
REMARKS	

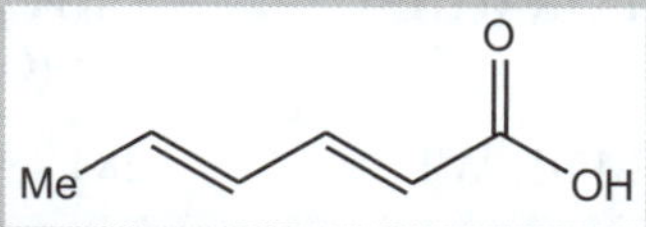

Natural organic compound used as a food preservative.

Examples of co-crystals:
Theophylline sorbic acid 1:1.[80]
2,4-Diamino-6-phenyl-1,3,5-triazine sorbic acid 1:1.[81]
AMG670129 sorbic acid 1:1.[82]
AMG831663 sorbic acid 1:1.[83]
AMG678809 sorbic acid 1:1.[84]

STEARIC ACID	
SYNONYMS	*Octadecanoic acid, C18:0*
CAS NO.	57-11-4
FORMULA	$C_{18}H_{36}O_2$; $CH_3(CH_2)_{16}COOH$ 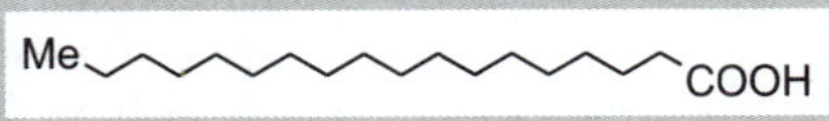
MOL WT.	284.48 g/mol
PHYSICAL STATE	White waxy solid
MELTING POINT	69.6 °C
SOLUBILITY	Insoluble in water, 1g dissolves in 26 mL acetone or 21 mL EtOH. *c*Log*P*: 7.5; *c*Log*S*: –4.7
pK_a	~4.9
HB DONORS	1
HB ACCEPTORS	2
POLYMORPHISM	Three (monoclinic) polymorphic forms
PXRD – 2θ(°)	Monoclinic polymorphs B (*STARAC*), C (*STARAC01*), and E (*STARAC05*). 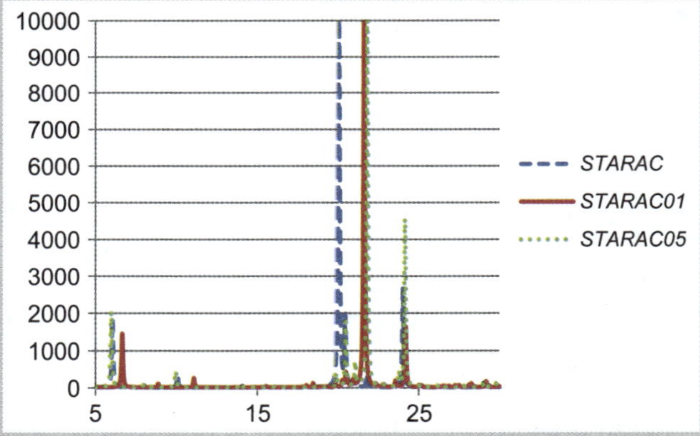
TOXICITY	Oral rat LD50: 4640 mg/kg
REMARKS	

Saturated fatty acid, occurring in animal (more common) and vegetable fats and oils.

Examples of co-crystals:
Deoxycholic acid stearic acid 1:1.[85]

SUCCINIC ACID		
SYNONYMS	*Butanedionic acid; dihydrofumaric acid; amber acid.*	
CAS NO.	110-15-6	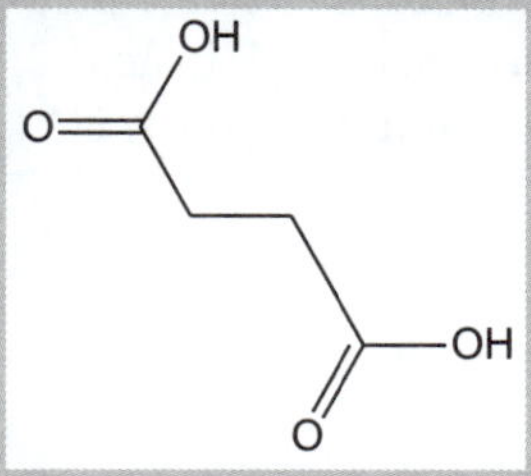
FORMULA	$C_4H_6O_4$; $HOOCCH_2CH_2COOH$	
MOL WT.	118.09 g/mol	
PHYSICAL STATE	Colourless crystals	
MELTING POINT	187–187 °C	
SOLUBILITY	Moderately soluble in water (100 g/100 mL at 100 °C). *c*Log*P*: –0.4; *c*Log*S*: –0.47	
pK_a	4.2; 5.6	
HB DONORS	2	
HB ACCEPTORS	4	
POLYMORPHISM	Two polymorphs are reported: the low temperature stable B form and the high temperature stable A form (metastable at room temperature).	
PXRD – 2θ(°)	Triclinic α (P_{-1}) *SUCACB07* Monoclinic β (P_{2_1}/c) *SUCACB02* 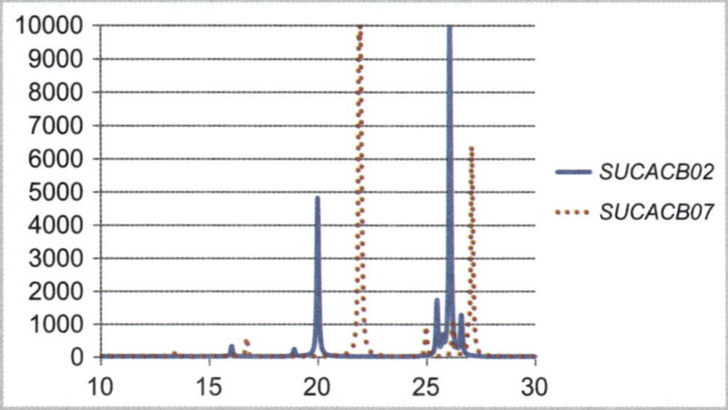	
TOXICITY	Oral rat LD50: 2260 mg/kg	
REMARKS		

In nutraceutical form as a food additive and dietary supplement, it is safe and approved by the FDA. As an excipient in pharmaceutical products it is used to control acidity and, more rarely, in effervescent tablets.

Examples of co-crystals:
Carbamazepine succinic acid 2:1.[86]

SUCCINIC ACID *(Continued)*
Caffeine succinic acid 2:1.[87] **Itraconazole** succinic acid 2:1.[88] **Piroxicam** succinic acid 2:1.[89] **Fluoxetine HCl** (Prozac®) succinic acid (2:1).[90]

(D, L or D/L) TARTARIC ACID	
SYNONYMS	*Dihydroxysuccinic acid; 2,3-dihydroxybutanedioic acid*
CAS NO.	526-83-0 87-69-4 L-(*R*,*R*)-(+)-tartaric acid 147-71-7 D-(*S*,*S*)-(–)-tartaric acid 147-73-9 (2*R*,3*S*)-tartaric (*meso*) 133-37-9 DL-(*S*,*S*/*R*,*R*)-(±)-tartaric acid 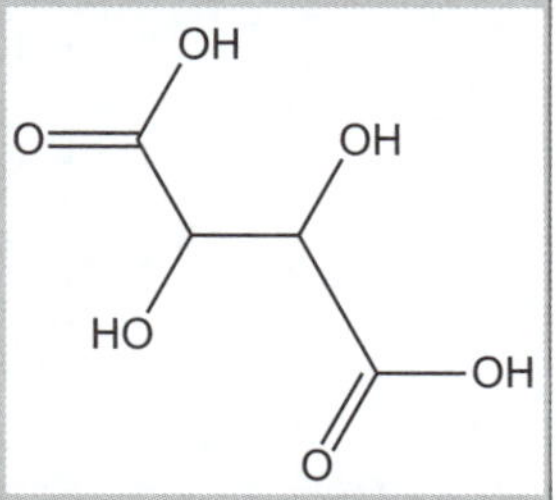
FORMULA	$C_4O_6H_6$; $HO_2CCH(OH)CH(OH)CO_2H$
MOL WT.	150.09 g/mol
PHYSICAL STATE	White crystalline powder
MELTING POINT	171–174 °C (L-tartaric) 206 °C (DL, racemic) 146–148 °C (*meso*)
SOLUBILITY	soluble in water (133 g/100 mL (20 °C)). *c*Log*P*: –2.4; *c*Log*S*: 0.32
pK_a	2.95; 4.25 (L(+) 25 °C) 3.22; 4.85 (*meso* 25 °C)
HB DONORS	4
HB ACCEPTORS	6
POLYMORPHISM	Two polymorphs of *meso* (2*R*,3*S*)-tartaric acid (triclinic (*P*1) *TARTAM* and orthorhombic *TARTAM01* (no coordinates)). Two polymorphs of *meso* (2*R*,3*S*)-tartaric monohydrate (triclinic (*P*1) *TARTMM* and monoclinic (*P*21/*c*) *TARTMM01*).
PXRD – 2θ(°)	L-(*R*,*R*)-(+)-tartaric acid: monoclinic (*P*21) *TARTAL* 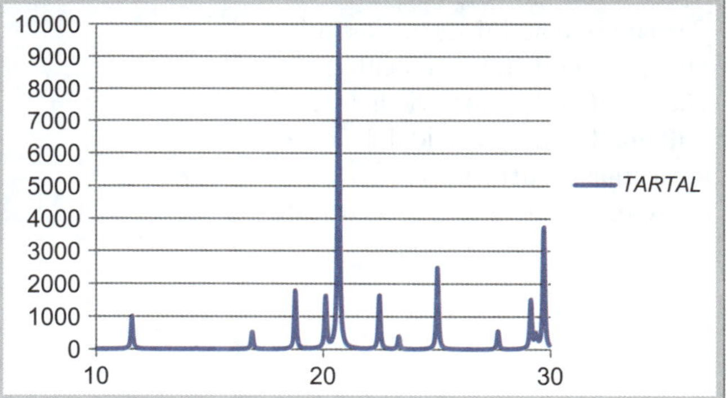

***(D, L or D/L) TARTARIC ACID** (Continued)*	
	Meso (2*R*,3*S*)-tartaric: triclinic (*P*1) *TARTAM* *Meso* (2*R*,3*S*)-tartaric (monohydrate): triclinic (*P*1) *TARTMM* and monoclinic (*P*21/*c*) *TARTMM01* 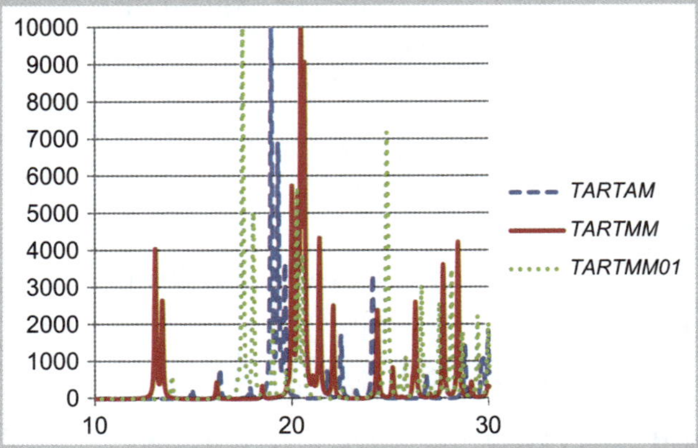 *Racemate* DL-(*S*,*S*/*R*,*R*)-(±)-tartaric acid: triclinic (P-1) *ZZZDUI01* *Racemate* DL-(*S*,*S*/*R*,*R*)-(±)-tartaric acid (monohydrate): triclinic (*P*-1) *TARTDL01*
TOXICITY	Mouse LD50: 485 mg/kg
REMARKS	

Occurs naturally in many plants (*e.g.* grapes, bananas) and is one of the main acids found in wine. It is added to other foods to give a sour taste, and is used as an antioxidant.

Hydrated forms of the *racemic* and *meso* forms of tartaric acid are known. The *meso* (2*R*,3*S*)-tartaric (anhydrous) exists under at least two polymorphic forms (triclinic (*P*1) and orthorhombic). The *meso* (2*R*,3*S*)-tartaric monohydrate exists under two polymorphic forms (triclinic (*P*1) and monoclinic (*P*21/*c*)).

Examples of co-crystals:
Itraconazole tartaric acid 2:1.[91]
Carbamazepine DL-tartaric acid.[92]
Theophylline L-tartaric acid 2:1.[93]
Theophylline DL-tartaric acid 2:1.[94]
Caffeine D-tartaric acid 1:1.[95]
Piracetam L-tartaric acid.[96]
Stanolone L-tartaric acid clathrate.[97]

TROMETHAMINE		
SYNONYMS	*2-Amino-2(hydroxymethyl)propane-1,3-diol, tris(hydroxymethyl) aminomethane TRIS, Trizma™, THAM.*	
CAS NO.	77-86-1	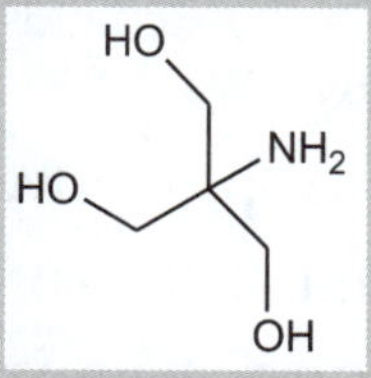
FORMULA	$C_4H_{11}NO_3$; $(HOCH_2)_3CNH_2$	
MOL WT.	121.14 g/mol	
PHYSICAL STATE	White crystalline powder	
MELTING POINT	171–172 °C	
SOLUBILITY	Solubility in water: 550 mg/mL (20°C).	
pK_a	8.04	
HB DONORS	4	
HB ACCEPTORS	4	
POLYMORPHISM	Two orthorhombic polymorphic forms: I *THXMAM11* and II *THXMAM13 (room temp.).* Third cubic (high temperature > 409 K) polymorph *THXMAM12 (no coordinates).*	
PXRD – 2θ(°)	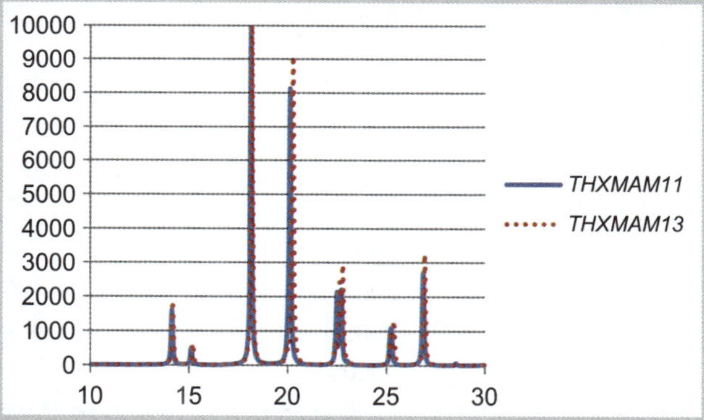	
TOXICITY	Oral rat LD50: 5900 mg/kg	
REMARKS		

TRIS is one of the most common buffers used in the biology/biochemistry labs. It is used as alternative to sodium bicarbonate in the treatment of metabolic acidosis. It is also an emulsifying agent and absorbent for acidic gases, alkalizer and osmotic diuretic.

Examples of co-crystals:
Glyburide–tromethamine 1:1.[98]

UREA		
SYNONYMS	*Carbamide, carbonyldiamine, diaminomethanal*	
CAS NO.	57-13-6	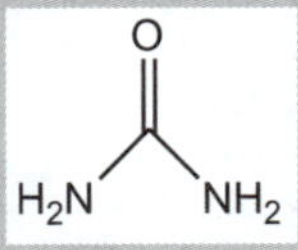
FORMULA	CH_4N_2O; $(NH_2)_2C{=}O$	
MOL WT.	60.06 g/mol	
PHYSICAL STATE	White solid	
MELTING POINT	133–135 °C	
BOILING POINT	Decomposes	
SOLUBILITY	54.5 g/100 mL (25 °C). *c*Log*P*: –1.25; *c*Log*S*: –0.80	
pK_a	0.18	
HB DONORS	2	
HB ACCEPTORS	1	
POLYMORPHISM	No polymorphism reported for urea but examples of polymorphism of co-crystals involving this co-former are known (*e.g.* with barbituric acid).	
PXRD – 2θ(°)	Tetragonal (*P*421*m*) *UREAXX* 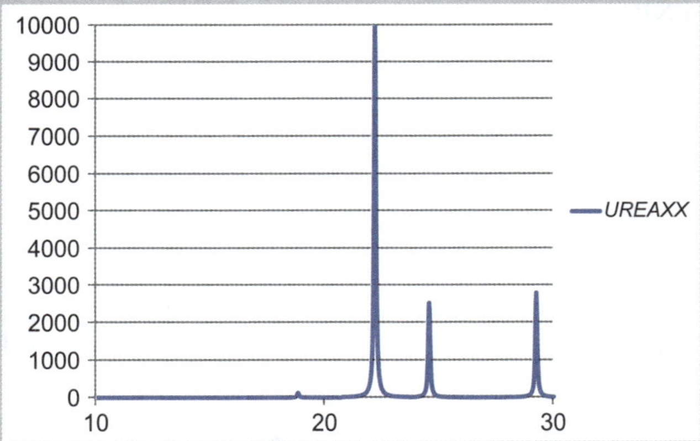	
TOXICITY	Rat oral LD50:8471 mg/kg	
REMARKS		

Urea is naturally produced when the body metabolizes proteins. It was the first natural compound to be synthesized using inorganic compounds

Examples of co-crystals:
Salicylic acid urea 1:1.[99]
Succinic acid urea 1:1 and 2:1.[100]

UREA (Continued)
Theophylline urea 1:1 monohydrate.[101] **Imidazolidone** urea 1:1.[102] **Barbituric** urea 1:1 *(three polymorphs reported)*.[103]

References

1. a) *The Merck Index*, Wiley, 2006; b) P. H. Stahl, C. G. Wermuth, *Handbook of Pharmaceutical Salts, Properties, Selection, and Use*, Wiley VCH, Weinheim, 2008.
2. Web Sources:
   http://www.inchem.org/
   http://www.cas.org/
   http://www.chemexper.com/
   http://webbook.nist.gov/chemistry/
   Safety data:
   http://msds.chem.ox.ac.uk/;
   http://www.msdssearch.com/;
   http://www.msdsonline.com/;
   Chemical Handbook:
   http://www.hbcpnetbase.com/ (login and password needed)
   p*K*a values:
   http://www.zirchrom.com/organic.htm;
   http://chemweb.unp.ac.za/chemistry/Physical_Data/pKa_compilation.pdf.
3. F. H. Allen, *Acta Cryst.*, 2002, **B58**, 380.
4. C. F. Macrae, I. J. Bruno, J. A. Chisholm, P. R. Edgington, P. McCabe, E. Pidcock, L. Rodriguez-Monge, R. Taylor, J. van de Streek and P. A. Wood, *J. Appl. Cryst.*, 2008, **41**, 466.
5. *XAQRAJ* in J. A. McMahon, J. A. Bis, P. Vishweshwar, T. R. Shattock, O. L. McLaughlin and M. J. Zaworotko, *Z. Kristallogr.*, 2005, **220**, 340.
6. *XAQREN* in J. A. McMahon, J. A. Bis, P. Vishweshwar, T. R. Shattock, O. L. McLaughlin and M. J. Zaworotko, *Z. Kristallogr.*, 2005, **220**, 340.
7. *NUHYEU* in G. Smith, K. E. Baldry, K. A. Byriel and C. H. L. Kennard, *Aust. J. Chem.*, 1997, **50**, 727.
8. *SORWIF* in M. R. Caira, *J. Crystallogr. Spectrosc. Res.*, 1991, **21**, 641.
9. *SORWEB* in M. R. Caira, *J. Crystallogr. Spectrosc. Res.*, 1991, **21**, 641.
10. *RUWFAR* in P. Kavuru, D. Aboarayes, K.K. Arora, H.D. Clarke, A. Kennedy, L. Marshall, Tien Teng Ong, J. Perman, T. Pujari, L. Wojtas and M. J. Zaworotko, *Cryst. Growth Des.*, 2010, **10**, 3568.
11. *RUWDUJ* in P. Kavuru, D. Aboarayes, K. K. Arora, H. D. Clarke, A. Kennedy, L. Marshall, Tien Teng Ong, J. Perman, T. Pujari, L. Wojtas and M.J. Zaworotko, *Cryst.Growth Des.*, 2010, **10**, 3568.

12. *RUWFEV* in P. Kavuru, D. Aboarayes, K. K. Arora, H. D. Clarke, A. Kennedy, L. Marshall, Tien Teng Ong, J. Perman, T. Pujari, L. Wojtas and M. J. Zaworotko, *Cryst. Growth Des.*, 2010, **10**, 3568.
13. *AQEXEZ* in M. Szyrszyng, E. Nowak, M. Gdaniec, M.J. Milewska and T. Polonski, *Tetrahedron: Asymm.*, 2004, **15**, 103.
14. V. Tomašić and Zoran Štefanić, *CrystEngComm*, 2007, **9**, 1124.
15. S. Ikonen, Nonappa and E. Kolehmainen, *CrystEngComm*, 2010, **12**, 4304.
16. *RUCFAX* in M. Viertelhaus, R. Hilfiker, F. Blatter and M. Neuburger, *Cryst. Growth Des.*, 2009, **9**, 2220.
17. *RUCFEB* in M. Viertelhaus, R. Hilfiker, F. Blatter and M. Neuburger, *Cryst. Growth Des.*, 2009, **9**, 2220.
18. *RUWGAS* in P. Kavuru, D. Aboarayes, K. K. Arora, H. D. Clarke, A. Kennedy, L. Marshall, Tien Teng Ong, J. Perman, T. Pujari, L. Wojtas, M. J. Zaworotko, *Cryst. Growth Des.*, 2010, **10**, 3568.
19. *CUYXUQ* in A. Lemmerer and J. Bernstein, *CrystEngComm*, 2010, **12**, 2029.
20. *KIGKER* in S. Karki, T. Friščić, W. Jones and W.D.S.Motherwell, *Mol. Pharmaceutics*, 2007, **4**, 347.
21. *KIGKAN* in S. Karki, T. Friščić, W. Jones and W.D.S.Motherwell, *Mol. Pharmaceutics*, 2007, **4**, 347.
22. S. L. Childs, N. Rodríguez-Hornedo, L. S. Reddy, D. Jayasankar, C. Maheshwari, L. McCausland, R. Shipplett and B. C. Stahly, *CrystEngComm*, 2008, **10**, 856.
23. S. L. Childs, L. J. Chyall, J. T. Dunlap, V. N. Smolenskaya, B. C. Stahly and G. P. Stahly, *J. Am. Chem. Soc.*, 2004, **126**(41), 13335.
24. *LUNNOX* in C. B. Aakeroy, A. M. Beatty and B. A. Helfrich, *J. Am. Chem. Soc.*, 2002, **124**, 14425.
25. *LGPYRG* in Z. Taira and W. H. Watson, *Acta Crystallogr., Sect.B, Struct. Crystallogr. Cryst. Chem.*, 1977, **33**, 3823.
26. *EXUQUJ* in A. V. Trask, W. D. S. Motherwell and W.Jones, *Chem. Commun.*, 2004, 890.
27. *XEJXIU* in A. V. Trask, W. D. S. Motherwell and W.Jones, *Int. J. Pharm.*, 2006, **320**, 114.
28. *MOXVOL* in S. L. Childs, P. A. Wood, N. Rodriguez-Hornedo, L. S. Reddy and K. I. Hardcastle, *Cryst. Growth Des.*, 2009, **9**, 1869.
29. *NUKYEY* in S. Karki, T. Friščić and W.Jones, *CrystEngComm*, 2009, **11**, 470.
30. *DAVPEW* in P. Vishweshwar, J. A. McMahon, M. L. Peterson, M. B. Hickey, T. R. Shattock and M. J. Zaworotko, *Chem. Commun.*, 2005, 4601.
31. *DEYREF* in Zi-Liang Wang and Lin-Heng Wei; *Acta Crystallogr., Sect.E, Struct. Rep. Online*, 2007, **63**, o1681.
32. *KIGLOC* in S. L. Childs, G. P. Stahly and A. Park, *Mol. Pharmaceutics*, 2007, **4**, 323.
33. *MOXVIF* in S. L. Childs, P. A. Wood, N. Rodriguez-Hornedo, L. S. Reddy, K. I. Hardcastle, *Cryst. Growth Des.*, 2009, **9**, 1869.

34. *XAQRAJ* in J. A. McMahon, J. A. Bis, P. Vishweshwar, T. R. Shattock, O. L. McLaughlin and M. J. Zaworotko, *Z. Kristallogr.*, 2005, **220**, 340.
35. *XAQREN* in J. A. McMahon, J. A. Bis, P. Vishweshwar, T. R. Shattock, O. L. McLaughlin and M. J. Zaworotko, *Z. Kristallogr.*, 2005, **220**, 340.
36. *MOZCUA* in D. K. Bucar, R. F. Henry, Xiaochun Lou, R. W. Duerst, L. R. McGillivray and G. G. Z. Zhang, *Cryst. Growth Des.*, 2009, **9**, 1932.
37. *GANYEA* in A. V. Trask, W. D. S. Motherwell and W.Jones, *Cryst. Growth Des.*, 2005, **5**, 1013.
38. *GANYIE* in A. V. Trask, W. D. S. Motherwell and W.Jones, *Cryst. Growth Des.*, 2005, **5**, 1013.
39. *MOXWOM* in S. L. Childs, P. A. Wood, N. Rodriguez-Hornedo, L. S. Reddy and K. I. Hardcastle, *Cryst. Growth Des.*, 2009, **9**, 1869.
40. *XEJXEQ* in A. V. Trask, W. D. S. Motherwell and W. Jones, *Int. J. Pharm.*, 2006, **320**, 114.
41. *CIZTAH* in T. Friščić, L. Fabian, J. C. Burley, D. G. Reid, M. J. Duer and W. Jones, *Chem. Commun.*, 2008, 1644.
42. *CODCOO* in T. Friščić, L. Fabian, J. C. Burley, D. G. Reid, M. J. Duer and W. Jones, *Chem. Commun.*, 2008, 1644.
43. T. Friščić, S. L. Childs, S. A. A. Rizvi and W. Jones, *CrystEngComm*, 2009, **11**, 418.
44. *NIVYOG* in C. B. Aakeroy, T. I. Cooke and M. Nieuwenhuyzen, *Supramol. Chem.*, 1996, **7**, 153.
45. *XASDIE* in C. B. Aakeroy, A. M. Beatty, M. Nieuwenhuyzen and M. Zou, *Tetrahedron* 2000, **56**, 6693.
46. M. Cheney, D. Weyna, Ning Shang, M. Hanna, L. Wojtas, M. Zaworotko, F. Meldrum and R. Ristic, *Crystal growth & design A*, 2010, **10**, 4401.
47. *MOXVUR* in S. L. Childs, P. A. Wood, N. Rodriguez-Hornedo, L. S. Reddy and K. I. Hardcastle, *Cryst. Growth Des.*, 2009, **9**, 1869.
48. *GANYAW* in A. V. Trask, W. D. S. Motherwell and W. Jones, *Cryst. Growth Des.*, 2005, **5**, 1013.
49. *HIJYEF* in S. Karki, L. Fabian, T. Friščić and W. Jones, *Org. Lett.*, 2007, **9**, 3133.
50. *RUCFIF* in M. Viertelhaus, R. Hilfiker, F. Blatter and M. Neuburger, *Cryst. Growth Des.*, 2009, **9**, 2220.
51. *XOZSOV* in M. Viertelhaus, R. Hilfiker, F. Blatter and M. Neuburger, *Cryst. Growth Des.*, 2009, **9**, 2220.
52. *LUNPAL* in C. B. Aakeroy, A. M. Beatty, B. A. Helfrich, *J. Am. Chem. Soc.*, 2002, **124**, 14425.
53. *JILZOU* in T. Friščić and W. Jones, *Faraday Discuss.*, 2007, **136**, 167.
54. *SODDOF* in D. J. Berry, C. C. Seaton, W. Clegg, R. W. Harrington, S. J. Coles, P. N. Horton, M. B. Hursthouse, R. Storey, W. Jones, T. Friščić and N. Blagden, *Cryst. Growth Des.*, 2008, **8**, 1697.
55. *SOGLAC* in D. J. Berry, C. C. Seaton, W. Clegg, R. W. Harrington, S. J. Coles, P. N. Horton, M. B. Hursthouse, R. Storey, W. Jones, T. Friščić and N. Blagden, *Cryst. Growth Des.*, 2008, **8**, 1697.

56. *UNEZES* in S. G. Fleischman, S. S. Kuduva, J. A. McMahon, B. Moulton, R. D. B. Walsh, N. Rodriguez-Hornedo and M. J. Zaworotko, *Cryst. Growth Des.*, 2003, **3**, 909.
57. *VIGDAR* in J. F. Remenar, M. L. Peterson, P. W. Stephens, Zhong Zhang, Yu Zimenkov and M. B. Hickey, *Mol. Pharmaceutics*, 2007, **4**, 386.
58. *MUPMOA* in H. D. Clarke, K. K. Arora, H. Bass, P. Kavuru, Tien Teng Ong, T. Pujari, L. Wojtas and M. J. Zaworotko, *Cryst. Growth Des.*, 2010, **10**, 2152.
59. *MUPNIV*, in H. D. Clarke, K. K. Arora, H. Bass, P. Kavuru, Tien Teng Ong, T. Pujari, L. Wojtas and M. J. Zaworotko, *Cryst. Growth Des.*, 2010, **10**, 2152.
60. *GOGQID* in S. Nicoli, S. Bilzi, P. Santi, M. Caira, J. Li and R. Bettini, *J. Pharm. Sci.*, 2008, **97**, 4830.
61. *NUKZAV* in S. Karki, T. Friščić and W. Jones, *CrystEngComm*, 2009, **11**, 470.
62. *NUKZEZ* in S. Karki, T. Friščić and W. Jones, *CrystEngComm*, 2009, **11**, 470.
63. *XAQPUB* in C. B. Aakeroy, J. Desper, E. Elisabeth, B. A. Helfrich, B. Levin and J. F. Urbina, *Z. Kristallogr.*, 2005, **220**, 325.
64. S. L. Childs, N. Rodríguez-Hornedo, L. Sreenivas Reddy, A. Jayasankar, C. Maheshwari, L. McCausland, R. Shipplett and B. Stahly, *CrystEngComm*, 2008, **10**, 856.
65. *GANXUP* in A. V. Trask, W. D. S. Motherwell and W. Jones, *Cryst. Growth Des.*, 2005, **5**, 1013.
66. *XOBCUN* in Enxian Lu, N. Rodriguez-Hornedo and R. Suryanarayanan, *CrystEngComm*, 2008, **10**, 665.
67. *UFERED* in S. P. Velaga, S. Basavoju and D. Bostrom, *Pharm. Res.*, 2008, **25**, 530.
68. *UNEZAO* in S. G. Fleischman, S. S. Kuduva, J. A. McMahon, B. Moulton, R. D. B. Walsh, N. Rodriguez-Hornedo and M. J. Zaworotko, *Cryst.Growth Des.*, 2003, **3**, 909.
69. *HORNEI* in K. Shiraki, N. Takata, R. Takano, Y. Hayashi and K. Terada, *Pharm. Res.*, 2008, **25**, 2581.
70. *ADECIW* in B. K. Saha, R. Banerjee, A. Nangia and G. R. Desiraju, *Acta Crystallogr., Sect. E, Struct. Rep. Online*, 2006, **62**, o2283.
71. *VUHFIO* in S. Aitipamula, Pui Shan Chow and R. B. H. Tan, *CrystEngComm*, 2009, **11**, 889.
72. Y. Gao, H. Zu and J. Zhang, *J. Pharm. Pharmacol.*, 2011, **63**, 483.
73. *MOXWAY* in S. L. Childs, P. A. Wood, N. Rodriguez-Hornedo, L. S. Reddy and K. I. Hardcastle, *Cryst. Growth Des.*, 2009, **9**, 1869.
74. *XOBCAT* in Enxian Lu, N. Rodriguez-Hornedo and R. Suryanarayanan, *CrystEngComm*, 2008, **10**, 665.
75. *KIGLES* in S. L. Childs, G. P. Stahly and A. Park, *Mol. Pharmaceutics*, 2007, **4**, 323.

76. *XAQQEM* in J. A. McMahon, J. A. Bis, P. Vishweshwar, T. R. Shattock, O. L. McLaughlin and M. J. Zaworotko, *Z. Kristallogr.*, 2005, **220**, 340.
77. *GEYSAE* in U. Patel, M. Haridas and T. P. Singh, *Acta Crystallogr.,Sect. C, Cryst. Struct. Commun.*, 1988, **44**, 1264.
78. *ODOHEV & ODOHIZ* in C. Lopez, R. M. Claramunt, M. A. Garcia, E. Pinilla, M. R. Torres, I. Alkorta and J. Elguero, *Cryst. Growth Des.*, 2007, **7**, 1176.
79. *NUKXAT* in S. Skovsgaard and A. D. Bond, *CrystEngComm*, 2009, **11**, 444.
80. *KIGLAO* in S. L. Childs, G. P. Stahly and A. Park, *Mol. Pharmaceutics*, 2007, **4**, 323.
81. *WINMIQ* in K. Thanigaimani, P. T. Muthiah and D. E. Lynch, *Acta Crystallogr., Sect. E, Struct. Rep. Online*, 2007, **63**, o4450.
82. *MOXTAV* in M. K. Stanton, S. Tufekcic, C. Morgan and A. Bak, *Cryst. Growth Des.*, 2009, **9**, 1344.
83. *MOXTID* in M. K. Stanton, S. Tufekcic, C. Morgan and A. Bak, *Cryst. Growth Des.*, 2009, **9**, 1344.
84. *MOXTUP* in M. K. Stanton, S. Tufekcic, C. Morgan and A. Bak, *Cryst. Growth Des.*, 2009, **9**, 1344.
85. *ZZZQHO* in G. Kratky; *Z. Phys. Chem.*, 1934, **B26**, 439.
86. *XOBCIB* in E. Lu, N. Rodriguez-Hornedo and R. Suryanarayanan, *CrystEngComm*, 2008, **10**, 665.
87. *SOCHOI* in T. Friščić, A. V. Trask, W. D. S. Motherwell and W. Jones, *Cryst. Growth Des.*, 2008, **8**, 1605.
88. *IKEQEU* in J. F. Remenar, S. L. Morissette, M. L. Peterson, B. Moulton, J. M. MacPhee, H. R. Guzman and O. Almarsson, *J. Am. Chem. Soc.*, 2003, **125**, 8456.
89. *DIKCIK* in S. L. Childs and K. I. Hardcastle, *Cryst. Growth Des.*, 2007, **7**, 1291.
90. S. L. Childs, L. Chyall, J. Dunlap, V. Smolenskaya, B. Stahly and G. Stahly, *J. Am. Chem. Soc.*, 2004, **126**, 13335.
91. J. F. Remenar, S. L. Morissette, M. L. Peterson, B. Moulton, J. M. MacPhee, H. R. Guzman and O. Almarsson, *J. Am. Chem. Soc.*, 2003, **125**, 8456.
92. *MOXWIG* in S. L. Childs, P. A. Wood, N. Rodriguez-Hornedo, L. S. Reddy and K. I. Hardcastle, *Cryst. Growth Des.*, 2009, **9**, 1869.
93. *NEXWOD* in T. Friščić, L. Fabian, J. C. Burley, W. Jones and W. D. S. Motherwell, *Chem. Commun.*, 2006, 5009.
94. *NEYCIE* in T. Friščić, L. Fabian, J. C. Burley, W. Jones and W. D. S. Motherwell, *Chem. Commun.*, 2006, 5009.
95. *NEXWUJ* in T. Friščić, L. Fabian, J. C. Burley, W. Jones and W. D. S. Motherwell, *Chem. Commun.*, 2006, 5009.
96. *RUCDUP* in M. Viertelhaus, R. Hilfiker, F. Blatter and M. Neuburger, *Cryst. Growth Des.*, 2009, **9**, 2220.
97. *YOFWEW* in N. Takata, K. Shiraki, R. Takano, Y. Hayashi and K. Terada, *Cryst. Growth Des.*, 2008, **8**, 3032.

98. J. McMahon, M. Peterson, M. J. Zaworotko, T. Shattock and M. Bourghol Hickey, WO2006/007448.
99. *SLCADC* in R. W. Gellert and I. N. Hsu, Acta Crystallogr., *Sect. A, Cryst. Phys., Diffr., Theor. Crystallogr.*,1981, **37**, C93.
100. A. Alhalaweh, S. George, D. Bostrom and S. Velaga, *Cryst. Growth Des*, 2010, **10**, 4847.
101. *DUXZAX* in H.Wiedenfeld and F.Knoch, *Arch. Pharm.*, 1986, **319**, 654.
102. Z. Deutsch and J. Bernstein, *Crystal Growth & Design*, 2008, **8**, 3537.
103. *EFOZAB* in M. Gryl, A. Krawczuk, K. Stadnicka, *Acta Crystallogr., Sect.B, Struct. Sci.*, 2008, **64**, 623.

Subject Index

References to tables are given in **bold** type. References to figures are given in *italic* type.